T0323532

# HANDBOOK OF LOW CARBON CONCRETE

# HANDBOOK OF LOW CARBON CONCRETE

**Ali NAZARI**
Swinburne University of Technology, Hawthorn, VIC, Australia

**Jay G. SANJAYAN**
Swinburne University of Technology, Hawthorn, VIC, Australia

AMSTERDAM • BOSTON • HEIDELBERG • LONDON
NEW YORK • OXFORD • PARIS • SAN DIEGO
SAN FRANCISCO • SINGAPORE • SYDNEY • TOKYO
Butterworth-Heinemann is an imprint of Elsevier

Butterworth-Heinemann is an imprint of Elsevier
The Boulevard, Langford Lane, Kidlington, Oxford OX5 1GB, United Kingdom
50 Hampshire Street, 5th Floor, Cambridge, MA 02139, United States

**Notices**
Knowledge and best practice in this field are constantly changing. As new research and experience broaden our understanding, changes in research methods, professional practices, or medical treatment may become necessary.

Practitioners and researchers must always rely on their own experience and knowledge in evaluating and using any information, methods, compounds, or experiments described herein. In using such information or methods they should be mindful of their own safety and the safety of others, including parties for whom they have a professional responsibility.

To the fullest extent of the law, neither the Publisher nor the authors, contributors, or editors, assume any liability for any injury and/or damage to persons or property as a matter of products liability, negligence or otherwise, or from any use or operation of any methods, products, instructions, or ideas contained in the material herein.

**British Library Cataloguing-in-Publication Data**
A catalogue record for this book is available from the British Library

**Library of Congress Cataloging-in-Publication Data**
A catalog record for this book is available from the Library of Congress

ISBN: 978-0-12-804524-4

For Information on all Butterworth-Heinemann publications
visit our website at https://www.elsevier.com

Working together
to grow libraries in
developing countries

www.elsevier.com • www.bookaid.org

*Publisher:* Joe Hayton
*Acquisition Editor:* Ken McCombs
*Editorial Project Manager:* Peter Jardim
*Production Project Manager:* Kiruthika Govindaraju
*Cover Designer:* Christian Bilbow

Typeset by MPS Limited, Chennai, India

# CONTENTS

## 11. An Overview on the Influence of Various Factors on the Properties of Geopolymer Concrete Derived From Industrial Byproducts    263

W.K. Part, M. Ramli and C.B. Cheah

## 12. Performance on an Alkali-Activated Cement-Based Binder (AACB) for Coating of an OPC Infrastructure Exposed to Chemical Attack: A Case Study    335

W. Tahri, Z. Abdollahnejad, F. Pacheco-Torgal and J. Aguiar

# LIST OF CONTRIBUTORS

**Z. Abdollahnejad**
University of Minho, Guimarães, Portugal

**J. Aguiar**
University of Minho, Guimarães, Portugal

**A. Castel**
The University of New South Wales, Sydney, NSW, Australia

**C. Baek**
Korea Institute of Civil Engineering and Building Technology, Goyang, Republic of Korea

**C.K. Chau**
The Hong Kong Polytechnic University, Hung Hom, Kowloon, Hong Kong

**C.B. Cheah**
Universiti Sains Malaysia, Penang, Malaysia

**M.-S. Cho**
Korea Hydro & Nuclear Power Co., Ltd, Seoul, Republic of Korea;
Hyundai Engineering & Construction, Seoul, Republic of Korea

**D.J.M. Flower**
Swinburne University of Technology, Melbourne, Australia

**A.M. Grabiec**
Poznan University of Life Sciences, Poznań, Poland

**C.-A. Graubner**
Technische Universität Darmstadt, Darmstadt, Germany

**S. Hainer**
Technische Universität Darmstadt, Darmstadt, Germany

**W.K. Hui**
The Hong Kong Polytechnic University, Hung Hom, Kowloon, Hong Kong

**E. Jamieson**
Alcoa of Australia, Kwinana, WA, Australia; Curtin University, Perth, WA, Australia

**Y.-B. Jung**
Korea Hydro & Nuclear Power Co., Ltd, Seoul, Republic of Korea; Hyundai Engineering & Construction, Seoul, Republic of Korea

**C. Kealley**
Curtin University, Perth, WA, Australia

**M. Kheradmand**
University of Minho, Guimarães, Portugal

**T.M. Leung**
The Hong Kong Polytechnic University, Hung Hom, Kowloon, Hong Kong

**A. Maghsoudpour**
WorldTech Scientific Research Center (WT-SRC), Tehran, Iran

**B. McLellan**
Kyoto University, Kyoto, Japan

**S. Miraldo**
University of Coimbra, Coimbra, Portugal

**A. Nazari**
Swinburne University of Technology, Hawthorn, VIC, Australia

**W.Y. Ng**
The Hong Kong Polytechnic University, Hung Hom, Kowloon, Hong Kong

**H. Nikraz**
Curtin University, Perth, WA, Australia

**F. Pacheco-Torgal**
University of Minho, Guimarães, Portugal; University of Sungkyunkwan, Suwon, Republic of Korea

**W.K. Part**
Universiti Sains Malaysia, Penang, Malaysia

**B. Penna**
Alcoa of Australia, Kwinana, WA, Australia; Curtin University, Perth, WA, Australia

**T. Proske**
Technische Universität Darmstadt, Darmstadt, Germany

**M. Ramli**
Universiti Sains Malaysia, Penang, Malaysia

**M. Rezvani**
Technische Universität Darmstadt, Darmstadt, Germany

**S. Roh**
Hanyang University, Ansan, Republic of Korea

**J.G. Sanjayan**
Swinburne University of Technology, Hawthorn, VIC, Australia

**J.-K. Song**
Chonnam National University, Gwangju, Republic of Korea

**K.-I. Song**
Chonnam National University, Gwangju, Republic of Korea

**J. Szulc**
Poznan University of Life Sciences, Poznań, Poland

**S. Tae**
Hanyang University, Ansan, Republic of Korea

**S.-H. Tae**
Hanyang University, Ansan, Republic of Korea

**W. Tahri**
University of Sfax, Sfax, Tunisia

**J.S.J. Van Deventer**
University of Melbourne, Melbourne, VIC, Australia; Zeobond Pty. Ltd., Melbourne, VIC, Australia

**A. van Riessen**
Curtin University, Perth, WA, Australia

**H. Wang**
University of Southern Queensland, Toowoomba, Australia

**J.M. Xu**
The Hong Kong Polytechnic University, Hung Hom, Kowloon, Hong Kong

**K.-H. Yang**
Kyonggi University, Suwon, Republic of Korea

**T. Yang**
China Oilfield Services Limited, Yanjiao, China

**D. Zawal**
Poznan University of Life Sciences, Poznań, Poland

**Z. Zhang**
University of Southern Queensland, Toowoomba, Australia

# PREFACE

Manufacture of ordinary Portland cement (OPC) requires the mining of limestone and releasing of carbon dioxide. For each ton of limestone mined, one-third is released as carbon dioxide that has been locked beneath the surface of the earth for millions of years. Emissions of greenhouse gases through industrial activities have a major impact on global warming and it is believed that at least 5–7% of $CO_2$ released to the atmosphere is due to the production of OPC. This has led to significant research on eco-friendly construction materials such as geopolymers and binary- and ternary-blended OPC concretes. *Handbook of Low-Carbon Concrete* is a collection of high-quality technical papers to provide the reader a comprehensive understanding of the ways in which carbon reductions can be achieved by careful choices of construction materials.

The demand for worldwide cement production is increasing by approximately 30% per decade as of 2016. The need for new infrastructure construction in developing nations is projected to force the demand up for cement in the coming years.

Manufacture of Portland cement is the fourth largest contributor to worldwide carbon emissions and is only behind petroleum, coal, and natural gas in releasing carbon dioxide that has been locked beneath the earth's surface for millions of years. The new cement factories that are being built mostly in developing nations to meet this forthcoming demand are unsustainable in the long term for the following reasons:

1. *Capital Intensive*: Cement factories are extremely capital-intensive developments. Once the capital is invested, the investor is committed to cement-production tonnages to recoup their capital investments. Cement manufacturers are notoriously well connected in the construction industry and are resistant to any new low-carbon technologies to protect their investments. Once capital investments are locked into new cement factories, there is little incentive for the cement manufacturers to embrace low-carbon technologies. For example, according to a Lafarge report, a new dry process cement line producing 1 million tons annually can cost up to $240 million.

2. *Low Employment/Capital Dollars Invested*: Cement manufacture is largely automated with low labor intensity. Despite large capital

investments, it offers very few employment opportunities. A modern plant usually employs less than 150 people.

3. *Energy Intensive*: Each ton of cement produced requires 60–130 kg of fuel oil or its equivalent, depending on the cement variety and the process used, and about 110 kWh of electricity.

Alternative low-carbon technologies utilize fly ash, slag, and other materials instead of calcination of limestone. Worldwide, there are 780 million tons of fly ash, only half of which is utilized in some form. The worldwide production of blast furnace slag is about 400 million tons per year and steel slag is about 350 million tons per year. In Australia, for example, there are 14 million tons of fly ash and about 3 million tons of slag produced per annum in comparison with the 9 million tons of cement demand per annum.

There is a serious case for the construction industry to utilize low-carbon concrete to meet the additional demand rather than investing in new cement factories.

This book has collected some of the most recent advances in low-carbon concrete technologies. The first six chapters are related to low-carbon OPC concretes, where other reactive cementitious materials are substituted for OPC. The last nine chapters are related to geopolymer concrete, eco-friendly materials with much lower $CO_2$ emissions that are produced from industrial byproducts such as fly ash, slag, or metakaolin, which are considered as the main possible low-carbon alternatives to Portland cement concrete.

# CHAPTER 1

# Greenhouse Gas Emissions Due to Concrete Manufacture

**D.J.M. Flower[1] and J.G. Sanjayan[2]**
[1]Swinburne University of Technology, Melbourne, Australia
[2]Swinburne University of Technology, Hawthorn, VIC, Australia

## 1.1 INTRODUCTION

Concrete is the most widely used construction material. Current average consumption of concrete is about 1 t/year per every living human being. Human beings do not consume any other material in such tremendous quantities except for water. Due to its large consumption, even small reductions of greenhouse gas emissions per ton of manufactured concrete can make a significant global impact. This chapter presents a systematic approach to estimate carbon dioxide ($CO_2$) emissions due to the various components of concrete manufacture. Reliable estimates of greenhouse gas emission footprints of various construction materials are becoming important, because of the environmental awareness of the users of construction material. Life cycle assessment of competing construction materials (e.g., steel and concrete) [1] can be conducted before the type of material is chosen for a particular construction. This chapter provides greenhouse gas emissions data collected from typical concrete manufacturing plants for this purpose.

The basic constituents of concrete are cement, water, coarse aggregates, and fine aggregates. Extraction of aggregates has considerable land use implications [2]. However, the major contributor of greenhouse emissions in the manufacture of concrete is Portland cement. It has been reported that the cement industry is responsible for 5% of global anthropogenic $CO_2$ emissions [3]. As a result, emissions due to Portland cement have often become the focus when assessing the greenhouse gas emissions of concrete. However, as demonstrated by the data presented in this chapter, there are also other components of concrete manufacture that are responsible for greenhouse gas emissions that need consideration. With users beginning to require detailed estimates of the environmental impacts of the materials in new construction projects, this study was intended to provide the basis for a rating tool for concrete, based on $CO_2$ emissions.

*Handbook of Low Carbon Concrete.*
DOI: http://dx.doi.org/10.1016/B978-0-12-804524-4.00001-4

1

Other cementitious components considered include ground granulated blast furnace slag (GGBFS), a byproduct of the steel industry, and fly ash, a byproduct of burning coal. These two materials are generally used to replace a portion of the cement in a concrete mix. The use of water in concrete leads to minimal $CO_2$ emissions, which leaves cement, coarse and fine aggregates, GGBFS, and fly ash as the main material contributors to the environmental impacts of concrete. In addition to the production of materials, the processing components of concrete production and placement were considered. Transport, mixing, and in situ placement of concrete all require energy input leading to $CO_2$ emissions. Fig. 1.1 shows the system that was considered for this research.

The $CO_2$ emissions from most of the activities involved in concrete production and placement result from the energy consumed to accomplish them. Hence, to find the $CO_2$ emissions associated with an activity, the energy consumption per unit of material produced had to be audited. The exception to this rule is cement, where approximately 50% of the emissions are process based, due to the decomposition of limestone in the kiln with the remainder associated with kiln fuels and electricity [3,4]. Previous

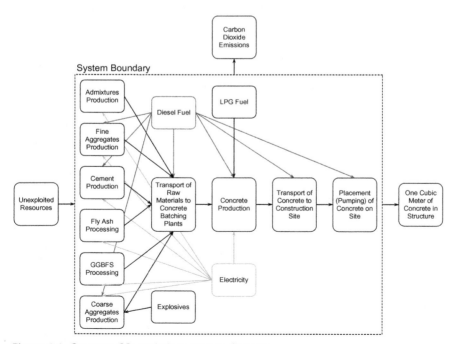

**Figure 1.1** Concrete $CO_2$ emissions system diagram.

research into the environmental impacts of cement production has already yielded several estimates of the $CO_2$ emissions per ton of cement produced. Similarly, fly ash and GGBFS have also both been investigated previously, and their emissions quantified. So the research that was conducted for this chapter covered the production of coarse and fine aggregates and admixtures, raw materials transport, concrete batching and transport, and onsite placement activities.

## 1.2 METHODOLOGY

The procedures used to calculate $CO_2$ emissions due to various energy sources in this study were obtained from the Australian Greenhouse Office Factors and Methods Workbook [5]. Table 1.1 shows the emission factors that were sourced from this publication. It should be noted that $CO_2$-e ($CO_2$ equivalents) are used as the unit, which is adjusted to include the global warming effects of any $CH_4$ or $N_2O$ emitted from the same fuel or process. These figures are appropriate for Melbourne, Australia, and may vary elsewhere around the world, due to differences in energy or fuel production methods. In 2004–05 the electricity mix in Melbourne was generated from brown coal (91.3%), oil (1.3%), gas (5.4%), hydro (1.4%), wind (0.5%), and biogas (0.1%) [6].

## 1.3 EMISSIONS DUE TO COARSE AGGREGATES

Data to estimate the $CO_2$ emissions due to the production of coarse aggregates was gathered from two quarries. The first produced granite and hornfels aggregates, and the second produced basalt aggregates. Note that the two quarries that were chosen for analysis were considered to be

**Table 1.1** Full fuel cycle $CO_2$ emission factors [5]

| Energy source | Emission factor | Unit |
|---|---|---|
| Diesel | 0.0030 | t $CO_2$-e/L |
| Electricity | 0.001392 | t $CO_2$-e/kWh |
| Riogel[a] | 0.1439 | t $CO_2$-e/ton product |
| Bulk emulsion[a] | 0.1659 | t $CO_2$-e/ton product |
| Heavy ANFO[a] | 0.1778 | t $CO_2$-e/ton product |
| LPG[b] | 0.0018 | t $CO_2$-e/L |

[a]Explosives.
[b]Liquefied petroleum gas.

typical examples. The production of both these types of coarse aggregates commences with the use of explosives to blast the rock from the quarry faces into medium-size boulders and rocks. Diesel-powered excavators and haulers then remove the rubble and dump it into electric crushing and screening equipment. Finally diesel-powered haulers move the final graded products into stockpiles. As part of this study two coarse aggregates quarries (basalt and granite/hornfels) were audited for energy consumption and total productivity over a 6-month period. This information was taken from fuel, electricity and explosives invoices, and site sales figures. The fuel, electricity, and explosives data was used to calculate the amount of $CO_2$ produced per ton of aggregate produced at each site. Using the emission factors presented in Table 1.1, $CO_2$ emissions per ton of granite/hornfels was found to be 0.0459 t $CO_2$-e/ton. $CO_2$ emissions per ton of basalt were found to be 0.0357 t $CO_2$-e/ton. These figures include the average contribution from transport from the quarry to the concrete-batching plants. Fig. 1.2 shows the contribution of each energy source.

Electricity is responsible for the majority of $CO_2$ emissions for each type of aggregate. This labels the crushing process as the most significant part of the coarse aggregates production process from an environmental perspective. Onsite blasting, excavation, and hauling, in addition to off-site transport, comprise less than 25% of the total emissions for coarse aggregates. It should be noted that while the explosives have very high emission factors, they contribute very small amounts (<0.25%) to the overall emissions, since such small quantities are used. To achieve significant environmental improvements in the production of aggregates, the crushing process needs to be targeted. Intelligent placement of explosives during the initial blasting process can reduce the demand on the electrical crushing equipment by blasting the rock into smaller fragments prior

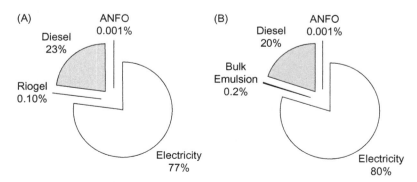

**Figure 1.2** (A) Basalt and (B) granite/hornfels $CO_2$ emissions breakdowns.

to crushing. Maintenance of crushing equipment is another way to lower electricity demands. Clearly the replacement of old, inefficient machinery will lead to lower energy demands.

## 1.4 EMISSIONS DUE TO FINE AGGREGATES

The fine aggregates investigated in this study begin as raw sand, which is strip mined by excavators and loaded into haulers. The haulers dump the sand where it is washed into a pumpable slurry that is piped to the grading plant. Electric vibrating screens filter the sand into standard grades, which are then stockpiled. One fine aggregates quarry was audited for energy consumption and total productivity over a 6-month period. The amount of $CO_2$ released during the production and subsequent transport of 1 t of concrete-sand was found to be 0.0139 t $CO_2$-e/ton. This is 40% of the figure for basalt coarse aggregate, and 30% of the figure for granite coarse aggregate. The lack of a crushing step explains the difference between the emissions of fine and coarse aggregates. Fig. 1.3 shows the contribution of each energy source to the $CO_2$ emissions associated with fine aggregates.

Diesel and electricity contribute almost equally to the $CO_2$ emissions from the production and transport of fine aggregates. The diesel is nearly all consumed by the strip mining and on-/offsite transport operations. The efficiency of these processes is largely dictated by the quality of the machinery being used. The replacement of aging excavators and haulers will lead to greater fuel efficiency, and hence lower $CO_2$ emissions. Electricity is consumed by the pumping and grading equipment. The emissions associated with these processes are largely fixed. Savings could be made by periodically relocating the screening plant closer to the source of the slurry, but the emissions associated with moving the equipment would need to be assessed before this course of action was taken. In general, the sand-mining process is fairly well established, and intentionally or otherwise, is already organized to generate minimal $CO_2$.

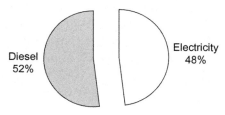

**Figure 1.3** Fine aggregates $CO_2$ emissions breakdown.

## 1.5 EMISSIONS DUE TO CEMENT, FLY ASH, GGBFS, AND ADMIXTURES

The environmental impacts associated with cement production have been investigated thoroughly in recent times [3,4,7–9]. Decomposition of limestone is an essential process in Portland cement production that takes place in the cement kiln. The chemical reaction for this process is $CaCO_3 \rightarrow CaO + CO_2$. This process releases 0.5 t of $CO_2$ for every ton of CaO produced. The high-energy consumption of the kiln produces additional $CO_2$ emissions, which are added to obtain the total emissions due to Portland cement manufacture. All of the figures for cement production in Australia lie around 0.8 t $CO_2$-e/ton, which is within the range of the other figures from around the world, which vary from approximately 0.7 to 1.0 t $CO_2$-e/ton [3,10,11]. The most recent and extensively researched figure was found to be that presented by Heidrich et al. [8], which was adopted for this project. The final emission factor that was used for cement in this project was 0.82 t $CO_2$-e/ton, which includes transport of cement to concrete-batching plants.

A part of the $CO_2$ emissions due to decomposition of limestone is reabsorbed from the atmosphere by concrete due to a chemical reaction called carbonation. The free lime, $Ca(OH)_2$, in the pores of the concrete, reacts with the atmospheric $CO_2$ and produces $CaCO_3$. This chemical reaction, $Ca(OH)_2 + CO_2 \rightarrow CaCO_3 + H_2O$, is what is commonly described as the carbonation of concrete. Sometimes it is mistakenly referred to as the reaction process involved in the hardening of concrete. Hardening of concrete is an entirely different reaction involving hydration of cement, which does not have any $CO_2$ implications. The carbonation of concrete structures only occurs near the surface of concrete. For a typical concrete structure, the carbonation depth would be about 20 mm from the surface after 50 years. Further, the major part of the CaO in cement is tied up as part of the hardened concrete in the form of calcium silicate hydrates that are not available for carbonation. Therefore, reabsorption of $CO_2$ by concrete during its lifetime would only be a very small proportion, and is not considered in the calculations in this chapter. Further discussions and estimates of $CO_2$ uptake by concrete can be found in Ref. [12].

The figures for the two supplementary cementitious materials (SCMs) considered in this study were also sourced from Heidrich et al. [8]. The emission factor adopted for fly ash was 0.027 t $CO_2$-e/ton. The emission factor adopted for GGBFS was 0.143 t $CO_2$-e/ton. Both fly ash and

**Table 1.2** $CO_2$ emissions associated with admixture manufacture

| Admixture type | Primary raw material | Production energy (kWh/L) | $CO_2$ emissions (t $CO_2$-e/L) |
|---|---|---|---|
| Superplasticizer | Polycarboxylate | 0.0037 | $5.2 \times 10^{-6}$ |
| Set accelerating | Calcium nitrate | 0.0380 | $53 \times 10^{-6}$ |
| Mid-range water reducing | Calcium nitrate | 0.0290 | $40 \times 10^{-6}$ |
| Water reducing | Lignin | 0.0016 | $2.2 \times 10^{-6}$ |

GGBFS are byproducts of industries (burning coal and producing steel, respectively) that would operate regardless of the production of these useful materials. So the emissions quoted here are based purely on activities conducted subsequent to initial production, including capture, milling, refining, and transport (100 km) processes.

Concrete often contains admixtures to enhance early age properties, such as the workability and strength-development characteristics. In this study, four different admixture types were considered, for which a large manufacturer supplied the typical figures presented in Table 1.2.

It can be seen that the $CO_2$ emissions associated with the manufacture of concrete admixtures are very small. The total volume of admixtures included in a typical mix design is generally less than $2 L/m^3$. Hence, the contribution to the total emissions per cubic meter of concrete is negligible. As a result of this, the $CO_2$ emissions generated by admixtures can justifiably be omitted from the calculations of the total $CO_2$ emissions of concrete.

## 1.6 EMISSIONS DUE TO CONCRETE BATCHING, TRANSPORT, AND PLACEMENT

Concrete batching is generally conducted at plants located at various strategic positions around a city or town to minimize transport time. Raw materials are mixed in elevated bins and placed directly into concrete trucks for final transport. This process is primarily powered by electricity, with small amounts of other fuels used on each site by small excavators used to move raw materials, etc. Over a 6-month summer/autumn period, the energy consumption and production levels of six different concrete-batching plants were audited. The average $CO_2$ emissions due to batching per cubic meter of concrete produced were found to be $0.0033 t$ $CO_2$-e/$m^3$. Fig. 1.4 shows the contributions of each energy source to the total $CO_2$ emissions.

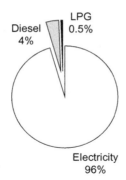

**Figure 1.4** Concrete-batching $CO_2$ emissions breakdown.

The electric mixing equipment is the most significant contributor to the emissions generated by concrete batching. It has been demonstrated in an internal review by a large concrete manufacturer that substantial improvements can be made to the efficiency of batching equipment by commissioning independent electrical contractors to report on the efficiency of batching equipment and making improvements. Aging equipment often contains inefficient wiring and switches. Often equipment is left running by old switching equipment during zero-load cycles when it could be paused. Thermal losses in poorly planned or low-quality wiring can be reduced by replacement. Installation of high-efficiency motors can reduce energy demands substantially. However, it should be noted that relative to other components of the concrete production process, the amount of $CO_2$ released through batching activities is fairly low, so it may be more critical to spend money on upgrading other more critical processes.

The transport of batched concrete consumes diesel fuel. Through trucking records taken over a 5-month period, the average amount of fuel consumed per cubic meter of concrete transported was found to be $3.1 \, \mathrm{L/m^3}$, which was found to be responsible for $0.009 \, \mathrm{t} \, CO_2\text{-e/m}^3$. Note that this figure includes empty return trips, since the total fuel consumption for the entire fleet of trucks was used. Since the trucking records included trucking to and from a wide range of construction sites and batching plants, it was assumed that the distances traveled were average for metropolitan concrete-transport activities.

Onsite placement activities such as pumping, vibrating, and finishing concrete consume liquid fuels. The amount of diesel consumed to pump $1 \, \mathrm{m}^3$ of concrete was found to be approximately $1.5 \, \mathrm{L/m^3}$, found by a survey of local pumping companies. The quantities of fuel consumed by other placement activities were impossible to accurately quantify, due to a lack of records and consistency between sites. Occasionally, concrete is craned into place instead of pumped, and this was also impossible to quantify. Hence, the original figure of $1.5\,L/m^3$ was doubled to account for all other placement activities. The final figure of $3\,L/m^3$ was assumed to be purely diesel fuel, and was found to be responsible for emissions of $0.009\,t$ $CO_2$-e/m$^3$. This is a conservative figure that is important since in very tall buildings, for example, the amount of fuel consumed by pumping could be higher than the average estimate, and the slack in this estimate allows room for such anomalies.

## 1.7 SUMMARY OF CO$_2$ EMISSIONS

The emissions associated with each activity in the concrete production and placement process were combined into a total figure based on mix design. The factors that were found are summarized in Table 1.3.

## 1.8 EMISSIONS GENERATED BY TYPICAL COMMERCIALLY PRODUCED CONCRETES

To investigate two of the methods by which the amount of $CO_2$ generated by concrete can be reduced, four mixes were selected with binders including SCMs. The first two mixes (25 and 32 MPa) have 25% of the general purpose (GP) cement replaced by fly ash. The second two mixes (25 and 32 MPa) have 40% of the GP cement replaced by GGBFS. These percentages are chosen because they are commonly used in construction

**Table 1.3** Final CO$_2$ emission factors

| Activity | Emission factor | Unit |
|---|---|---|
| Coarse aggregates: granite/hornfels | 0.0459 | t CO$_2$-e/ton |
| Coarse aggregates: basalt | 0.0357 | t CO$_2$-e/ton |
| Fine aggregates | 0.0139 | t CO$_2$-e/ton |
| Cement | 0.8200 | t CO$_2$-e/ton |
| Fly ash (F-type) | 0.0270 | t CO$_2$-e/ton |
| GGBFS | 0.1430 | t CO$_2$-e/ton |
| Concrete batching | 0.0033 | t CO$_2$-e/m$^3$ |
| Concrete transport | 0.0094 | t CO$_2$-e/m$^3$ |
| Onsite placement activities | 0.0090 | t CO$_2$-e/m$^3$ |

**Table 1.4** $CO_2$ emissions generated by typical commercially produced concretes

| Strength (MPa) | 100% GP cement | | 25% Fly ash | | 40% GGBFS | |
|---|---|---|---|---|---|---|
| | 25 | 32 | 25 | 32 | 25 | 32 |
| Emissions (t $CO_2$-e/m$^3$) | 0.290 | 0.322 | 0.253 | 0.273 | 0.225 | 0.251 |
| % $CO_2$ reduction | 0 | 0 | 13 | 15 | 22 | 22 |

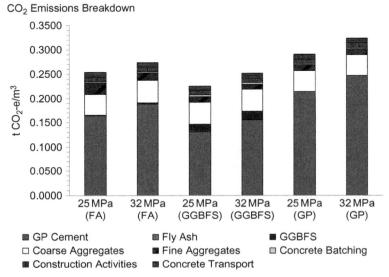

**Figure 1.5** $CO_2$ emissions generated by typical commercially produced concretes.

projects. It is noted that large cement replacements in lower-grade concretes such as these will not result in efficient construction because properties such as set time and early strength development can be affected. In addition, 25 and 32 MPa concretes are commonly used standard strengths. Table 1.4 and Fig. 1.5 show the results of this analysis, with two Type GP cement concretes as a benchmark.

Type GP cement is the dominant source of emissions in all of the concretes, blended or otherwise. The fly ash–blended concretes show reduced $CO_2$ emissions (13–15%), but it is the GGBFS-blended concretes that show more substantial reductions (22%). This is because more GGBFS is typically included in a blended mix without significantly changing the engineering properties of the concrete, due to its natural cementitious properties. So while GGBFS has a higher material-emission factor than fly ash, it can replace more cement, which leads to lower total emissions.

## 1.9 CASE STUDY: THE ROLE OF CONCRETE IN SUSTAINABLE BUILDINGS

The result of a design competition held in 2001 by the Victorian Office of Housing, the K2 public housing project in Melbourne, Australia, is an excellent example of innovative sustainable building design. The competition required the core structure to have a 200-year lifespan, generate renewable energy onsite, consume no nonrenewable energy, and halve mains water consumption [13]. In the final design, currently under construction, these requirements have been subjected to some interpretation, but generally they have all been achieved in some capacity. The winning design, by architects DesignInc Melbourne Pty Ltd, supported by engineering firm Arup, features four medium-rise buildings, with a total of 96 apartments suitable for public housing.

The main environmentally sustainable design (ESD) features of the design are: (1) maximized incident solar energy through building orientation, (2) passive ventilation through building orientation and apartment design, (3) photovoltaic cells for onsite renewable energy generation, (4) strategic placement of thermally massive materials for energy storage, (5) strategic placement of insulation to prevent unwanted energy migration, (6) use of low embodied-energy materials (structural and otherwise), (7) solar-powered hydronic heating for extreme winter weather, (8) gray- and storm-water recycling, and (9) water-efficient appliances and fittings.

Based on predictive models compiled by the design team, it can be estimated that the probable annual operational energy consumption at K2 (lighting, elevators, hot water, and appliances) will be approximately 1000 MWh, depending on ongoing tenant education and choice of appliances [14]. This energy is sourced from both electricity and natural gas. Note that depending on the uptake of tenant education, annual operational energy could be substantially lower than this, under the predicted best-case scenario. The most probable energy-consumption scenario was used for this case study.

When predicting the total operational energy consumption over the lifespan of the structure, it is appropriate to consider only 100 years of operation. Due to the demographic changes expected over 100 years, the purpose of the structure may change after that period of time. Hence, it is expected that a major refit will be required after 100 years. According to the design team's probable development scenario, 1000 MWh/year equates to approximately 850 t $CO_2$-e/year including contributions from both gas

**Table 1.5** Concrete volumes and target embodied-energy $CO_2$ emissions

| Strength (MPa) | Structural element | Quantity (m³) | Emission factor (t $CO_2$-e/m³) | Emissions (t $CO_2$-e) |
|---|---|---|---|---|
| 15 | Blinding | 589 | 0.20 | 119 |
| 32 | Footings | 489 | 0.24 | 119 |
| 32 | Slabs | 1948 | 0.27 | 533 |
| 40 | In situ columns and walls | 235 | 0.27 | 63 |
| 40 | Precast walls | 1067 | 0.33 | 351 |
| | | | | 1185 |

and electricity. Hence, over 100 years, building operations will generate a total of approximately 85,000 t $CO_2$-e.

This figure is based on the current electricity and gas emission factors. However, the methods of electricity generation in Melbourne may change substantially over the next 100 years, from burning brown coal to more sustainable techniques.

It is now interesting to investigate the initial material-based $CO_2$ emissions associated with concrete. There are a range of other sources of initial $CO_2$ emissions at K2, including glass, steel, aluminum, photovoltaics, and fitout materials; however, this investigation will focus on concrete alone.

Based on the K2 bill of quantities and the component emission factors outlined earlier, the volumes of concrete and associated target $CO_2$-equivalent emissions shown in Table 1.5 were found. On average across the whole structure, the design target is to replace 30% of total Portland cement with fly ash. Note that this target has not yet been achieved, since construction is not complete.

To quantify the target $CO_2$ savings that will be made by substituting fly ash for some of the cement content of the concrete, a similar investigation was performed using mix designs containing only pure GP cement. The total $CO_2$-equivalent emissions generated by the pure GP cement–based concretes were found to be 1391 t $CO_2$-e. Hence the target savings that will be made by replacing a portion of the GP cement with fly ash are approximately 206 t $CO_2$-e.

According to the design team's estimated energy consumption as described earlier, the yearly $CO_2$ emissions associated with building operations will be approximately 850 t $CO_2$-e/year. Over the 100-year building lifespan, the $CO_2$ emissions generated by the structural concrete will be less than 1.4% of the emissions associated with operation, assuming all design targets are met.

Furthermore, as a result of the energy-efficient design of K2, the most probable estimated energy consumption is already expected to be reduced by 57% [14], or approximately $1125\,t\ CO_2$-e/year. So by designing the building with passive energy measures and educating the tenants to minimize energy consumption, the target tonnage of $CO_2$-e that will be saved per year will be over five times greater than that predicted to be saved initially by the use of fly ash in the structural concrete. Hence, over 100 years, the tonnage of $CO_2$-e that will be saved due to the efficient building design will be approximately 500 times greater than that estimated to be saved by the use of fly ash in the structural concrete, again assuming all design targets are met.

This case study shows that passive design measures, which enhance the operational energy performance of a building, have the potential to make a greater impact on the overall greenhouse gas emissions of a building than using fly ash substitution in concrete mix designs. However, the short-to-medium-term benefits of using low embodied-energy concretes are still significant and valuable. It is worth noting that using fly ash in structural concrete results in accurately quantifiable capital $CO_2$ savings. Passive energy measures have the capacity to be more effective in the long term, but depend on a large number of variables, such as tenant behavior, which can be difficult to control.

This case study also shows that for comparison of $CO_2$ emissions of alternative construction materials such as steel with concrete, the emissions associated with concrete should be considered rather than just the cement component alone, since emissions due to cement are only part of the concrete emissions, albeit a significant part.

## 1.10 CONCLUSIONS

While there have been many studies conducted to estimate the $CO_2$ emissions due to Portland cement manufacture, very few reliable estimates are available for the emissions due to concrete manufacture. The figures for the emissions for two types of coarse aggregates, fine aggregates, cement, fly ash, slag, concrete batching, and transport have been developed based on a large number of records obtained from aggregates quarries, concrete-batching plants, and other sources. Although the data presented above was collected from locations around Melbourne, Australia, it can be used as a guide to estimate the emissions due to concrete production and placement in other parts of the world with similar production methods.

The following conclusions can be drawn from the data collected in this study:

1. The equivalent $CO_2$ emissions generated by a particular concrete with known mix proportions can be estimated using the emissions contributions from the constituents of concrete.

2. Portland cement was found to be the primary source of $CO_2$ emissions generated by typical commercially produced concrete mixes, being responsible for 74–81% of total $CO_2$ emissions.

3. The next major source of $CO_2$ emissions in concrete was found to be coarse aggregates, being responsible for 13–20% of total $CO_2$ emissions.

4. The majority contribution of $CO_2$ emissions in coarse aggregates production was found to from electricity, typically about 80%. Blasting, excavation, hauling, and transport comprise less than 25%. While the explosives have very high emissions, they contribute very small amounts (<0.25%) to coarse aggregate production, since only small quantities are used.

5. Production of a ton of fine aggregates was found to generate 30–40% of the emissions generated by the production of a ton of coarse aggregates. Fine aggregates generate less equivalent $CO_2$ since they are not crushed.

6. Diesel and electricity were found to contribute almost equally to the emissions due to fine aggregates.

7. Emission contributions due to admixtures were found to be negligible.

8. Concrete-batching, transport, and placement activities were all found to contribute very small amounts of $CO_2$ to total concrete emissions.

9. The $CO_2$ emissions generated by typical normal strength concrete mixes using Portland cement as the only binder were found to range between 0.29 and 0.32 t $CO_2$-e/m$^3$.

10. GGBFS was found to be capable of reducing concrete $CO_2$ emissions by 22% in typical concrete mixes.

11. Fly ash was found to be capable of reducing concrete $CO_2$ emissions by 13–15% in typical concrete mixes.

12. The target $CO_2$ emissions due to the structural concrete at the sustainable apartment complex considered as a case study will form less than 1.4% of the estimated probable total lifetime $CO_2$ emissions generated by the building. Note that the award-winning design of this particular building is estimated to reduce operational energy consumption by 57% under the most probable operational scenario

compared to a typical conventional apartment building of comparable size designed without any ESD features.

13. The case study showed that passive design measures, which enhance the operational energy performance of a building, have the potential to make a greater impact on the overall greenhouse gas emissions of a building than using fly ash substitution in concrete mix designs.

## 1.11 RECOMMENDATIONS AND PERSPECTIVES

The various rating schemes used to compare alternative construction materials should use models that are based on hard data, such as the one presented in this chapter, so that reliable comparisons can be made. A case study is presented in the chapter demonstrating how the results may be utilized.

## ACKNOWLEDGMENTS

Work on this project was conducted at Monash University with support from Rinker Australia under R&D Project RD849. Thanks are expressed to Dr. Daksh Baweja, Jacques Teyssier, Damian Hope, Paul Rocker, and Joshua Choong from Readymix Holdings for their valuable assistance during the data-collection phase of this project. Thanks are also expressed to John MacDonald and Jennifer Dudgeon from DesignInc Melbourne, and Malcolm Barr and Kate West from Arup for their valuable assistance during the preparation of the case study.

## REFERENCES

[1] Jönsson Å, Björklund T, Tillman A-M. LCA of concrete and steel building frames. Int J Life Cycle Assess 1998;3(4):216–24.

[2] Schuurmans A, Rouwette R, Vonk N, Broers J, Rijnsburger H, Pietersen H. LCA of finer sand in concrete. Int J Life Cycle Assess 2005;10(2):131–5.

[3] Humphreys K, Mahasenan M. Toward a sustainable cement industry, substudy 8, climate change. : World Business Council for Sustainable Development; 2002.

[4] CIF Cement industry environment report. : Cement Industry Federation; 2003.

[5] Commonwealth of Australia. Australian Greenhouse Office factors and methods workbook. Australian Greenhouse Office; 2004.

[6] Cuevas-Cubria C, Riwoe D. Australian energy: national and state projections to 2029–30. : Australian Bureau of Agricultural and Resource Economics; 2006.

[7] Collins F, Sanjayan JG. The challenge of the cement industry federation towards the reduction of greenhouse emissions, towards a better built environment—innovation, sustainability, information technology. Proceedings of the international association of bridge and structural engineers symposium. Melbourne; 2002.

[8] Heidrich C, Hinczak I, Ryan B. SCM's potential to lower Australia's greenhouse gas emissions profile Iron and steel slag products: a significant time of scarcity. Sydney: Australasian Slag Association Conference; 2005.

[9] John VM. On the sustainability of concrete. UNEP Ind Environ 2003:62–3.

[10] Gartner E. Industrially interesting approaches to "low $CO_2$" cements. Cement Concrete Res 2004;34:1489–98.

[11] Josa A, Aguado A, Heino A, Byars E, Cardim A. Comparative analysis of available life cycle inventories of cement in the EU. Cement Concrete Res 2003;34:1313–20.

[12] Pade C, Guimaraes M. The $CO_2$ uptake of concrete during a 100 year perspective Proceedings of advances in cement and concrete X–sustainability. Davos, Switzerland: Engineering Conferences International; 2006:114–19, July, 2006.

[13] Office of Housing internet site, http://hnb.dhs.vic.gov.au/ooh/oohninte.nsf/frameset/Ooh?Opendocument [accessed 09.08.05], last updated 20 July 2005.

[14] Personal communication with John MacDonald, DesignInc Melbourne Pty Ltd; 15 July 2005.

# CHAPTER 2

# Life Cycle $CO_2$ Evaluation on Reinforced Concrete Structures With High-Strength Concrete

**S. Tae[1], C. Baek[2] and S. Roh[1]**
[1]Hanyang University, Ansan, Republic of Korea
[2]Korea Institute of Civil Engineering and Building Technology, Goyang, Republic of Korea

## 2.1 INTRODUCTION

Global warming, resource depletion, and pollution are causing many countries to adopt environmentally friendly policies. According to the report of the Environmental Protection Agency, energy consumption and $CO_2$ emissions of buildings in the United States are responsible for 70% of the entire energy consumption and 38% of the entire $CO_2$ emissions of the country [1,2]. Construction is an environmentally demanding industry requiring mass consumption and disposal. Architectural production activities should focus on sustainable development to reduce the environmental load of design, construction work, maintenance, and disposal [3–5]. Under the World Trade Organization system, international organizations such as the United Nations, Organization for Economic Cooperation and Development, and International Organization for Standardization have considered techniques to reduce global warming, create environmentally sound and sustainable practices, and set up compulsory regulations for environmental load reduction [6,7].

Skyscrapers have been constructed more frequently since the early 2000s due to their increased land efficiency and recent progress in modern construction techniques, and recently a considerable amount of attention has been paid to environmentally sound and sustainable "green" buildings. Skyscrapers are advantageous for supporting broad greens and open space, and reducing the building-to-land ratio. Their weak points include lack of social contact and ground connections, and difficulty with natural ventilation. Research and development under the principle of environmentally sound and sustainable development is now firmly established as an international paradigm [8–10].

*Handbook of Low Carbon Concrete.*
DOI: http://dx.doi.org/10.1016/B978-0-12-804524-4.00002-6

Hence, the purpose of this study is to evaluate the environmental performance of buildings by the application of high-strength concrete, mainly used in supertall buildings as a material of environmental stress reduction (hereinafter "high-strength concrete building").

This study proposed a plan for the evaluation of energy consumption amount and $CO_2$ emission amount throughout the life cycle of building, and calculated energy consumption amount and $CO_2$ emission amount throughout the life cycle of a tall apartment building actually constructed (hereinafter "existing building") by using this plan.

Thereafter, this study evaluated energy consumption and $CO_2$ emission–reduction performance for the life cycle of the building by the decrease of concrete and reinforcing-bar quality obtained through conversion from the existing building's concrete compressive strength to 40 MPa high-strength concrete.

## 2.2 METHOD OF EVALUATING ENVIRONMENTAL LOAD FOR THE LIFE CYCLE OF BUILDING

This study assessed the environmental load of a structure through its life cycle with stages classified into construction, use/maintenance, and removal/disposal. Construction included material production, transportation, and construction work on the site. Interindustry relations analyses were carried out to measure the $CO_2$ released during material production. Use/maintenance was divided into use of a building and its maintenance steps, and it was analyzed by considering the assessment period and the life of the building, based on the annual energy consumption. Removal/disposal was divided into removal of a structure and disposal of the removed wastes [11]. Fig. 2.1 and Table 2.1 show the classification of buildings by the method of environmental load evaluation and evaluation items for the life cycle of the building proposed in this study, respectively. In addition, Eqs. (2.1) and (2.2) show the calculations of energy consumption and $CO_2$ emission for the life cycle of the building [12].

$$LCE = \sum E_{ij} \tag{2.1}$$

$$LCCO_2 = \sum CO_{2ij} \tag{2.2}$$

where, LCE is life cycle energy consumption $(MJ/m^2)$, $LCCO_2$ is life cycle $CO_2$ emission $(kg\text{-}CO_2/m^2)$, $E_{ij}$ is life cycle energy consumption $(MJ/m^2)$ for each stage (i) and material (j), $CO_{2ij}$ is life cycle $CO_2$

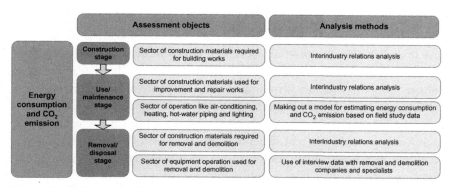

**Figure 2.1** Method of environmental load assessment.

**Table 2.1** Classification of environmental load assessment

| Stage | Classification | Subclassification |
|-------|----------------|-------------------|
| 1. Construction stage | 1. Material production step | ① Construction work ② Public work ③ Facility work |
| | 2. Transportation step | ① Transportation |
| | 3. Construction work step | ① Construction work ② Public work ③ Landscaping work ④ Power consumption |
| 2. Use/maintenance stage | 1. Use step | ① Power consumption ② Heating energy ③ City gas consumption |
| | 2. Maintenance step | ① Improvement and repair |
| 3. Removal/disposal stage | 1. Removal step | ① Removal |
| | 2. Disposal step | ① Loading ② Returning |

emission for each stage (i) and material (j), and $E = 1$: construction stage (1-1 = material production step, 1-2 = transportation step, 1-3 = construction work step), 2: use/maintenance stage (2-1 = use step, 2-2 = maintenance step), 3: removal/disposal stage (3-1 = removal step, 3-2 = disposal step).

## 2.2.1 Construction Stage

The construction stage, which generally makes up 30% of the building LCCO$_2$ emissions, was divided into three steps: material production, transportation, and construction work [13]. The material production step ranged from gathering raw materials to producing building materials to be used in the construction work stage. The transportation step refers to the energy consumption by freight vehicles transporting building materials to the construction sites. The construction work step refers to the energy consumption by construction machinery, field offices, and other facilities from starting construction to construction completion. The construction stage was also divided into three kinds of work: construction work, public work, and facility work. The energy consumption and $CO_2$ emissions were determined for each kind of work. The construction work included 17 types of subwork, including temporary, pile, reinforced concrete, masonry, waterproofing, tile, stone, and steel works. The public work was composed of three types of subwork: a retaining wall and waterproofing, pile, and appurtenant public works. The facility work included 17 types of subworks including facility, piping of machine rooms, and gas piping works.

The energy consumption and $CO_2$ emissions during construction material production were calculated by applying a unit of a construction material, which was drawn by interindustry relations analyses, to the material volume to be used for buildings [14].

### 2.2.1.1 Material Production Step

Energy consumption and $CO_2$ emission for the production of each construction material are computed, as described above, based on the interindustry relation analysis.

The material production step is the stage of calculating the $CO_2$ emission and energy amount consumed to produce the construction materials used in building construction. The energy consumption and $CO_2$ emission to produce each construction material are based on interindustry relation analysis as mentioned above. Therefore, through identification of the material quantity put into the construction of buildings, the energy consumption and $CO_2$ emission in the production of the construction materials used in the building intended for evaluation can be calculated.

$$E_{C\text{-}M} = \sum M_{ij} \cdot COST_m \cdot U_{M,E} \qquad (2.3)$$

$$CO_{2C\text{-}M} = \sum M_{ij} \cdot COST_m \cdot U_{M,CO2} \qquad (2.4)$$

where $E_{C\text{-}M}$ (MJ/m$^2$) is energy consumption of the material production step, $M_{ij}$ (Unit/m$^2$) is the amount of construction material (j) used for the construction type (i), $COST_m$ (Won/Unit) is the cost of construction material (m), $U_{M,E}$ (Mj/Won) is the energy consumption factor for construction material (m), $CO_{2C\text{-}M}$(kg-$CO_2$/m$^2$) is $CO_2$ emission of the material production step, and $U_{M,CO2}$(kg-$CO_2$/Won) is the $CO_2$ emission factor for construction material (m).

### 2.2.1.2 Transportation Step

Energy consumption and CO$_2$ emission of the transportation step can be computed based on transportation method, transportation distance, load on the transportation vehicle, and expenditure of oil and power used for transportation. However, records on equipment use are often omitted in construction diaries furnished at construction sites, and in many cases it is difficult to secure sufficient data because of conditions at the site [15]. Therefore, this study used Eqs. (2.5) and (2.6) to compute energy consumption and CO$_2$ emission for transportation. Eqs. (2.5) and (2.6) calculate energy consumption and CO$_2$ emission for the transportation of materials used in general construction work. Transportation distance is based on a travel distance of 30 km, a value suggested by the Korean Ministry of Environment for an energy consumption factor and CO$_2$ emission factor for transportation equipment.

$$E_{C\text{-}T} = 104.6 \, \text{MJ/m}^2 \tag{2.5}$$

$$CO_{2C\text{-}T} = 7.4 \, \text{kg-CO}_2/\text{m}^2 \tag{2.6}$$

where $E_{C\text{-}T}$ is energy consumption (MJ/m$^2$) of transportation step and $CO_{2C\text{-}T}$ is CO$_2$ emission (kg-CO$_2$/m$^2$) of the transportation step.

### 2.2.1.3 Construction Work Step

This study is programmed to use Eqs. (2.7)–(2.12) for the computation of energy consumption and CO$_2$ emission of the construction work step by classifying construction work, engineering work, landscape work, and electricity use. Existing studies that proposed Eqs. (2.7)–(2.12) classified the energy source of construction work into oil and electricity [14]. In addition, oil consumption for each construction item was computed by analyzing the data on oil consumption per unit of machine time. Power consumption, as shown in Eq. (2.12), was computed by investigating the actual power expended during construction. Energy consumption computation results can be substituted into Eqs. (2.7) and (2.8) to compute the CO$_2$ emission of the construction work step.

$$E_{C\text{-}C} = E_{ca} + E_{cc} + E_{cl} + E_{ce} \tag{2.7}$$

$$CO_{2C\text{-}C} = (E_{ca} + E_{cc} + E_{cl}) \times 3.06 + E_{ce} \times 1.64 \tag{2.8}$$

$$E_{ca} = 95.13 \, \text{MJ/m}^2 \tag{2.9}$$

$$E_{cc} = 15.29 \, \text{MJ/m}^2 \tag{2.10}$$

$$E_{cl} = 3.04\,\text{MJ}/\text{m}^2 \tag{2.11}$$

$$E_{ce} = 100.71\,\text{MJ}/\text{m}^2 \tag{2.12}$$

where $E_{C\text{-}C}$ is the energy consumption ($\text{MJ}/\text{m}^2$) of the construction work step and $CO_{2C\text{-}C}$ is the $CO_2$ emission (kg-$CO_2$/$\text{m}^2$) of the construction work step, $E_{ca}$ is the energy consumption factor ($\text{MJ}/\text{m}^2$) for the construction type during construction work, $E_{cc}$ is the energy consumption factor ($\text{MJ}/\text{m}^2$) for civil construction, $E_{cl}$ is the energy consumption factor ($\text{MJ}/\text{m}^2$) for landscape construction, $E_{ce}$ is the electric power consumption factor ($\text{MJ}/\text{m}^2$) during construction work.

## 2.2.2  Use/Maintenance Stage

This stage, which makes up about 70% of the building LCCO$_2$ emissions, considers the $CO_2$ emissions due to energy consumed during the service life of the building. Energy sources used for air conditioning, lighting, and cooking were classified into electric, heating, and gas energy. Heating energy was divided into district, central, and individual heating in terms of heating method and into LPT, heavy oil, light oil, and kerosene by heating source type. The total emission of $CO_2$ was calculated by adding the $CO_2$ emission from each type of source. This study presently sets the heating type ratios as follows: city gas (65%), heavy oil (25%), incineration heat (9%), and light oil (1%). The ratio can be configured by users. Eqs. (2.13) and (2.14) show the energy consumption and CO2 emission in the use step.

Energy consumption and $CO_2$ emission in the maintenance step can be calculated, based on the life of a structure, by using data on repaired and replaced volumes of building materials due to wear-out, damage, and destruction, and by using breakdowns of oil and electric power for repair and replacement work. This study sets the energy consumption and $CO_2$ emission used in the maintenance step as 6.24 MJ/$\text{m}^2$/year and 0.59 kg-$CO_2$/$\text{m}^2$/year, respectively.

$$E_U = (E_{ue} + E_{uh} + E_{ug}) \cdot Y \tag{2.13}$$

$$CO_{2U} = (E_{ue} + E_{uo} + E_{ug}) \cdot Y \cdot U_{E,CO2} \tag{2.14}$$

$$E_M = 6.24\,\text{MJ}/\text{m}^2/\text{year} \cdot Y \tag{2.15}$$

$$CO_{2M} = 0.59\,\text{kg-}CO_2/\text{m}^2/\text{year} \cdot Y \tag{2.16}$$

Here, $E_U$ and $CO_{2U}$ indicate the energy consumption amount (MJ/m$^2$) and $CO_2$ emission (kg-CO$_2$/m$^2$) of each use step. $E_{ue}$ indicates the amount of electric energy used (MJ/m$^2$/year), $E_{uh}$ the amount of heating energy used (MJ/m$^2$/year), $E_{ug}$ the amount of city gas used (MJ/m$^2$/year), $U_{E,CO2}$ the $CO_2$ emission factor by the consumed energy sources (kg-CO$_2$/MJ), and $Y$ the number of years the building is used (year). $E_M$ and $CO_{2M}$ indicate energy consumption (MJ/m$^2$) and $CO_2$ emission (kg-CO$_2$/m$^2$).

### 2.2.3 Removal/Disposal Stage

The removal/disposal stage included the removal and disposal of buildings at the end of their life cycle, or replacement and transportation to handle building waste. The recycling of building wastes was left for future consideration. The objects of analyses were materials and equipment used for removal, vehicles, oil and electric power required for the transportation of waste. Disposal included loading and returning, and the energy consumption of a vehicle's return trip to a waste-generating place from a disposal area was assumed to require half of the energy of the loading case.

$$E_R = 1.07\,\text{MJ/m}^2 \tag{2.17}$$

$$CO_{2R} = 0.0734\,\text{kg-CO}_2/\text{m}^2 \tag{2.18}$$

$$E_D = O_i \cdot U_{O,E} \tag{2.19}$$

$$CO_{2D} = O_i \cdot U_{O,CO2} \tag{2.20}$$

Here, $E_R$ and $CO_{2R}$ respectively indicate the energy consumption (MJ/m$^2$) and $CO_2$ emission (kg-CO$_2$/m$^2$) of the removal step. $E_D$ and $CO_{2D}$ respectively indicate the energy consumption (MJ/m$^2$) and $CO_2$ emission (kg-CO$_2$/m$^2$) of the disposal step. $O_i$ is the amount of oil used by transporting vehicle by (i) the wastes, and $U_{O,E}$ and $U_{O,CO2}$ the energy consumption factor (MJ/L) and $CO_2$ emission factor (kg-CO$_2$/L) of oil, respectively.

## 2.3 EVALUATING ENVIRONMENTAL LOAD BY THE APPLICATION OF HIGH-STRENGTH CONCRETE

This chapter evaluated energy consumption amount and $CO_2$ emission–reduction effects by the application of high-strength concrete on the tall apartment building actually constructed based on Section 2.2, Method of Evaluating Environmental Load for the Life Cycle of Building.

The target building for evaluation is a tall apartment building made of reinforced concrete with a total floor area of $14,424 \, m^2$, 35 stories above the ground, which had its construction completed in May 2004.

In this chapter, the life cycle of the case building was classified into construction stage, use/maintenance stage, and removal/disposal stage to calculate energy consumption and $CO_2$ emission, and the energy consumption and $CO_2$ emission of each stage were calculated.

In addition, we selected the concrete compressive strength that can support more than a 100-year lifespan in an urban environment (carbonized environment) and applied it to the case building, and evaluated the effects of the reinforcing bar and concrete reduction and lifespan extension on the environmental load.

Table 2.2 shows the evaluation conditions of the existing building and high-strength concrete building. In Table 2.2, the evaluation period was set to 100 years, which is the lower limit of the lifespan of the high-strength concrete structure set in this study. The evaluation condition is at level 3 of cases 1–3. Case 1 is the case in which the 100-year evaluation period is reached through the reconstruction work step, after reaching 50 years of its lifespan without a repair process on the carbonization of the existing building. On the other hand, case 2 is the case where the lifespan of the existing building is extended up to 100 years through maintenance

**Table 2.2** Overview of the building assessed

| | |
|---|---|
| Building overview | • Apartment with 35 stories above ground<br>• RC structure<br>• Total area: $14,424 \, m^2$<br>• Building-to-land ratio: 59.22% |
| Assessment conditions | • Assessment period: 100 years<br>• Service life: Case 1: Existing building (50 years)<br>Case 2: Existing building renovation (50 years)<br>Case 3: High-strength concrete building (100 years) |
| Concrete compressive strength | • 27–35 stories: 24 MPa<br>• 20–26 stories: 27 MPa<br>• 10–19 stories: 30 MPa<br>• 1–9 stories: 35 MPa |
| Minimum concrete-cover thickness | • 20 mm |
| Deterioration environment | • Carbonation environment |

of structure members that were deteriorated by carbonization, and case 3, the case of the high-strength concrete building, is the case in which the 100-year evaluation period can be reached without a maintenance of structural members by carbonization.

### 2.3.1 Evaluation Method

#### 2.3.1.1 Materials Production Step

The energy consumption and $CO_2$ emission of the material production stage were calculated as the sum of the amount of energy consumed and that of $CO_2$ emitted in the process of production of the construction materials used in building construction. The construction materials used in building construction were identified by obtaining the actual quantity sheet of existing apartment houses, and for the energy consumption factor and $CO_2$ emission factor of each construction material, the energy consumption factor and $CO_2$ emission factor calculation results deduced with the use of the 2003 interindustry relation analysis of Korea were applied [13]. Table 2.3 shows the energy consumption factor and $CO_2$ emission factor for the main construction material applied in this chapter.

Specifically, case 1, the case where reconstruction is done after removal/disposal without maintenance when the building reaches 50 years of its limit lifespan by the carbonization phenomenon, calculated the $CO_2$ of the construction work step provided that construction is done twice within 100 years of the average period. Also, case 2, the case where lifespan is extended up to 100 years through maintenance of the concrete deteriorated by carbonization, gains one session of construction work step within 100 years of the average period, but with the amount of $CO_2$ by the use of additional construction materials used in the maintenance stage added to the amount of $CO_2$ emitted during the maintenance stage. On the other hand, case 3, the high-strength concrete building having 100 years of lifespan, gains one session of construction work step within 100 years of the average period, and was evaluated as not having any additional $CO_2$ emission by maintenance.

#### 2.3.1.2 Transportation Step

The energy consumption and $CO_2$ emission during transportation step were calculated with the use of Eqs. (2.5) and (2.6). While Eqs. (2.5) and (2.6) are calculation equations that can evaluate the amount of energy consumption and that of $CO_2$ emitted during the transportation step with the use of the total floor area of the case building for evaluation, cases 1–3

**Table 2.3** Energy consumption factor and $CO_2$ emissions factor for main construction materials

| No. | Article | Interindustry analysis (domestic and overseas) | |
|-----|---------|---------------------------------------|----|
| | | **Amount of energy consumption** | **Amount of $CO_2$ emission** |
| 1 | REMICON | 2420.993 MJ/m$^3$ | 186.493 $CO_2$-kg/ m$^3$ |
| 2 | Deformed iron bar | 35.300 MJ/kg | 3.052 $CO_2$-kg/kg |
| 3 | Waterproof plywood | 22.574 MJ/kg | 1.516 $CO_2$-kg/kg |
| 4 | Rectangular lumber | 17.885 MJ/kg | 1.216 $CO_2$-kg/kg |
| 5 | Wire | 90.953 MJ/kg | 6.813 $CO_2$-kg/kg |
| 6 | Nail | 61.512 MJ/kg | 4.607 $CO_2$-kg/kg |
| 7 | Concrete brick | 2.679 MJ/each | 0.206 $CO_2$-kg/ each |
| 8 | Concrete block | 21.886 MJ/kg | 1.683 $CO_2$-kg/kg |
| 9 | Concrete tile | 15.575 MJ/kg | 1.197 $CO_2$-kg/kg |
| 10 | Cement | 6.916 MJ/kg | 0.556 $CO_2$-kg/kg |
| 11 | Sand | 72.936 MJ/m$^3$ | 5.033 $CO_2$-kg/m$^3$ |
| 12 | Gravel | 70.537 MJ/m$^3$ | 4.868 $CO_2$-kg/m$^3$ |
| 13 | Foamed polystyrene board | 140.014 MJ/kg | 10.229 $CO_2$-kg/kg |
| 14 | PVC ceiling panel | 99.947 MJ/kg | 7.302 $CO_2$-kg/kg |
| 15 | PVC rain leader pipe | 708.421 MJ/kg | 50.872 $CO_2$-kg/kg |
| 16 | Water-based paint | 652.690 MJ/kg | 48.017 $CO_2$-kg/kg |
| 17 | Ceramic paint | 489.419 MJ/kg | 36.005 $CO_2$-kg/kg |

applied the same total floor area of 14,424 m$^2$. However, since there are two sessions of construction work steps, case 1 was evaluated as having double the energy consumed and $CO_2$ emitted during the transportation stage in comparison with cases 2 and 3 during one session of construction work step.

### 2.3.1.3 Construction Work Step

The energy consumption and $CO_2$ emission during the construction work step were calculated with the use of Eqs. (2.7)–(2.12). Eqs. (2.7)–(2.11) calculate the amount of oil used in the construction work step with the use of the total floor area and apartment complex area of the case building for evaluation, and Eq. (2.12) means the amount of electric energy used during the construction work step with the use of total floor area. The energy consumption and $CO_2$ emission were calculated by substituting such deduced amounts of oil and electric energy used into Eqs. (2.7) and (2.8). The total floor areas and apartment complex areas of cases 1–3, 14,424 and 3291 m$^2$, respectively, were evaluated to be the same. However, since case 1 involves two sessions of construction steps just as with the construction work step, this case was evaluated to have double the energy consumption and CO$_2$ emission of the construction work step in comparison with cases 2 and 3, which include only one session of the construction work step.

### 2.3.1.4 Use Step

The energy consumption and CO$_2$ emission during the use step of cases 1–3 were calculated by obtaining the data on the amount of energy used for years in practice in existing apartment houses. The energy sources of existing apartment houses were classified into electric power, heating energy, and city gas, and heating was evaluated provided that city gas, heavy oil, incineration heat, and light oil are used at 65%, 25%, 9%, and 1%, respectively, in the local heating method. For the amount of energy used for years in cases 1–3, the energy consumption for years taken above was applied in the same way, and the total amount of energy for the average 100-year period was evaluated assuming that the surveyed amount of energy per year used in the apartment house in practice has been the same for 100 years.

### 2.3.1.5 Maintenance Step

The energy consumption and CO$_2$ emission during the maintenance step in cases 1–3 were calculated based on Eqs. (2.13) and (2.14), respectively. However, in case 2, the amount of CO$_2$ emitted by concrete used in the maintenance step of the structure deteriorated by carbonization within the average 100-year period was added in the calculation. At this time, only the amount of CO$_2$ emitted in the concrete-production step was added in the calculation, and the amount of CO$_2$ occurring in the transportation and construction work steps, being judged as negligible, were excluded in this calculation process.

### 2.3.1.6 Removal Step and Disposal Step

The energy consumption and CO$_2$ emission during the removal steps of cases 1–3 were calculated with the use of Eqs. (2.15) and (2.16). If Eqs. (2.15) and (2.16) are used, the energy consumption and CO$_2$ emission during the removal process can be calculated by substituting the total floor area. In the meantime, the removal step can be divided into loading and returning, and in case the vehicle returns to the place where construction wastes occur to load again after transporting construction wastes to the landfill, it was assumed that half the amount of energy consumed during loading is consumed. The amount of energy used during loading in the construction step was calculated by deducing the amount of light oil consumed in transporting the construction wastes that occurred after removal to the landfill in a vehicle. At this time, the amount of construction wastes that occurred was set to $3.5\,m^3/m^2$, and one-way transportation distance to 30 km. In addition, with a transportation vehicle assumed to be a 16-t dump truck, fuel consumption rate and loading weight were respectively set to $2.74\,km/L$ and $20\,m^3$. Eqs. (2.17) and (2.18) calculate the energy consumption and $CO_2$ emission during loading.

## 2.3.2 Selection of High-Strength Concrete

The minimum value of concrete strength that can guarantee at least 100 years of lifespan in an urban environment (carbonation environment) was selected as the target. The use of high-strength concrete reduces the cross-section of structural members, and thus reduces the number of reinforcing bars and concrete used in those structural members. Accordingly, energy use and $CO_2$ emission during the production of reduced reinforcing bars and concrete are also reduced. By performing structural analysis on the building with high-strength concrete, the size of the cross-section of a structural member and the amount of reduction in reinforcing bars and concrete were computed.

## 2.3.3 Calculation of Quantity Reduction Effect by Application of High-Strength Concrete

Structural analysis was performed by replacing the four compressive strengths (24, 27, 30, and 35 MPa) with 40 MPa high-strength concrete. Based on this result, the quantity of concrete and reinforcing bars was computed and compared with existing designs. Reductions in cross-sections by the application of high-strength concrete were limited to vertical members (column, wall, core wall, and wall column). As a result, the reduction in concrete and reinforcing bars of vertical members was 8.8% and 30.3%, respectively, and such reduction is converted to a 5.7% and 19.7% reduction rate for the entire concrete and reinforcing bars on the entire building. Computed reduction rate was used to decrease the quantity of concrete and reinforcing bars. Such a decrease in quantity is caused by a reduction in the cross-section of vertical members from using high-strength concrete. Table 2.4 shows the reduction of the volume of concrete and reinforcing bars by applying high-strength concrete.

Relatively more $CO_2$ is emitted when high-strength concrete is used because the amount of cement used is increased compared to

**Table 2.4** The reduction of the volume of concrete and reinforcing bars by applying high-strength concrete (40 MPa)

|  |  | Previous design | Design of high-strength concrete application | Reduction rate (%) |
|---|---|---|---|---|
| Vertical members | Concrete (m$^3$) | 11,377.30 | 10,374.10 | 8.8 ↓ |
|  | Steel (ton) | 545.33 | 379.94 | 30.3 ↓ |
| Total members | Concrete (m$^3$) | 17,596.06 | 16,589.57 | 5.7% ↓ |
| Horizontal + vertical | Steel (t) | 1667.94 | 1339.44 | 19.7% ↓ |

normal-strength concrete. In order to solve this problem, methods such as substitution of a portion of cement with industrial wastes like blast furnace slag are being proposed [16–18]. This study assumed a mixture with 20% blast furnace slag in the cement. If a different mixture composition is used, an increase or decrease in concrete composition materials results, and such changes must be taken into consideration when calculating the amount of CO$_2$. Fig. 2.2 shows the changes in the amount of cement, coarse aggregate, and fine aggregate with application of 40 MPa pressure. According to Fig. 2.2, the amount of cement increased by 1156 t, coarse aggregate increased by 490 t, and fine aggregate decreased by 1649 t when the high-strength concrete mixture composition was used. CO$_2$ emission from the production of 1 kg of blast furnace slag is 0.0263 kg-CO$_2$/kg, which is about 4% of the CO$_2$ emission by 1 kg of cement at 0.7466 kg-CO$_2$/kg. Table 2.5 shows the mixture compositions of concrete for each strength.

## 2.3.4 Calculation of Building Lifespan

Carbonation is a phenomenon in which CO$_2$ in the atmosphere propagates into concrete to react with calcium hydroxide to form calcium carbonate, reducing the pH of the concrete pore solution down to 8.3–10.0. Once the pH inside the concrete is reduced, the stability of reinforcing bars buried inside the concrete is lost, and they begin to corrode. Corrosion in reinforcing bars by carbonation is a representative deterioration phenomenon of reinforced concrete structures [19–21].

The infiltration rate of CO$_2$ into concrete must be computed in order to compute the life cycle of reinforced concrete in a carbonation environment, and the infiltration rate of CO$_2$ in general can be expressed as the square root of time, as shown in Eq. (2.21). In addition, the velocity coefficient $A$ used in Eq. (2.21) is calculated from Eq. (2.22), and $A$ is

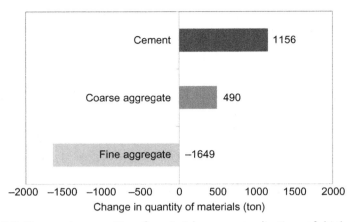

**Figure 2.2** Change in quantity of materials upon application of high-strength concrete.

**Table 2.5** Mixture compositions of concrete for each strength

| Strength (MPa) | Water/ cement ratio (%) | S/A[a] (%) | Unit weight (kg/m³) | | | | |
| | | | Water (kg/m³) | Cement (kg/m³) | Fine aggregate (kg/m³) | Coarse aggregate (kg/m³) | Slag (kg/m³) |
|---|---|---|---|---|---|---|---|
| 24 | 50 | 48 | 169 | 337 | 859 | 919 | 0 |
| 27 | 46 | 47 | 167 | 362 | 833 | 924 | 0 |
| 30 | 43 | 46 | 169 | 392 | 796 | 931 | 0 |
| 35 | 38 | 44 | 165 | 435 | 752 | 945 | 0 |
| 40 | 33 | 39 | 165 | 400 | 629 | 983 | 99 |

[a]Sand (fine aggregate) to aggregate (fine aggregate + coarse aggregate) ratio.

a coefficient dependent on (1) type of concrete, (2) type of cement, (3) water–cement ratio, and (4) temperature and humidity. The coefficient A for this study was determined using the method proposed by an existing study [22], and carbonation depth versus time was computed. Table 2.6 shows the values of variables that determine the velocity coefficient of carbonation. This study used the values for each variable shown in Table 2.6 to compute carbonation velocity.

$$C = A\sqrt{t} \qquad (2.21)$$

$$A = \alpha_1 \cdot \alpha_2 \cdot \alpha_3 \cdot \beta_1 \cdot \beta_2 \cdot \beta_3 \qquad (2.22)$$

where $C$ is the carbonation depth (cm), $A$ is the carbonation velocity coefficient, $t$ is the time (year).

**Table 2.6** Variables of carbonation velocity coefficient $A$

| Variable | Details | Applied value |
|---|---|---|
| $\alpha_1$ | Concrete type | Normal concrete $=$ $>1$ |
| $\alpha_2$ | Cement type | Normal concrete $=$ $>1$ |
| $\alpha_3$ | Water-to-binder ratio | W/B $= 0.6 =$ $>0.22$ |
| $\beta_1$ | Temperature | Annual average temperature $15.9°C =$ $>1$ |
| $\beta_2$ | Humidity | Annual average humidity $63\% =$ $>1$ |
| $\beta_3$ | Carbon dioxide concentration | $CO_2$ concentration $0.05\% =$ $>1$ |

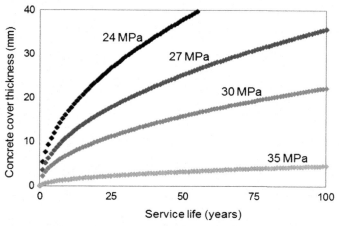

**Figure 2.3** Results of carbonation velocity for each strength.

Fig. 2.3 shows an estimation of carbonation velocity. Among the four types of concrete strength (24, 27, 30, and 35 MPa) in existing buildings, the time needed for carbonation to reach 20 mm of cover thickness (distance from concrete surface to buried reinforcing bar) is 14, 32, and 81 years for a building with compressive strengths of 24, 27, and 30 MPa, respectively. Carbonation reached the 20-mm cover thickness in less than 100 years in all cases. Based on such results, this study selected 40 MPa as the compressive strength of concrete with at least 100 years of durability in a carbonation environment. This value took into consideration various safety factors, such as flaws resulting from construction error. Such consideration corresponds to a 40% water–cement ratio (W/C) (corresponds to a compressive strength of about 40 MPa), generally known to be safe in a carbonation environment, and a 10% safety factor. Table 2.7 shows the compressive strengths of existing buildings and high-strength concrete building for each floor.

**Table 2.7** Redesign of high-strength concrete

|  | Previous design (MPa) | Redesign of high-strength concrete (MPa) |
|---|---|---|
| 27–top | 24 | 40 |
| 20–26 stories | 27 | 40 |
| 10–19 stories | 30 | 40 |
| 1–9 stories | 35 | 40 |

## 2.4 THE RESULTS OF ENVIRONMENTAL PERFORMANCE BY THE APPLICATION OF HIGH-STRENGTH CONCRETE

The results of environmental performance evaluation on the evaluation conditions cases 1–3 were shown in classes of construction stages and life cycle. That is because reinforcing bar and concrete quantity reduction and lifespan extension by the application of high-strength concrete occur in the form of energy consumption amount and $CO_2$ reduction in the construction stage.

### 2.4.1 Energy Consumption and $CO_2$ Emission in Construction Stage

Figs. 2.4 and 2.5 show the energy consumption and $CO_2$ emission of the construction stage respectively. According to Figs. 2.4 and 2.5, energy consumption and $CO_2$ emission appear in the sequential order of material production, transportation, and construction regardless of evaluation conditions, and energy consumption and $CO_2$ emission tended to be the highest in case 1 and the lowest in case 3, i.e., a high-strength concrete building.

In particular, the energy consumption of case 3 decreased 51.89% and 3.79%, respectively, compared with cases 1 and 2, and $CO_2$ emission also decreased 52.06% and 4.12%, respectively, compared with cases 1 and 2. The energy consumption and $CO_2$ emission–reduction effects of case 3 were analyzed due to the reduction of concrete and reinforcing bar quantity by the lifespan extension of the building by the application of high-strength concrete and the cross-section decrease of the vertical member. In particular, case 3 could obtain more than double the lifespan compared with cases 1 and 2 by the application of high-strength concrete, and through this, energy consumption and $CO_2$ emission were evaluated without one session of construction work step and separate maintenance stage by carbonization for a 100-year evaluation period.

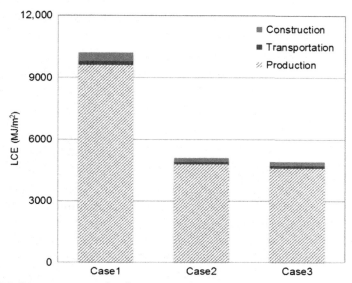

**Figure 2.4** Energy consumption for construction stage.

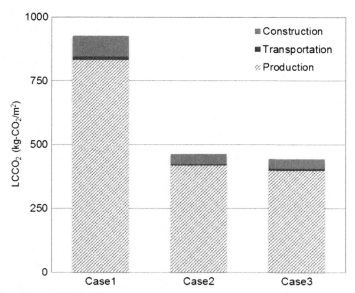

**Figure 2.5** CO$_2$ emission for construction stage.

However, in case 1, a relatively large amount of energy consumption and CO$_2$ are evaluated to be emitted in construction stage compared with cases 2 and 3 as two sessions of construction work step is done for a 100–year evaluation period. In practice, in Korea, many buildings have been reconstructed for 20–30 years since construction, although a lifespan of over 40 years is required by tax law in reinforced concrete structures. Through this study, buildings reconstructed after a short period of use are evaluated to emit a relatively large amount of energy consumption and $CO_2$ compared with long-lifespan buildings, and the rate of $CO_2$ reduction in the construction stage is evaluated to reach about 50% by doubling the lifespan.

On the other hand, case 2 is the case where lifespan is extended up to a 100-year evaluation period through sustainable maintenance of the structure members that were deteriorated by carbonization for an existing building. According to Figs. 2.4 and 2.5, case 2 was evaluated to consume more energy and emit more $CO_2$ compared with case 3, but showed a dramatically lower energy consumption amount and $CO_2$ emission amount compared with case 1. However, case 2 consumes energy and emits $CO_2$ in the process of maintenance of the structure members that were deteriorated by carbonization during the evaluation period. However, such additional energy consumption and $CO_2$ emission are added to the maintenance part during the life cycle stages, and are not expressed in the construction stage, so the energy consumption and $CO_2$ emission of the construction stage of case 2 is considered to be evaluated very low compared with case 1.

As a result of discussion in this chapter, application of high-strength concrete is evaluated to have an outstanding reduction effect of energy consumption and $CO_2$ emission by the lifespan extension of building, along with the effects of construction material reduction.

There is a method of extending building lifespan through maintenance up to a target lifespan level without applying lifespan extension technology in the initial construction stage. However, such a method would increase the $CO_2$ emission of the maintenance stage during the life cycle stages of a building, and involves apprehension of decrease of residential performance in the maintenance stage. Therefore it is desirable to sufficiently secure building lifespan through high-strength concrete and other lifespan extension technologies at the initial stage of construction.

## 2.4.2 Energy Consumption and CO₂ Emission for Life Cycle

Figs. 2.6 and 2.7 show the energy consumption and $CO_2$ emission for the building life cycle. According to Figs. 2.6 and 2.7, the energy consumption amount and $CO_2$ emission amount tended to be very high in the construction stage and using stage, which indicates that an effective practice for the reduction of environmental load is to apply energy consumption and $CO_2$ emission of reduction technology in the construction stage and using stage.

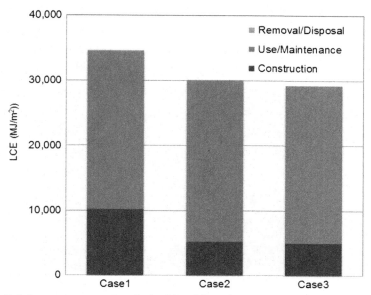

**Figure 2.6** Energy consumption for building life cycle.

In particular, the energy consumption of case 3, a high-strength concrete building, in total $34{,}617\,\mathrm{MJ/m^2}$, was evaluated at 29% in the construction stage, 70% in the using and maintenance stage, and 1% in the disposal stage. In addition, the stage that involves the largest energy consumption was the heating energy consumption sector of the using and maintenance stage followed by the electric energy consumption sector and building construction material production stage. Such a trend appeared in cases 1 and 2 as well.

In addition, $CO_2$ emission, in total $2{,}918\,\mathrm{kg\text{-}CO_2/m^2}$, was evaluated at 31% in construction, 68% in the using and maintenance stage, and 1% in the removal and disposal stage, and the results were similar in cases 1 and 2. Table 2.8 shows the energy consumption and $CO_2$ emission by each stage of cases 1–3.

According to Table 2.8, the amounts of energy consumption of case 3. a high-strength concrete building, decreased 15.53% and 2.95%, respectively, compared with cases 1 and 2, and $CO_2$ emission was also evaluated to show 16.70% and 3.37% reduction effects, respectively, compared with cases 1 and 2.

The reduction effects of case 3 are evaluated to be due to the reduction of concrete and reinforcing bar quantity by the lifespan extension of the building and the cross-section reduction of the vertical member by the application of high-strength concrete as is described in Section 2.4.1, Energy Consumption and $CO_2$ Emission in Construction Stage.

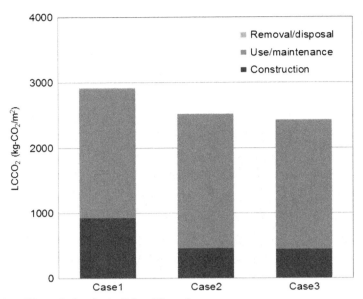

**Figure 2.7** $CO_2$ emission for building life cycle.

**Table 2.8** Life cycle energy consumption and $CO_2$ emission

| | LCE (MJ/m²) | | | LCCO₂ (kg-CO₂/m²) | | |
|---|---|---|---|---|---|---|
| | Case 1 | Case 2 | Case 3 | Case 1 | Case 2 | Case 3 |
| Production | 9608 | 4804 | 4611 | 833 | 416 | 397 |
| Transportation | 209 | 105 | 105 | 15 | 7 | 7 |
| Construction | 400 | 200 | 200 | 78 | 39 | 39 |
| Use | 24,250 | 24,250 | 24,250 | 1981 | 1981 | 1981 |
| Maintenance | 0 | 695 | 0 | 0 | 66 | 0 |
| Removal | 2 | 1 | 1 | 0 | 0 | 0 |
| Disposal | 148 | 74 | 74 | 11 | 5 | 5 |
| Total | 34,617 | 30,129 | 29,241 | 2918 | 2514 | 2429 |

On the other hand, case 2 was evaluated to consume less energy and emit less $CO_2$ compared with case 1, and that indicates that an effective method of reducing environmental load is to obtain a lifespan up to the target years through maintenance for a deteriorated environment. However, case 2 was evaluated to consume more energy and emit more $CO_2$ compared with cases 2 and 3, and that is because in case 3, concrete and reinforcing bar quantity decreased in the construction material stage by the application of high-strength concrete, which produced the effect of reducing environmental load by it, and in case 2, energy consumption and CO$_2$ emission of the maintenance stage were added through the maintenance repair process.

Therefore, according to the results of the above study, when high-strength concrete is applied to a building, energy consumption and CO$_2$ emission–reduction effects are evaluated to be outstanding for life cycle as well as for the construction stage. In addition, use of the method of evaluating environmental load for the life cycle of buildings proposed in this study can possibly evaluate energy consumption and CO$_2$ emission for the case where high-strength concrete is applied to various buildings.

## 2.5 CONCLUSIONS

This chapter proposed a plan for the evaluation of energy consumption and CO$_2$ emission for the life cycle of a building in order to evaluate the environmental performance of building by the application of high-strength concrete, and evaluated the energy consumption and CO$_2$ emission–reduction performance for the life cycle of a high-strength concrete building. Now, the following conclusions can be made:

1. We propose a plan for the evaluation of the energy consumption and CO$_2$ emission for the life cycle of a building.
2. The distribution of the energy consumption and CO$_2$ emission of concrete building was evaluated within 30% in the construction stage, 70% in the using and maintenance stage, and over 1% in the removal and disposal stage, roughly.
3. The energy consumption of a high-strength concrete building in the construction stage (case 3) decreased by 51.89% and 3.79%, respectively, compared with cases 1 and 2, which were general-strength concrete buildings, and CO$_2$ emission also decreased 52.06% and 4.12%, respectively, compared with cases 1 and 2.
4. The energy consumption of case 3, a high-strength concrete building, for the life cycle decreased 15.53% and 2.95%, respectively, compared with cases 1 and 2, which were general-strength concrete buildings, and CO$_2$ emission also decreased 16.70% and 3.37%, respectively, compared with cases 1 and 2.
5. Such reduction effects of energy consumption and CO$_2$ emission in case 3, a high-strength concrete building, are attributed to the reduction of concrete and reinforcing bar quantity by the lifespan extension of the building and cross-section reduction of vertical members by the application of high-strength concrete.

# REFERENCES

[1] Damtoft JS, Lukasik J, Herfort D, Sorrentino D, Gartner EM. Sustainable development and climate change initiatives. Cement and Concrete Research 2008;38:115–27.

[2] Sisomphon K, Franke L. Carbonation rates of concrete containing high volume of pozzolanic materials. Cement and Concrete Research 2007;37:1647–53.

[3] Zhang Z, Shu X, Yang X, Zhu Y. BEPAS-a life cycle building environmental performance assessment model. Building and Environment 1990;41:669–75.

[4] Kim DG. Master Thesis Basic Environmental & Economical Comparison Analyses of Reinforced vs. Concrete Bridges Using Life Cycle Assessment Methodology. Korea Advanced Institute of Science and Technology; 1995;23–5.

[5] Li Z. A new life cycle impact assessment approach for building. Building and Environment 2006;41:1414–22.

[6] Ardente F, Beccali M, Cellura M, Mistretta MA. LCA case study of kenaf-fibres insulation board. Building energy performance 2008;40:1–10.

[7] Forsberg A, Malmborg F. Tools for environmental assessment of the built environment. Building and Environment 2004;39:223–8.

[8] Gao NP, Niu JL, Perino M, Heiselberg P. The airborne transmission of infection between flats in high-rise residential building: particle simulation. Build Environ 2009;44:402–10.

[9] Giridharan R, Lau SSY, Ganesan S, Givoni B. Urban design factors influencing heat island intensity in high-rise high-density environments of Hong Kong. Build Environ 2007;42:3669–84.

[10] Lai JHK, Yik FWH. Perception of importance and performance of the indoor environmental quality of high-rise residential building. Build Environ 2009;44:352–60.

[11] Roh SJ, Tae SH, Shin SW, Woo JH. Development of an optimum design program (SUSB-OPTIMUM) for the life cycle CO2 assessment of an apartment house in Korea, Build Environ 2014;73:40–54.

[12] Shin SW. Sustainable building technology. Seoul: Kimoondang; 2007. p. 124–47.

[13] Shin SW, Tae SH, Woo JH, Roh SJ. The development of environmental load evaluation system of a standard Korean apartment house. Renewable and Sustainable Energy Reviews 2011;15:1239–49.

[14] Lee KH, Tae SH, Shin SW. Development of a Life Cycle Assessment Program for building(SUSB-LCA) in South Korea. Renew Sust Energ Rev 2009;13:1994–2002.

[15] Tae SH, Shin SW, Woo JH, Roh SJ. The development of apartment house life cycle CO2 simple assessment system using standard apartment houses of South Korea. Renewable and Sustainable Energy Reviews 2011;15:1454–67.

[16] Robeyst N, Gruyaert E, Grosse CU, Belie ND. Monitoring the setting of concrete containing blast-furnace slag by measuring the ultrasonic p-wave velocity. Cement Concrete Res 2008;38:1169–76.

[17] Yuksel I, Bilir T, Ozkan O. Durability of concrete incorporating non-ground blast furnace slag and bottom ash as fine aggregate. Build Environ 2007;42:2676–85.

[18] Lee KM, Lee HK, Lee SH, Kim GY. Autogenous shrinkage of concrete containing granulated blast-furnace slag. Cement Concrete Res 2006;36:1304–11.

[19] Sisomphon K, Franke L. Carbonation rates of concrete containing high volume of pozzolanic materials. Cement Concrete Res 2007;37:1647–53.

[20] Chang CF, Chen JW. The experimental investigation of concrete carbonation depth. Cement and Concrete Research 2006;36:1760–7.

[21] Tae SH, Ujiro T. A study on the corrosion resistance of Cr-bearing rebar in mortar in corrosive environments involving chloride attack and carbonation. ISIJ International 2007;47:715–22.

[22] AIJ (Architectural Institute of Japan). Recommendations for durability design and construction practice of reinforced concrete. Tokyo: Architectural Institute of Japan; 2004;92 (In Japanese).

# CHAPTER 3

# Assessment of $CO_2$ Emissions Reduction in High-Rise Concrete Office Buildings Using Different Material-Use Options

**C.K. Chau, W.K. Hui, W.Y. Ng, T.M. Leung and J.M. Xu**
The Hong Kong Polytechnic University, Hung Hom, Kowloon, Hong Kong

## 3.1 INTRODUCTION

Buildings account for 40% of the world's energy sources and 36% of the energy-related carbon emissions in industrialized countries [1]. In the United States, buildings account for around 40% of energy use, which is equivalent to 7.7% of global carbon emissions [2]. The building sector in the European Union accounted for 40% of total energy use, and the building sector in the United Kingdom accounted for 50% of its total $CO_2$ emissions [3]. The building sector in Hong Kong consumes even more, i.e., more than 80% of total electricity and fuel energy, due to the absence of large industrial bases in Hong Kong [4]. In response to imminent climate change issues, substantial efforts have been spent on reducing the operating energy consumption of buildings. High-efficiency lighting installations and appliances, high-efficiency ventilation and cooling systems, waste heat recovery, smart glass, smart meters, advanced insulation, reflective building materials, and multiple glazing systems have been incorporated into many new buildings [5–10].

On the other hand, substantial attention has also been diverted to lowering the embodied energy contents of buildings so as to reduce the total carbon emissions associated with buildings. As buildings become more energy efficient and their functional obsolescence becomes more rapid, the relative importance of energy embodied in, or carbon

*Handbook of Low Carbon Concrete.*
DOI: http://dx.doi.org/10.1016/B978-0-12-804524-4.00003-8

emissions associated with, building materials on the overall life-cycle energy use or carbon emissions will become higher [11–13]. Building materials play a more and more important role in $CO_2$ emissions. Different types of buildings have different structures, which will influence the distribution of building materials. Previous studies showed a maximum of 30% carbon emissions can be reduced through careful selection of low-environmental-impact materials in residential houses in Valladolid [14]. Constructing a steel-framed and concrete-framed office building would incur similar energy and $CO_2$-emission implications over a 50-year lifespan [15]. Using 40-MPa high-strength concrete in supertall buildings could reduce their total $CO_2$ emissions up to 17% [16]. Increasing the recycling rate of concrete from deconstructed buildings from 27% to 50% could yield a 2–3% reduction in buildings' greenhouse gas emissions [17].

In this chapter, we intended to determine the carbon footprint of materials and building elements for the superstructure of a high-rise concrete-framed office building. The $CO_2$ emissions reduction resulting from the implementation of various material-use options were also evaluated. Accordingly, the major building superstructure elements were identified before examining the impacts of different material-use options on the $CO_2$ emissions reduction.

## 3.2 SYSTEM DEFINITIONS AND BOUNDARIES

Given no consensus on the scope and boundaries for the $CO_2$ emissions study, it is vital to define them clearly at the outset. The scope of the emission-impacts study covers the emissions associated with extraction and production of materials and components, their transportation from countries of origin to ports in Hong Kong, construction of building elements onsite, and replacement of materials. For concrete products, net $CO_2$ emissions due to calcination and carbonation are also considered. Although calcination reactions of concrete products only occur during the production of cement in the kiln, they take significant impacts on the net $CO_2$ emissions of concrete products [18]. Carbonation is also considered since it occurs throughout the life cycle of concrete products. Fig. 3.1 shows the boundaries defined for this study.

Carbonation has also been accounted for over the lifetime of the concrete products but has been omitted from the figure for clarity.

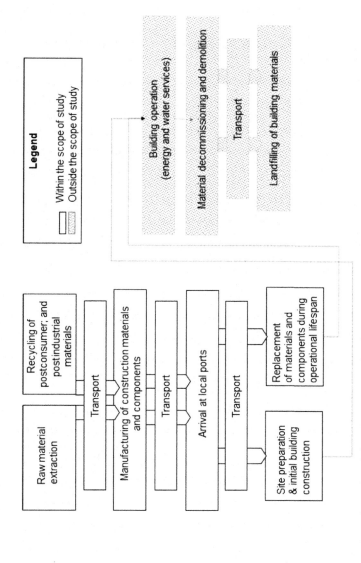

**Figure 3.1** Boundaries that define the processes for which their $CO_2$ emissions impacts have been examined for this study.

## 3.3 METHODOLOGY

Thirteen Grade A[1] high-rise concrete-framed office buildings completed between 1995 and 2005 have been selected for studying the $CO_2$ emissions impacts of their constituent building materials. The total number of stories of the studied buildings varied between 16 and 62.

### 3.3.1 Identify the Types and Quantities of Materials for Building Elements

Information such as construction floor area, as well as types and quantities of building materials used for individual building, was extracted from the bills of quantities, which are the tender documents of building projects. Given certain elements are needed to be replaced over the building life, it is necessary to estimate the number of times for which the elements are needed to be replaced before the end of their life cycles. Data on the life expectancies of different building elements from the report published by the Building Cost Information Service of the Royal Institution of Chartered Surveyors [19] was used (see Table 3.1).

The replacement factors for the various building elements were computed using the following formula with the assumption of a 60-year lifespan for a building [20].

$$\text{Replacement factor} \quad 60/\text{expected life span (years)} \qquad (3.1)$$

The replacement factor quantifies the number of times that resource input is needed for construction/installation of the element within the lifespan of a building. Accordingly, the total weight of a building element used in its life cycle will be the impact of the first installation scaled up by the replacement factors.

Given that a majority of the currently available inventory databases express the gaseous emissions of a material in terms of unit mass, i.e., kilogram of $CO_2$ emissions per kilogram, it is necessary to convert the quantities of various building materials into their respective masses prior to estimating their emissions impacts. The masses of building materials were

---

[1] Flexible layout; spacious circulation areas; larger floor plates (area per floor in the office tower around $1600\,m^2$); more window area (window to wall area ratio around 0.4); many with reflective glazing; smaller floor plan aspect ratio (around 1.6); the majority would be equipped with variable air volume air conditioning systems; lower central plant capacity per unit floor area (around $0.17\,kW/m^2$); good lift services zoned for passengers and delivery of goods; car parking facilities normally available.

**Table 3.1**  Life expectancies of building elements

| Element | Typical life expectancy (years) |
|---|---|
| *Frame* | |
| Concrete frame | 81 |
| *Upper floors* | |
| Reinforced concrete floor | 71 |
| Precast concrete slab | 78 |
| *Roof* | |
| Asphalt covering to flat roof | 36 |
| PVC covering to flat roof | 27 |
| Ethylene Propylene Diene Monomer (EPDM) covering to flat roof | 25 |
| *Stairs* | |
| Concrete stairs | 74 |
| Steel stairs | 50 |
| Aluminum stair nosings | 21 |
| Plastic stair nosings | 15 |
| *External walls* | |
| Aluminum curtain walling | 43 |
| *Windows* | |
| Aluminum windows | 44 |
| *Internal doors* | |
| Internal softwood door | 42 |
| *Wall finishes* | |
| Plasterboard to wall | 39 |
| Clay tiling to wall | 37 |
| *Floor finishes* | |
| Vinyl sheet floor covering | 17 |
| Vinyl tile to floor covering | 18 |
| Carpet floor covering | 13 |
| *Ceiling finishes* | |
| Suspended ceilings | 24 |

*Source*: From reference Building Cost Information Service. Life expectancy of building components, 2006.

estimated based on the data available in textbooks and handbooks on construction materials, published trade literature, product technical datasheets, product catalogs (including e-catalogs), data collected from contractors, reference information disseminated from trade organizations and professional bodies, as well as published information from suppliers/specialist contractors. Concrete and reinforcing bars for fabric elements, and glass, aluminum, and sealant for windows, had been identified as far as possible in respect to the types and quantities so that the environmental impacts incurred due to the production of the components could be adequately accounted for.

## 3.3.2 $CO_2$ Emissions Associated with Building Materials

It is found that the amount of embodied energy and $CO_2$ emissions associated with building materials is highly dependent on the type and amount of energy used in their manufacturing processes [14]. To convert embodied energy to $CO_2$ emissions requires information on the amount of $CO_2$ emitted during the production of different types of energy (such as oil, wind, solar, nuclear). The amount of $CO_2$ emitted from individual building materials was estimated by multiplying the mass of materials with the corresponding embodied energies and $CO_2$ emissions factors [14]. Besides, the materials were grouped and aggregated under a building element format as shown in Table 3.2, developed by the Building Research Establishment (BRE) in the United Kingdom [21]. Therefore, the amount of $CO_2$ emitted by the $i$th building element ($Q_{\text{Element, i}}$, in kg $CO_2$) is estimated by summing up the amount of $CO_2$ emitted from all its constituent materials, i.e.,

$$Q_{\text{Element, i}} = \sum_{1}^{i} e_i \, \beta_i \, m_i \qquad (3.2)$$

Or

$$q_i = e_i \, \beta_i \qquad (3.3)$$

where $q_i$ is the $CO_2$ emissions per kilogram of $i$th type building material (in kg $CO_2$/kg); $e_i$ is the embodied energy intensity of the $i$th type of building material (in MJ/kg); $\beta_i$ is the $CO_2$ emission factor for the $i$th type of building material (in kg $CO_2$/MJ); and $m_i$ is the mass of the $i$th type of building material (in kg).

As the magnitudes of embodied energy intensities reported in different studies vary considerably, ranges of the embodied energies for various building materials were used in the estimation (see Table 3.3). In addition, adjustments were made for different types of fuel mixes employed in different countries during the manufacturing processes. Table 3.4 lists the

**Table 3.2** Classification of building elements as suggested by BRE in the United Kingdom

| Class (BRE) | Major material group |
|---|---|
| Doors | Plastic |
| | Plywood |
| | Stainless steel |
| External walls | Aluminum |
| | Concrete |
| | Reinforcing bar |
| | Stainless steel |
| | Stone |
| Floor surfacing and finishes | Plaster |
| | Galvanized steel |
| | Stone |
| | Tile |
| Internal walls and partitioning | Aluminum |
| | Bricks and blocks |
| | Concrete |
| | Galvanized steel |
| | Glass |
| | Reinforcing bar |
| | Stainless steel |
| Paint system | Paint |
| Roof construction | Concrete |
| | Galvanized steel |
| | Plaster |
| | Stone |
| | Tile |
| Roof insulation | Asphalt and bitumen |
| | Plaster |
| | Thermal insulation |
| Suspended ceilings and ceilings finishes | Aluminum |
| | Galvanized steel |
| | Plaster |
| Upper-floor construction | Concrete |
| | Galvanized steel |
| | Plaster |
| | Reinforcing bar |
| | Structural steel |
| Wall finishes | Aluminum |
| | Galvanized steel |
| | Plaster |
| | Stone |
| | Tile |
| Wall insulation | Plaster |
| | Thermal insulation |
| Windows/curtain wall | Aluminum |
| | Glass |

**Table 3.3** Embodied energy intensities for different types of building materials

| Type of building material | Embodied energy intensities[a] (in MJ/kg) |
|---|---|
| Aluminum | 166.0–312.7 |
| Bitumen and asphalt | 3.4–50.2 |
| Bricks and blocks | 0.5–3.3 |
| Concrete | 0.7–1.6 |
| Galvanized steel | 30.6–34.8 |
| Glass | 6.8–25.8 |
| Stone, gravel, and aggregate | 0.1–0.8 |
| Purified fly ash (PFA) | <0.1 |
| Paint | 60.2–144 |
| Plaster, render, and screed | 0.1–2.0 |
| Plastic, rubber, and polymer | 70.0–116.0 |
| Plywood | 3.1–18.9 |
| Precast concrete element | 2.0 |
| Reinforcing bar and structural steel | 6.2–42.0 |
| Stainless steel | 8.2–13.3 |
| Thermal and acoustic insulation | 1.2–17.6 |
| Ceramic and tile | 2.2–5.5 |

[a]Embodied energy values are extracted from studies [14,22–25].

**Table 3.4** Average $CO_2$ emission factor values for electricity generation in different countries

| Country | Emission factor,[a] $\beta$ (in kg $CO_2$/MJ) |
|---|---|
| Australia | 0.02294 |
| Belgium | 0.00775 |
| Brazil | 0.00186 |
| China | 0.02176 |
| France | 0.00148 |
| Germany | 0.01253 |
| Hong Kong | 0.01655 |
| India | 0.02165 |
| Indonesia | 0.01911 |
| Italy | 0.01460 |
| Japan | 0.01261 |
| Korea | 0.01473 |
| Malaysia | 0.01781 |
| Romania | 0.01677 |
| Russian | 0.01658 |
| Singapore | 0.01755 |
| South Africa | 0.02358 |
| Spain | 0.01129 |
| Taiwan | 0.01479 |
| Thailand | 0.01641 |
| United Kingdom | 0.01453 |
| United States and Canada | 0.01583 |
| Vietnam | 0.00817 |

[a]Emission factors extracted from the reports issued by Refs. [26,27].

average values of $CO_2$ emission factors for electricity generation in different countries.

However, for concrete-framed buildings, the impact of calcination and carbonation processes imposed on the life-cycle evaluation of carbon emissions of concrete was so significant that we should not overlook it. Calcination-process emissions occur during concrete manufacturing when limestone is decomposed to calcium oxide (CaO) and carbon dioxide at high temperatures. For the remainder of their life cycle, concrete products absorb carbon dioxide from the atmosphere through carbonation, a chemical process in which the calcium oxide present in hardened cement products binds with $CO_2$ in the atmosphere to form carbonate [18].

$CO_2$ emissions due to calcination were estimated using the model by Ref. [28]. The model estimates the calcium oxide content of clinker at 65%, which agrees with the reported range of CaO in clinker of 64–67% by weight [29]. The clinker-to-cement ratio was estimated at 65% [30] and we assumed an average concrete mix ratio of cement:sand:aggregate of 1:2:5 by weight. The average moisture content in cured concrete was assumed to be 3.5%, being midway in the normal range of 2–5% [31].

With the aid of Fig. 3.1 given in Ref. [18], over a 60-year timespan, it was estimated that approximately 17% of the calcination $CO_2$ emissions will be reabsorbed through carbonation. The net effect of calcination and carbonation is about 0.033 kg of $CO_2$ emitted per kilogram concrete based on the above assumptions.

Considering that the embodied energy and $CO_2$ emission data values vary considerably, the Monte Carlo method was applied in this study for handling the inherent uncertainties and variations arisen from the collected data. The procedure is discussed in details in the following section.

### 3.3.3 Applying the Monte Carlo Method for CO$_2$ Emission Prediction

The Monte Carlo method was used for generating probability distributions to define the boundaries for the $CO_2$ emissions from various materials [32]. We determined the 5th and 95th percentile levels using the computer software called EasyFit [33] together with MATLAB [34]. EasyFit was initially employed for developing model distributions that closely resemble the real scenario. Subsequently, it was used for defining the distributions for the input variables that contain uncertainties. The fitness or correctness of these distributions was checked with aid of goodness-of-fit (GOF) statistics, which are the statistical measures used

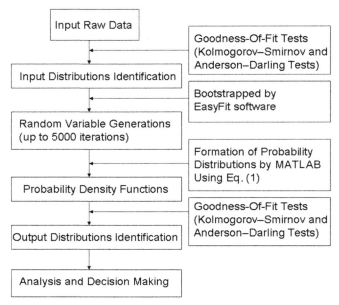

**Figure 3.2** Monte Carlo process protocol.

for describing the validity of fitting a dataset to distributions. Other than visualizing through graphs, like $p$–$p$ plots or $q$–$q$ plots [35], EasyFit can automate the decision of choosing the best-fitting distributions. Among many GOF statistics available in EasyFit, Kolmogorov–Smirnov statistic and Anderson–Darling statistic tests have been selected to determine the best-fitting distributions [36]. Once the underlying distributions were determined, EasyFit was applied to generate random values based on a maximum number of 5000 iterations. This propagation results in probability density functions (PDFs) for estimating the emissions. In the second stage, MATLAB was employed to aggregate the PDFs using Eq. (3.1) and construct the PDFs for estimating the output values. The output values can not only be used for constructing empirical distributions but also for deriving the percentiles and other statistics for the distribution. The validity of the PDFs was further examined by the two above tests. Fig. 3.2 shows our Monte Carlo process protocol.

## 3.3.4 Material-Use Options

Many different material-use options are available for implementation in the design stage for reducing the $CO_2$ emissions from buildings. For instance, the $CO_2$ emissions can be reduced by reducing material use, minimizing waste, and specifying localized materials, recycled materials, or alternative low-carbon materials. In general, there are five options for material use discussed in the study: (1) importing regional materials, (2) maintaining the existing structural and nonstructural building elements, (3) reusing existing resources, (4) diverting construction wastes to recycling, and (5) using prefabricated materials.

## 3.3.5 Calculation Methods for Different Material-Use Options

### 3.3.5.1 Importing Regional Materials

Construction of buildings usually requires a large amount of materials to be transported to construction sites from different countries in different geographical regions. A change in the origins of a material source will induce changes in the amount of transportation energy use. Therefore, it is worthwhile to investigate how much energy use and emissions can be reduced by importing more materials from nearby countries, and preferably from those within the same geographical region. Statistics of the breakdown of materials imported from different countries, and distance as well as the CO$_2$ emission intensities associated with transportation of the imported materials, comes from the Census and Statistics Department of Hong Kong [37]. Table 3.5 listed the major original countries for building materials, and distance as well as the CO$_2$ emission intensities associated with transportation for imported materials. The differences in the transportation energy use were determined by using the embodied energy intensities data for different modes of transportation shown in Table 3.6. With this information, the change in the amount of CO$_2$ emissions, $\Delta Q_R$ (in kg CO$_2$), was determined from the following:

$$\Delta Q_R = \sum_1^i [m_{R,i}(1 + w_i) \star \lambda_{R,i} \star (\Delta q_{R,T,i} + \Delta q_{R,\text{origin},i})] \quad (3.4)$$

where $\Delta Q_R$ is the change in the CO$_2$ emissions due to importing regional materials (in kg CO$_2$), and the positive sign denotes an increase in the CO$_2$ emissions; $m_R \star (1+w_i)$ is the mass of materials originally imported from the other geographical regions, which includes the wastages generated during construction (in kg); $\lambda_R$ is the fraction of materials imported from the same geographical region; $\Delta q_{R,T}$ is the difference in the CO$_2$ emissions per kilogram of material associated with transportation of imported regional material (in kg CO$_2$/kg); $\Delta q_{R,\text{origin}}$ is the difference in the CO$_2$ emissions per kilogram of material associated with manufacturing of imported regional material (in kg CO$_2$/kg). $i$, pertains to the $i$th type of building material; R, pertains to regional use of material.

**Table 3.5** Major material source profiles for different types of building materials

| Type of building material | Country of origin | Percentage by weight of imported material (in %) | Transportation distance (in km) Land | Transportation distance (in km) Sea | $CO_2$ emission intensity associated with transportation (in kg $CO_2$/kg) |
|---|---|---|---|---|---|
| Asphalt and bitumen | Korea | 85 | 561.5 | 2246.7 | 0.0238 |
| Aluminum | China | 84 | 250.0 | 150.0 | 0.0070 |
| Blocks and bricks | China | 83 | 250.0 | 150.0 | 0.0070 |
| Building stones | China | 75 | 250.0 | 150.0 | 0.0070 |
| Cement | China | 42 | 250.0 | 150.0 | 0.0070 |
| Float glass | China | 61 | 250.0 | 150.0 | 0.0070 |
| Galvanized steel | Australia | 56 | 4946.3 | 7152.0 | 0.1903 |
| Gravels | China | 100 | 250.0 | 150.0 | 0.0070 |
| Granite tiles | China | 48 | 250.0 | 150.0 | 0.0070 |
| Paints | China | 51 | 250.0 | 150.0 | 0.0070 |
| Plasters | China | 40 | 250.0 | 150.0 | 0.0070 |
| Plywood | United Kingdom | 30 | 877.9 | 18,240.0 | 0.1368 |
| Prefabricated structural components | Malaysia | 87 | 104.5 | 2122.0 | 0.0195 |
| Reinforcing bars | China | 100 | 250.0 | 150.0 | 0.0070 |
| Sand | China | 60 | 250.0 | 150.0 | 0.0070 |
| Stainless steel | China | 100 | 250.0 | 150.0 | 0.0070 |
| Structural steel | India | 72 | 3077.9 | 6797.0 | 0.1355 |
|  | Romania | 28 | 1000.0 | 13,907.0 | 0.1259 |

*N.B. Source:* From reference Census and Statistics Department. Hong Kong merchandise trade statistics, 2002.

**Table 3.6** Embodied energy intensities associated with different modes of transportation

| Mode of transportation | Embodied energy/kg-km[a] |
|---|---|
| Rail | 0.0003 |
| Tanker | 0.0001 |
| Truck | 0.0027 |

[a]Embodied energy values are extracted from studies [38,39].

### 3.3.5.2 Maintaining the Existing Structural and Nonstructural Building Elements

Maintaining the existing building elements can reduce the CO$_2$ emissions through reducing the amount of material use and construction wastes. Theoretically, these benefits can be maximized by reusing the entire building through maintaining the existing walls, floors, and roof. It was assumed based on the Hong Kong Building Environment Assessment Method that 15–30% of the superstructure, e.g., a portion of the lower floors, would be retained in a new building design. The change in the amount of CO$_2$ emissions, $\Delta Q_M$ (in kg CO$_2$), due to maintaining the existing elements was determined from the following:

$$\Delta Q_M = -\sum_{1}^{i} (m_M \star (1 + w_i) \star (q_{M,i} + q_{M,T,i})) \qquad (3.5)$$

where $\Delta Q_M$ is the change in the CO$_2$ emissions related to maintaining the existing building element (in kg CO$_2$); $q_M$ is the CO$_2$ emissions per kilogram of material related to maintaining the material in the existing building element (in kg CO$_2$/kg); $q_{M,T}$ is the CO$_2$ emissions per kilogram of material related to transportation of the material if it is no longer maintained (in kg CO$_2$/kg); $w_i$ is the fraction of material wastes generated during construction; $m_M\star(1+w_i)$ is the mass of materials (which include the amount of material wastages generated during construction) that can be saved as a result of maintaining the existing element (in kg). $i$, pertains to the $i$th type of building material; M, pertains to maintaining the existing building element.

Table 3.7 listed the percentages of construction wastages used in calculation.

### 3.3.5.3 Reusing Existing Resources

Building materials and components, such as flooring panels, doors, cabinetry, bricks, concrete, suspended ceilings, and decorative items, can be

**Table 3.7** Percentages of construction wastages for different types of building material

| Types of materials | Wastage (in %) |
|---|---|
| Aluminum | 5 |
| Bricks and blocks | 3 |
| Cast iron | 5 |
| Concrete | 3 |
| Copper | 5 |
| Durasteel | 3 |
| Fiberglass | 8 |
| Galvanized steel | 5 |
| Glass | 5 |
| Precast concrete elements | 2.5 |
| Precast structural concrete element | 2.5 |
| Reinforcing bar | 5 |
| Special aggregates (Dynagrip, in nonskid finish) | 10 |
| Stainless steel | 5 |
| Structural steel | 5 |
| Stone | 5 |

*Source*: Local data from reference Poon CS, Ann TW, Ng LH. On-site sorting of construction and demolition waste in Hong Kong. Resour Conserv Recycl 2001;32(2):157–172.

salvaged from previously demolished other sites for uses in current project sites if they are properly stored and maintained. Hence, the amount of embodied energy and $CO_2$ emissions can be reduced together with the total quantities of materials use. The change in the amount of $CO_2$ emissions, $\Delta Q_{reuse}$ (in kg $CO_2$) due to reusing existing resources was determined from the following:

$$\Delta Q_{reuse} = -\sum_{1}^{i} m_{reuse} \star (1 + w_i) \star (q_{reuse,i} + q_{reuse,T,i}) \qquad (3.6)$$

where $\Delta Q_{reuse}$ is the change in the $CO_2$ emissions due to reusing the existing resources (in kg $CO_2$); $m_{reuse}(1+w_i)$ is the mass of materials that can be saved due to reuse of the material (which include the amount of material wastages occurred during construction) (in kg); $q_{reuse}$ is the $CO_2$ emission intensity per kilogram of material associated with the reuse of the material (in kg $CO_2$/kg); $q_{reuse,T}$ is the $CO_2$ emissions per kilogram of material related to transportation of the material if it is no longer reused (in kg $CO_2$/kg). $i$, pertains to the $i$th type of building material; reuse, pertains to reusing the existing building material.

### 3.3.5.4 Diverting Construction Wastes to Recycling

Broadly speaking, building construction wastes can be classified under structural wastes or finishing wastes. Structural wastes, which embrace ferrous and nonferrous metals and concrete fragments, have higher recycling potential. For instance, concrete fragments can be reused for land reclamation, while metals can be recycled in construction sites. By contrast, finishing wastes, like surplus cement mortar, broken mosaic, tiles, ceramics, paints, and plastering materials, are usually contaminated with a high portion of organic matters and debris and have little or zero recycling potential. Accordingly, only the structural wastes are taken into consideration in estimating the amount of construction wastes to be diverted.

The change in CO$_2$ emissions due to diverting construction wastes to recycling, $\Delta Q_D$ (in kg CO$_2$), was determined from the following:

$$\Delta Q_D = \sum_1^i m_{D,i} \star w_i \star (q_{D,i} \star \alpha_{D,i}) \tag{3.7}$$

where $\Delta Q_D$ is the change in the CO$_2$ emissions due to diverting construction wastes (in kg CO$_2$); $q_D$ is the CO$_2$ emissions per kilogram of material for diverting the material (in kg CO$_2$/kg); $\alpha_D$ is the increase in percentage of the CO$_2$ emissions of the recycled material compared to its virgin material; $m_D \star w$ is the mass of wastes generated during construction (in kg). $i$, pertains to the $i$th type of building material; D, pertains to diverting construction wastes.

The maximum amount of construction wastes that can be recycled was estimated by multiplying the total quantity of a specific type of building materials with the percentage of construction material wastages listed in Table 3.7. On the other hand, the differences in percentage of CO$_2$ emissions between recycled and virgin materials were extracted from Table 3.8.

### 3.3.5.5 Offsite Fabricated Materials

Prefabrication techniques have been increasingly applied in building construction in Hong Kong in response to high demands for improvement in overall quality and reduction in wastage of materials, and speedy erection processes. Nowadays, a majority of precast concrete suppliers have set up their fabrication yards in remote areas to take advantage of cheap labor and land costs. As a result, higher mileages and energy are needed for transporting fabricated materials from manufacturing yards to construction sites. Of paramount interest is whether an offsite prefabricated element emits less CO$_2$ than a cast–in–place element. In order to examine

**Table 3.8** Increase in percentage of the $CO_2^-$ emissions of the recycled material compared to its virgin material

| Type of materials | Percentage change in $CO_2$ emissions over the virgin materials (%) |
|---|---|
| Recycled concrete | 5 |
| Recycled plasterboard | 48 |
| Recycled aluminum | −80 |
| Recycled steel | −40 |
| Recycled wood | −22 |

*Source*: From reference Gao W, Ariyama T, Ojima T, Meier A. Energy impacts of recycling disassembly material in residential buildings. Energy Build 2001;33(6):553–562.

this, it is necessary to take both embodied energy and transportation energy into account during the $CO_2$ evaluation. Assuming that 50–80% of offsite fabricated building materials were used for a new office building, the corresponding change in $CO_2$ emissions, $\Delta Q_p$ (in kg $CO_2$), was determined from the following:

$$\Delta Q_p = \sum_1^i m_{p,i} \star (q_{p,i} + q_{p,T,i} - q_{instiu,i}) \tag{3.8}$$

where $\Delta Q_p$ is the change in the $CO_2$ emissions due to prefabrication (in kg $CO_2$); $q_p$ is the $CO_2$ emissions per kilogram of material associated with the manufacturing of the material constituting the prefabricated building element (in kg $CO_2$/kg); $q_{p,T}$ is the $CO_2$ emissions per kilogram of material related to transportation of the building material constituting the prefabricated building element from the manufacturing yard to the construction site (in kg $CO_2$/kg); $q_{insitu}$ is the $CO_2$ emissions per kilogram of material used for casting the building element in place (in kg $CO_2$/kg). $i$, pertains to the $i$th type of building material; p, pertains to prefabrication.

In estimating the change in the amount of $CO_2$ emissions, facades, staircases, slabs, external elements, and partition walls were assumed to be prefabricated within a yard, and the amount of energy required for assembling offsite fabricated building elements was assumed to be the same as that required for assembling the building elements on site.

## 3.4 RESULTS AND ANALYSIS

Tables 3.9 and 3.10 list the basic parameter values that portray the distribution profiles of weights and $CO_2$ emission factors respectively for

**Table 3.9** Mass distributions of different building elements and types of materials

| Building element | Major material group | Mass per construction floor area (in kg/m²) |
|---|---|---|
| Doors | Plastic | 0.03–0.3 |
| | Plywood | 0.1–0.9 |
| | Stainless steel | 0.04–0.7 |
| External walls | Aluminum | 1.7–13.7 |
| | Concrete | 41.4–628.2 |
| | Reinforcing bar | 2.3–68.4 |
| | Stainless steel | 0.2–1.8 |
| | Stone | 0.2–2.5 |
| Floor surfacing and finishes | Galvanized steel | 0.6–4.7 |
| | Plaster | 0.02–0.4 |
| | Stone | 1.4–9.8 |
| | Tile | 0.7–9.4 |
| Internal walls and partitioning | Bricks and blocks | 0.5–3.3 |
| | Concrete | 0.7–1.6 |
| | Galvanized steel | 0.2–8.5 |
| | Glass | 0.04–2.5 |
| | Reinforcing bar | 2.0–7.1 |
| | Stainless steel | 0.01–0.8 |
| Paint system | Paint | 0.09–0.9 |
| Roof construction | Galvanized steel | 0.3–2.4 |
| | Concrete | 0.3–6.1 |
| | Plaster | 5.2–11.7 |
| | Stone | 0.3–2.4 |
| | Tile | 0.7–1.0 |
| Roof insulation | Asphalt and bitumen | 0.1–1.9 |
| | Plaster | 0.3–19.5 |
| | Thermal insulation | 0.1–0.3 |
| Suspended ceilings and finishes | Acoustic insulation | 0.1–3.8 |
| | Aluminum | 0.1–1.2 |
| | Galvanized steel | 0.1–7.6 |
| | Plaster | 0.7–4.9 |
| | Thermal insulation | 0.1–3.0 |
| Upper-floor construction | Concrete | 490.4–1271.8 |
| | Galvanized steel | 2.0–51.0 |
| | Plaster | 0.1–6.5 |
| | Reinforcing bar | 24.5–237.3 |
| | Structural steel | 1.0–130.0 |
| | Tile | 0.1–1.8 |
| Wall finishes | Aluminum | 0.01–0.8 |
| | Galvanized steel | 0.1–1.7 |
| | Plaster | 10.1–43.3 |
| | Stone | 2.3–13.8 |
| | Tile | 0.4–10.3 |
| Wall insulation | Plaster | 0.5–8.4 |
| | Thermal insulation | 0.3–21.9 |
| Windows/curtain wall | Aluminum | 0.1–0.6 |
| | Glass | 2.8–32.3 |

**Table 3.10** $CO_2$ emission factor distributions of different types of building materials

| Type of building material | $CO_2$ emission factors (in kg $CO_2$/kg) |
|---|---|
| Aluminum | 3.49–6.58 |
| Bitumen and asphalt | 0.074–1.09 |
| Brick/block | 0.011–0.072 |
| Concrete | 0.045–0.06 |
| Galvanized steel | 0.63–0.72 |
| Glass | 0.078–0.29 |
| Stone/gravel/aggregate | 0.0020–0.016 |
| PFA | 0.0017 |
| Paint | 0.98–2.35 |
| Plaster, render, and screed | 0.0022–0.044 |
| Plastic, rubber, and polymer | 1.52–2.52 |
| Plywood | 0.054–0.33 |
| Precast concrete element | 0.033 |
| Reinforcing bar/structural steel | 0.12–0.80 |
| Stainless steel | 0.16–0.27 |
| Thermal/acoustic insulation | 0.022–0.33 |
| Ceramic/tile | 0.048–0.12 |

different building elements and materials based on the Monte Carlo method implied with PDFs. According to the results, the average embodied energy content for the superstructure of a new concrete-framed office building is $10.3 \, GJ/m^2$ of construction floor area. This is higher than those reported by some earlier studies, which found that the initial embodied energy values for office buildings lie between 4 and $12 \, GJ/m^2$ [42,43]. Concrete invoked an extremely large mass in external walls and upper-floor construction. For the entire the high-rise concrete-framed office building, the mass of concrete also ranked high. However, the carbon emission of concrete was not as significant as the mass compared with other materials.

## 3.4.1 $CO_2$ Emissions from Building Elements

Fig. 3.3 shows the ranges of $CO_2$ emission values for different building elements. Among all the superstructure elements, upper-floor construction was the highest $CO_2$ emissions contributor with average emissions of $75.7 \, kg \, CO_2/m^2$. External wall, and suspended ceilings and finishes were the next two highest-impact elements. On average, emissions associated with external walls were $75.3 \, kg \, CO_2/m^2$, whereas emissions associated with suspended ceilings and finishes were $30.1 \, kg \, CO_2/m^2$. Taken

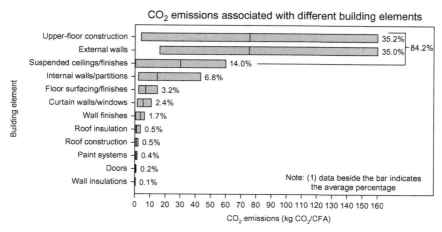

**Figure 3.3** Range of $CO_2$ emission values for different building elements.

**Table 3.11** Average contribution of the materials constituting the three major building elements

| Building element | Material | Average contribution of $CO_2$ emissions from the building element (in %) |
|---|---|---|
| Upper-floor construction | Concrete | 18.3 |
| | Galvanized steel | 1.4 |
| | Plaster | 0.1 |
| | Reinforcing bar | 68.9 |
| | Structural steel | 11.3 |
| | Tile | 0.1 |
| External walls | Aluminum | 69.9 |
| | Concrete | 18.6 |
| | Reinforcing bar | 11.1 |
| | Stainless steel | 0.4 |
| | Stone | 0.1 |
| Suspended ceilings and finishes | Acoustic insulation | 3.6 |
| | Aluminum | 39.7 |
| | Galvanized steel | 50.0 |
| | Plaster | 0.7 |
| | Thermal insulation | 6.0 |

together, emissions associated with these three elements were 181.1 kg $CO_2/m^2$. In view of their significant impacts, we will only focus on these three elements in our subsequent analysis.

Table 3.11 shows a breakdown in the average percentage contribution to the $CO_2$ emissions by the types of materials constituting the three

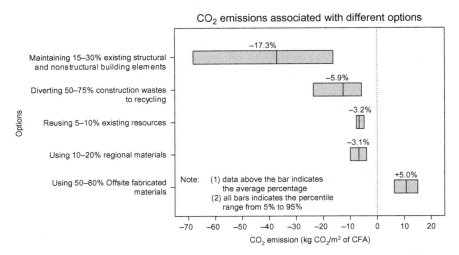

**Figure 3.4** $CO_2$ emissions reduction resulting from the implementation of different material-use options.

major elements. Although concrete contributes the largest mass of upper-floor construction and external walls, the $CO_2$ emission from concrete was not so high compared with aluminum and reinforcing bars.

### 3.4.2 Impact of Different Material-Use Options

Fig. 3.4 shows the $CO_2$ emissions reduction resulting from the implementation of different material-use options. Maintaining or reusing 15–30% of the existing structural and nonstructural elements can significantly reduce the $CO_2$ emissions by 37.1 kg $CO_2/m^2$ or 17.3% of the total. Diverting 50–75% of construction wastes to recycling is the second most effective option. Additionally, reusing 5–10% of the existing resources can reduce existing total emissions by 3.2% or 6.8 kg $CO_2/m^2$. An additional 10.7 kg of $CO_2$ emissions per meter square or 5.0% of the $CO_2$ emissions will be emitted if 50–80% offsite fabricated materials are used in facades and concrete elements.

### 3.5 DISCUSSIONS AND CONCLUSIONS

This study successfully developed Monte Carlo method for portraying the $CO_2$ emission profile for the superstructure of a new high-rise concrete office building in Hong Kong and for evaluating the impacts of various material-use options. Our findings indicate that the average $CO_2$ emissions due to the use of materials in the superstructure of current office buildings are $215.1\,kg\ CO_2/m^2$. External walls and upper-floor construction accounted for the highest $CO_2$ emissions from the superstructure of an office building, followed by suspended ceilings and finishes. These three elements became the major focus in our evaluation of impacts of different material-use options as they together already accounted for 84.2% of the $CO_2$ emissions on average. Concrete, reinforcing bars, aluminum, and galvanized steel were the major materials for $CO_2$ emissions.

The amount of emissions reduction greatly depends on the quantities of materials to be maintained or reused for the existing elements. On the other hand, the $CO_2$ emissions will even be increased if offsite prefabricated materials are used in a building, i.e., it will emit an additional $6.3–15.1\,kg$ of $CO_2$ per $m^2$, or 5% ($10.7\,kg$ of $CO_2$ per $m^2$) on average. However, this should be weighed against the benefit gains reaped by an increase in speed of construction, improvement in quality of products, and reduction in material wastage if prefabricated materials are used.

This study provides a general view of the $CO_2$ emission of concrete-framed high-rise office buildings. Through improving the accuracy of estimating or manufacturing data for material quantities; the embodied energy collected in different countries; and waste management methods, human behaviors, and government policies, the $CO_2$ emissions results will make a greater difference [40,44,45]. Calcination and carbonation of concrete should not be ignored for the study of concrete framed high-rise office buildings. Besides, recycling content can also be considered in future studies, especially for concrete, reinforcing bars, aluminum, and galvanized steel.

## REFERENCES

[1] Change IC. Impacts, adaptation and vulnerability. Contribution of working group iii to the fourth assessment report of the intergovernmental panel on climate change. New York: Cambridge University Press; 2007.
[2] U.S. Department of Energy. Building energy data book, 2010.
[3] Dowden M. Climate change and sustainable development: law, policy and practice: Taylor & Francis; 2008.
[4] Electrical and Mechanical Services Department. Hong Kong energy end-use data 2010, 2010.
[5] Joshi AS, Dincer I, Reddy BV. Performance analysis of photovoltaic systems: a review. Renew Sust Energy Rev 2009;13(8):1884–97.
[6] Jiang P, Tovey NK. Opportunities for low carbon sustainability in large commercial buildings in China. Energy Policy 2009;37(11):4949–58.
[7] Knudstrup MA, Hansen HTR, Brunsgaard C. Approaches to the design of sustainable housing with low $CO_2$ emission in Denmark. Renew Energy 2009;34(9):2007–15.
[8] Gratia E, De Herde A. Design of low energy office buildings. Energy Build 2003;35(5):473–91.

[9] Karlsson JF, Moshfegh B. A comprehensive investigation of a low-energy building in Sweden. Renew Energy 2007;32(11):1830–41.

[10] Sartori I, Hestnes AG. Energy use in the life cycle of conventional and low-energy buildings: a review article. Energy Build 2007;39(3):249–57.

[11] Cole RJ, Wong KS. Minimizing environmental impact of high-rise residential buildings. Proceedings of housing for millions: the challenge ahead 1996:262–5.

[12] Reddy BV, Jagadish KS. Embodied energy of common and alternative building materials and technologies. Energy Build 2003;35(2):129–37.

[13] Junnila S, Horvath A, Guggemos AA. Life-cycle assessment of office buildings in Europe and the United States. J Infrastruct Syst 2006;12(1):10–17.

[14] González MJ, Navarro JG. Assessment of the decrease of $CO_2$ emissions in the construction field through the selection of materials: practical case study of three houses of low environmental impact. Build Environ 2006;41(7):902–9.

[15] Guggemos AA, Horvath A. Comparison of environmental effects of steel-and concrete-framed buildings. J Infrastruct Syst 2005;11(2):93–101.

[16] Tae S, Baek C, Shin S. Life cycle $CO_2$ evaluation on reinforced concrete structures with high-strength concrete. Environ Impact Assess Rev 2011;31(3):253–60.

[17] Vieira PS, Horvath A. Assessing the end-of-life impacts of buildings. Environ Sci Technol 2008;42(13):4663–9.

[18] Dodoo A, Gustavsson L, Sathre R. Carbon implications of end-of-life management of building materials. Resour Conserv Recycl 2009;53(5):276–86.

[19] Building Cost Information Service. Life expectancy of building components, 2006.

[20] Chau CK, Yik FWH, Hui WK, Liu HC, Yu HK. Environmental impacts of building materials and building services components for commercial buildings in Hong Kong. J Clean Prod 2007;15(18):1840–51.

[21] Anderson J, Shiers D, Sinclair M. The green guide to specification: an environmental profiling system for building materials and components. : John Wiley & Sons; 2002.

[22] Scheuer C, Keoleian GA, Reppe P. Life cycle energy and environmental performance of a new university building: modeling challenges and design implications. Energy Build 2003;35(10):1049–64.

[23] Chen TY, Burnett J, Chau CK. Analysis of embodied energy use in the residential building of Hong Kong. Energy 2001;26(4):323–40.

[24] Huberman N, Pearlmutter D. A life-cycle energy analysis of building materials in the Negev desert. Energy Build 2008;40(5):837–48.

[25] Kofoworola OF, Gheewala SH. Life cycle energy assessment of a typical office building in Thailand. Energy Build 2009;41(10):1076–83.

[26] Hong Kong: Environmental Protection Department and Electrical and Mechanical Services Department. Guidelines to account for and report on greenhouse gas emissions and removals for buildings (commercial, residential or institutional purpose) in Hong Kong, 2010.

[27] Electrical and Mechanical Services Department of Hong Kong. Life cycle energy analysis of building construction; final report—consultancy agreement, 2005.

[28] Pommer K, Pade C. Guidelines: uptake of carbon dioxide in the life cycle inventory of concrete. : Nordic Innovation Centre; 2006;1–82.

[29] Worrell E, Price L, Martin N, Hendriks C, Meida LO. Carbon dioxide emissions from the global cement industry 1. Annu Rev Energy Environ 2001;26(1):303–29.

[30] Force, A.C.C.T. Status report of China cement industry. In 8th CTF meeting, vol. 24. Vancouver; 2010.

[31] Cement, Concrete and Aggregates Australia. Moisture in concrete and moisture sensitive finishes and coatings, 2007.

[32] Harris RH, Burmaster DE. Restoring science to superfund risk assessment. Toxics Law Reporter 1992;6:1318–23.

[33] Easy–fit. Easy–fit software, 2008.

[34] Mat–lab. Mat–lab software, 2007.

[35] Kelton WD, Law AM. Simulation modeling and analysis. Boston: McGraw Hill; 2000.

[36] Raychaudhuri, S. Introduction to Monte Carlo simulation. In simulation conference, 2008. WSC 2008. Winter, IEEE; 2008. p. 91–100.

[37] Census and Statistics Department. Hong Kong merchandise trade statistics, 2002.

[38] Davis, S.C., Diegel, S.W., & Boundy, R.G.. Transportation energy data book, 2008.

[39] Corbett JJ, Koehler HW. Updated emissions from ocean shipping. J Geophys Res Atmos 2003;108(D20).

[40] Poon CS, Ann TW, Ng LH. On-site sorting of construction and demolition waste in Hong Kong. Resour Conserv Recycl 2001;32(2):157–72.

[41] Gao W, Ariyama T, Ojima T, Meier A. Energy impacts of recycling disassembly material in residential buildings. Energy Build 2001;33(6):553–62.

[42] Cole RJ, Kernan PC. Life-cycle energy use in office buildings. Build Environ 1996;31(4):307–17.

[43] Yohanis YG, Norton B. Life-cycle operational and embodied energy for a generic single-storey office building in the UK. Energy 2002;27(1):77–92.

[44] Chung SS, Lo CW. Evaluating sustainability in waste management: the case of construction and demolition, chemical and clinical wastes in Hong Kong. Resour Conserv Recycl 2003;37(2):119–45.

[45] Tam VW, Tam CM, Chan JK, Ng WC. Cutting construction wastes by prefabrication. Int J Constr Manag 2006;6(1):15–25.

# CHAPTER 4

# Eco-Friendly Concretes With Reduced Water and Cement Content: Mix Design Principles and Experimental Tests

**T. Proske, S. Hainer, M. Rezvani and C.-A. Graubner**
Technische Universität Darmstadt, Darmstadt, Germany

## 4.1 CONCRETE FOR ECO-FRIENDLY STRUCTURES

To ensure the future competitiveness of concrete as a building material, it is essential to improve the sustainability of concrete structures. Great potential for reducing the environmental impact and consumption of scarce resources has been identified in the field of concrete construction, especially in the production of raw materials, concrete technology, and structures [1] (see Fig. 4.1). For concretes that are developed, produced, and used in an environmentally friendly manner the term "green concrete" [2] is commonly used.

The major environmental impact of concrete comes from $CO_2$ emissions during cement production as a result of the calcination and grinding process. The $CO_2$ emissions are mainly related to the decarbonation of limestone and the consumption of electricity and fuel [3]. Approximately 5% of global anthropogenic $CO_2$ emissions are connected with the production of 3.3 billion tons of cement per annum [4]. Therefore, reducing the cement clinker content might have positive effects on the environmental life–cycle assessment of concrete. Some research work on reducing the cement clinker content in concrete has already been carried out. However, there exist different research strategies. Often replacement of some clinker for large amounts of slag or fly ash was investigated [5–7] based on conventional concrete technology. This could lead to a waste of scarce raw materials such as slag and fly ash. The aim of other research activities is the efficient use of cement and reactive materials like slag and fly ash in concrete [1,2,8–12] based on a modified mix design approach.

*Handbook of Low Carbon Concrete.*
DOI: http://dx.doi.org/10.1016/B978-0-12-804524-4.00004-X

63

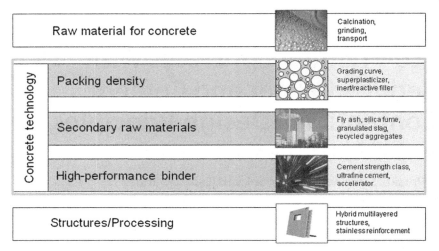

**Figure 4.1** Opportunities for ecological optimization in concrete construction [1].

In the following sections the general procedure for the development of structural concrete with low environmental impact and normal compressive strength, including the step-by-step development of the mix design, is outlined. The results of performance tests on clinker-reduced concretes conducted in laboratory conditions are also shown. In addition, the advantages with regard to the evaluation of environmental performance were verified. Finally, this chapter covers the application in the precast and ready-mix industry as well as the technical benefits of clinker-reduced eco-friendly concretes.

## 4.2 PRINCIPLES FOR THE DEVELOPMENT OF ECO-FRIENDLY CONCRETES WITH LOW ENVIRONMENTAL IMPACTS

### 4.2.1 Generals

In the following section an approach to reduce the environmental impacts connected to concrete production will be presented. This general approach can be applied on two different levels, the binder level and the concrete level.

The binder or cement level, focuses on the development of binder or cement with low environmental impacts in the range of current cement standards [13] or a totally new binder with very low or no clinker content. Such research was already conducted by the authors [13,14]. New cements with limestone content up to 50% were developed. However, the concrete technology has to be adapted in accordance with the principle mentioned later in Section 4.2.2.

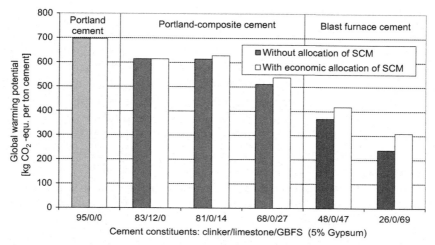

**Figure 4.2** GWP of different cements according to Ref. [16], without allocation and based on economic allocation coefficients [17].

This chapter is mainly focused on the concrete level, which aims to develop new eco-friendly concretes with reduced cement content. However, the cements used in such concretes are either in the range of conventional allowable cements or beyond the limits of national standards.

## 4.2.2 Low-Carbon Concretes With Reduced Cement Contents

Based on experimental results, a step-by-step procedure for the development of low-carbon concretes with efficient use of reactive materials was devised [15]. The following key steps are recommended:

1. Selection of cement of a high-strength class and eco-friendly constituents such as limestone, granulated blast furnace slag (GBFS), or fly ash.
2. Optimization of water content and cementitious material in the concrete paste.
3. Optimization of the paste volume.

The first step is the selection of cement. Preferable are cements with a low environmental impact, especially with a low global warming potential (GWP) (see Fig. 4.2), as well as a relatively high-strength performance such as Portland composite cements and blast furnace cements with a compressive strength of more than $42.5 \, N/mm^2$ based on a water–cement ratio of 0.5. However, the increased use of slag and blast furnace cements as well as fly ash is limited in several countries by the availability of these materials. In this case and if a high early strength concrete is required,

Portland cement with a strength of at least $52.5 \, \text{N/mm}^2$ is also an appropriate option. It must be mentioned that the values in Fig. 4.2 are not universally valid [16]. Kiln energy efficiency and $CO_2$ footprint of used fuel have a large influence. Furthermore, standards allow a large variation in supplementary cementing materials (SCMs) content for any given exposure class, which increases uncertainty in the evaluation. The recent tendency of allocating $CO_2$ to SCM, which are byproducts, will change the outcomes [17]. The effect of the allocation to the GWP of cements with GBFS is presented in Fig. 4.2. For the calculation of the $CO_2$ footprint, the coefficients based on the economic allocation [17] were used. It is shown that the impact of blast furnace cements is now significantly higher but still lower than that of Portland cement. Of course these results do not consider other impact categories (e.g., human toxicity or energy consumption), which have much higher allocation factors, and the limited availability of GBFS. For future application, in accordance with a modified low-water concrete technology, the development of cements with higher limestone content is in progress [18].

In the second step the volume of cement and cementitious materials should be minimized. To achieve a significant reduction, the concrete technology for ordinary concretes was modified based on the principles of high-performance concretes. The basic idea is the reduction of water so as to allow reducing the reactive components, i.e., clinker, slag, and fly ash in the concrete mixture. However, acceptable workability has to be ensured. This can be provided by sufficient paste content and surplus water based on increased powder content and a higher packing density of the granular mixture (Fig. 4.3).

Fig. 4.3 shows that the application of high-performance superplasticizers increases the dispersion of particles and allows a higher *actual* packing density of the solid powder particles (<0.125 mm) to be obtained. *Virtual* packing is the maximum achievable packing density and therefore not influenced by the superplasticizer or the particular packing process [19].

To achieve higher actual and virtual packing density, optimization of the particle-size distribution is recommended, using different approaches [19–21]. Of special importance for higher *actual* packing density is an increased grading span, which can be obtained by a certain amount of finer particles, the use of compact or rounded particles, and continuous grading with increased ratio of fractions of larger particle size [19].

The measures described lead to less water being required, which in turn allows the water–powder ratio in the mixture to be reduced while still

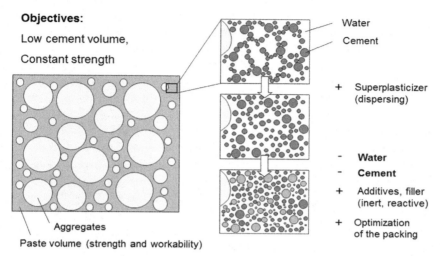

**Figure 4.3** Evolution from the conventional mixture proportion to cement-reduced eco-friendly concrete [8].

providing sufficient workability. Based on a decreased water content while achieving an increase in strength and durability, it is possible to reduce the cement content. The corresponding cement and water volume is replaced by environmentally friendly powders such as limestone. Investigations have shown that optimized limestone powders, though mostly inert, contribute considerably to strength development. The use of fly ash or slag is also possible. However, considering the limited availability of these reactive byproducts, widely available limestone should be preferred for the major part. It should also be mentioned that the current most-common practice of considering fly ash and slag as $CO_2$ neutral is under revision.

An additional reduction in cement content in step 3 can be realized based on a reduction in paste content. For the requisite optimization of aggregate grading, existing knowledge can be applied [19,21,22].

The principles for development of eco-friendly concretes are described and their effects on concrete strength, water content, and workability are presented qualitatively in Fig. 4.4. In this case the optimization is based on a conventional concrete mixture. The potential reduction in cement clinker conforms to the cement clinker quality, the contribution of additives to concrete performance, and the decrease in water. An almost linear correlation between cement clinker content and environmental impacts results in a better environmental performance of the concretes with reduced water and cement content.

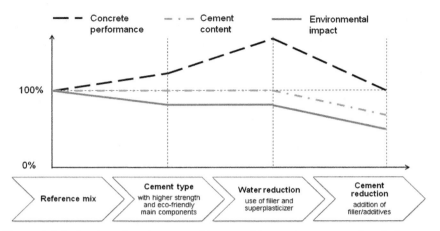

**Figure 4.4** Reduction of environmental impact of low-water concrete in relation to a conventional reference concrete (steps 1 and 2).

For the application of eco-friendly concretes in practice, the questions must be answered regarding strength development and durability as well as workability and pumpability. Changes to the conventional mixtures must not diminish the material performance. In particular, where eco-friendly concretes are to be exposed to severe conditions, durability is of great importance including performance in the presence of carbonation-induced corrosion, freeze–thaw cycling, and sulfate attack [23–25]. The viscosity of the eco-friendly concretes is higher compared to conventional concrete as a result of the required low water–powder ratio. The contractor should define the maximal acceptable viscosity depending on the concreting technology. This parameter controls the minimum water content and hence the potential for the clinker reduction.

## 4.3 LABORATORY TESTS

### 4.3.1 Overview and Targets

Different concrete mixtures with reduced cement content were developed especially for conventionally reinforced concrete structures. The initial investigations based on laboratory tests are described below.

A compressive strength of $10\,N/mm^2$ at an age of 24 h with a curing temperature of 20°C was targeted to enable demolding. After 28 days an average compressive strength of $38\,N/mm^2$ was desired to obtain a concrete strength sufficient for certain applications. The cement-reduced concretes must have sufficient workability. Therefore the table-test flow value according to DIN EN 12350-05:2009-08 was chosen to be 550 mm. The cement-reduced concretes are intended to be used for interior structures (exposure class XC1) as well as for exterior structures (exposure classes XC4, XF1, and XA1). The standard DIN 1045-2:2008-08 defines the German national requirements for concrete mix design depending on the exposure classes. For application in exterior structures (exposure classes XC4, XF1, and XA1) the minimum cement content is 270 kg/m³, for interior structures 240 kg/m³ (XC1). The water–cement ratio, including all cement constituents, may not exceed 0.60 and 0.75, respectively. In case of addition of fly ash an efficiency factor of 0.4 is considered, up to a weight of 33% of cement mass.

## 4.3.2 Constituents and Concrete Mix Design

To evaluate concrete performance, conventional reference concretes based on the concrete mix design according to DIN 1045-2:2008-08 were included in the test program. The mix design for the reference concretes with a cement content of 240 and 270 kg/m³ is shown in Table 4.1.

Starting with the reference concrete, the conventional cement content was progressively reduced from 270 to 100 kg/m³ (see Table 4.2). Additives were gradually substituted for cement. At the same time, the water volume was reduced. The lowest water content was 125 L/m³. To maintain sufficient workability the powder content (<0.125 mm) was increased up to 440 kg/m³ by the addition of fly ash and limestone powder. Concrete consistency was adjusted by changing the dosage of superplasticizer. Generally a Portland cement with nominal strength of 52.5 N/mm², high early strength, and a defined cement clinker content (Clinker 1) were used. In addition, two Portland cements with a lower strength class (Clinkers 2 and 3) were included. Subsequently, the influence of a blast furnace cement composed of 60% clinker (Clinker 4) and 40% GBFS was analyzed. Limestone powder and fly ash additives were used in ratios of 0, 0.5, and 1.0 by volume.

In an additional test series, the effect of the limestone powder fineness on the concrete properties was analyzed (see Table 4.3). The Blaine value of the standard Limestone 1 was 0.31 m²/g and of the fine Limestone 2 1.60 m²/g. The location parameters (particle diameter for accumulation by weight) are $d_{10} = 3.3 \, \mu m$, $d_{50} = 15.4 \, \mu m$, $d_{90} = 59.1 \, \mu m$ and $d_{10} = 0.7 \, \mu m$, $d_{50} = 1.8 \, \mu m$, $d_{90} = 3.9 \, \mu m$, respectively.

**Table 4.1** Mix design of the reference concretes

| Mix design | Mass per m³ | Reference concretes | | | | | | |
|---|---|---|---|---|---|---|---|---|
| | | C1-270-FA10-w165 | C2-270-FA10-w165 | C3-270-FA10-w165 | C1-240-w180 | C3-240-w180 | C1-240-FA160-w180 | C1-240-FA160-w145 |
| Clinker 1: CEM 1 52.5 R | kg | 270 | — | — | 240 | — | 240 | 240 |
| Clinker 2: CEM 1 42.5 R | kg | — | 270 | — | — | — | — | — |
| Clinker 3: CEM 1 32.5 R | kg | — | — | 270 | — | 240 | — | — |
| Fly ash (EN 450) | kg | 10 | 10 | 10 | — | — | 160 | 160 |
| Limestone powder 1 | kg | — | — | — | — | — | — | — |
| Water | kg | 162 | 162 | 162 | 180 | 180 | 179 | 142 |
| Superplasticizer | kg | 2.8 | 1.9 | 3.0 | — | 1.3 | 1.7 | 4.0 |
| River sand 0–2mm | kg | 597 | 603 | 603 | 601 | 601 | 569 | 509 |
| River gravel 2–8mm | kg | 446 | 446 | 446 | 444 | 444 | 394 | 446 |
| River gravel 8–16mm | kg | 847 | 847 | 847 | 842 | 842 | 748 | 846 |
| w/c | [–] | 0.61 | 0.61 | 0.61 | 0.75 | 0.76 | 0.75 | 0.60 |
| w/c$_{eq}$ | [–] | 0.60 | 0.60 | 0.60 | 0.75 | 0.76 | 0.66 | 0.53 |

**Table 4.2** Mix design of the cement-reduced concretes

| Mix design | Mass per m³ concrete | C1-200-FA 200-w145 | C1-175-FA225-w145 | C1-150-FA250-w145 | C1-150-FA 125-LS 145-w145 | C1-150-LS289-w145 | C1-150-FA250-w125 | C1-150-LS289-w125 | C4-90-GBFS60-FA 250-w145 | C4-90-GBFS60-FA 125-LS 145-w145 | C4-90-GBFS60-LS289-w145 | C1-125-FA275-w145 | C1-100-FA 300-w145 |
|---|---|---|---|---|---|---|---|---|---|---|---|---|---|
| | | | | | | | | Cement clinker-reduced concretes | | | | |
| Clinker 1: CEM 152.5 R | kg | 200 | 175 | 150 | 150 | 150 | 150 | 150 | — | — | — | 125 | 100 |
| Clinker 4: CEM 152.5 R | kg | — | — | — | — | — | — | — | 90 | 90 | 90 | — | — |
| GBFS | kg | — | — | — | — | — | — | — | 60 | 60 | 60 | — | — |
| Fly ash (EN 450) | kg | 200 | 225 | 250 | 125 | — | 250 | — | 250 | 125 | — | 275 | 301 |
| Limestone powder 1 | kg | — | — | — | 145 | 289 | — | 289 | — | 145 | 289 | — | — |
| Water | kg | 142 | 142 | 142 | 142 | 142 | 120 | 120 | 143 | 142 | 142 | 143 | 144 |
| Superplasticizer | kg | 4.1 | 3.9 | 3.1 | 4.0 | 5.1 | 6.9 | 6.5 | 2.4 | 3.3 | 4.5 | 3.1 | 1.9 |
| River sand 0–2mm | kg | 515 | 519 | 523 | 524 | 524 | 542 | 542 | 523 | 524 | 524 | 528 | 534 |
| River gravel 2–8mm | kg | 440 | 436 | 434 | 434 | 434 | 444 | 444 | 434 | 434 | 434 | 429 | 424 |
| River gravel 8–16mm | kg | 834 | 828 | 823 | 823 | 823 | 843 | 843 | 823 | 823 | 823 | 814 | 804 |
| w/c | [–] | 0.73 | 0.83 | 0.96 | 0.97 | 0.97 | 0.84 | 0.83 | 1.61 | 1.61 | 1.61 | 1.16 | 1.45 |
| $w/c_{eq}$ | [–] | 0.64 | 0.73 | 0.85 | 0.85 | 0.97 | 0.74 | 0.83 | 1.42 | 1.42 | 1.61 | 1.02 | 1.28 |

**Table 4.3** Mix design of cement-reduced concretes with different limestone fineness

| Mix design | Mass per m³ concrete | C5-150-LS289/0-w145 | C5-150-LS246/43-w145 | C5-150-LS202/87-w145 | C5-150-LS159/130-w145 | C5-150-LS116/173-w145 | C5-150-LS72/217-w145 | C5-150-LS0/289-w145 |
|---|---|---|---|---|---|---|---|---|
| | | Cement clinker-reduced concretes with different limestones | | | | | | |
| Clinker 5: CEM I 52.5 R | kg | 150 | 150 | 150 | 150 | 150 | 150 | 150 |
| Limestone powder 1[a] | kg | 289 | 246 | 202 | 159 | 116 | 72 | 0 |
| Limestone powder 2[b] | kg | 0 | 43 | 87 | 130 | 173 | 217 | 289 |
| Water | kg | 142 | 142 | 142 | 142 | 142 | 142 | 142 |
| Superplasticizer[c] | kg | 2.5 | 2.5 | 1.8 | 2.2 | 3.0 | 3.0 | 3.1 |
| River sand 0–2 mm | kg | 524 | 524 | 524 | 524 | 524 | 524 | 524 |
| River gravel 2–8 mm | kg | 434 | 434 | 434 | 434 | 434 | 434 | 434 |
| River gravel 8–16 mm | kg | 823 | 823 | 823 | 823 | 823 | 823 | 823 |
| w/c | [–] | 0.96 | 0.96 | 0.96 | 0.96 | 0.96 | 0.96 | 0.96 |

[a]Normal fineness.
[b]High fineness.
[c]Different producer.

### 4.3.3 Test Methods

After mixing, the table flow value of the concrete was determined and the dosage of superplasticizer adjusted as necessary to achieve a table flow of $550 \pm 20$ mm. Then, the specimens for the compressive strength (150 mm cubes) and carbonation tests (prisms $100 \times 100 \times 500$ mm) were produced. The samples for the compressive strength test were demolded after 1 day and stored in water at 20°C. Compressive strength was tested after 1, 3, 7, 28, and 91 days.

Resistance to carbonation was analyzed using the accelerated carbonation test method (ACC test method) according to Ref. [26]. After demolding the concrete prisms were stored until the age of 7 days in water at a temperature of $T = 20°C$. Subsequent to water storage the specimens were placed in a climate chamber for 21 days at $T = 20°C$, and RH = 65%. Then, the specimens were exposed for 28 days to an increased $CO_2$ concentration of 2%, $T = 20°C$, and RH = 65%. After this storage, the specimens were split and the carbonation depth $x_c$ was measured at the plane of rupture using an indicator solution (phenolphthalein). In addition to the ACC test method, selected concretes were stored in a normal $CO_2$ concentration ($T = 20°C$ and RH = 65%) for 2.5 years.

## 4.4 CONCRETE PROPERTIES

### 4.4.1 Workability and Strength Development

The requirements of table flow and compactibility were fulfilled by all mixtures, in spite of the significant reduction in water content. However, a higher demand of superplasticizer was necessary compared with the reference mixes. The minimum water volume for workability oriented on practical application was identified to be 145 L/m³. The plastic viscosity increased noticeably for concretes with 125 L/m³. Detailed studies on the viscosity of water-reduced eco-friendly concretes were conducted in additional test series [27].

The compressive strength measurements are presented in Table 4.4 and Fig. 4.5. It was noted that the loss of compressive strength, corresponding to the cement clinker reduction, can be compensated by decreased water volume and by using reactive powder such as fly ash and blast furnace slag. The application of cement clinker with higher strength is also advantageous.

**Table 4.4** Strength development and carbonation depth

| Concrete mix | | Compressive strength $f_{cm,cube}$ (N/mm$^2$) | | | | | Carbonation depth $X_c$ (mm) | |
| | | Concrete age | | | | | ACC test 28 days | Long-term test 2.5 years |
| Clinker additives | Water | 1 day | 3 days | 7 days | 28 days | 91 days | 2% CO$_2$ | Normal CO$_2$ |
|---|---|---|---|---|---|---|---|---|
| C1-270-FA10- | w165 | 34.6 | 42.0 | 49.4 | 53.9 | 61.2 | 2.9 | 2.7 |
| C2-270-FA10- | w165 | 15.9 | 30.1 | 38.5 | 40.8 | 52.3 | 2.8 | — |
| C3-270-FA10- | w165 | 10.0 | 20.9 | 28.9 | 34.7 | 39.6 | 5.6 | 8.1 |
| C1-240- | w180 | 15.1 | 22.7 | 30.5 | 33.6 | 34.3 | 6.1 | 5.4 |
| C3-240- | w180 | 4.1 | 11.9 | 18.0 | 24.0 | 25.6 | 11.1 | — |
| C1-240-FA160- | w180 | 22.8 | 31.6 | 38.2 | 46.8 | 59.7 | 5.4 | — |
| C1-240-FA160- | w145 | 34.5 | 42.9 | 56.0 | 69.4 | 82.0 | 1.1 | 0.1 |
| C1-200-FA200- | w145 | 21.8 | 36.5 | 45.1 | 57.0 | 72.9 | 2.3 | — |
| C1-175- FA225- | w145 | 21.3 | 32.6 | 38.5 | 55.0 | 66.5 | 4.1 | — |
| C1-150-FA250- | w145 | 11.7 | 24.4 | 29.2 | 42.7 | 54.6 | 9.8 | 6.0 |
| C1-150-FA125-LS145- | w145 | 15.0 | 25.6 | 30.7 | 38.8 | 52.9 | 9.4 | 7.3 |
| C1-150-LS289- | w145 | 11.6 | 21.8 | 22.2 | 27.6 | 30.4 | 12.3 | — |
| C1-150-FA250- | w125 | 14.3 | 28.4 | 32.2 | 55.3 | 69.6 | 3.1 | — |
| C1-150-LS289- | w125 | 18.3 | 31.9 | 32.6 | 37.8 | 38.4 | 8.4 | — |
| C4-90-GBFS60-FA250- | w145 | 7.2 | 17.2 | 30.9 | 50.9 | 62.9 | 6.3 | 8.1 |
| C4-90-GBFS60-FA125-LS145- | w145 | 6.2 | 18.1 | 29.7 | 45.4 | 50.0 | 7.0 | 8.7 |
| C4-90-GBFS60-LS289- | w145 | 6.9 | 15.0 | 29.9 | 40.8 | 42.4 | 7.4 | 9.4 |
| C1-125-FA275- | w145 | 8.2 | 17.3 | 26.4 | 39.3 | 44.0 | 14.2 | 8.9 |
| C1-100-FA300- | w145 | 4.8 | 10.7 | 16.9 | 26.3 | 31.9 | 21.0 | 14.1 |

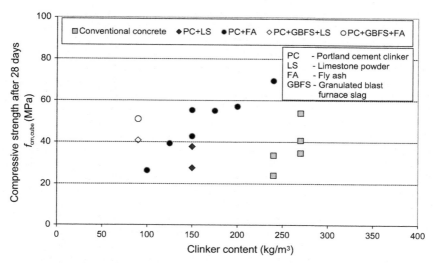

**Figure 4.5** Compressive strength versus cement clinker content.

Fig. 4.5 and Table 4.4 show that concretes with a low cement clinker content of 150 kg/m³ were able to meet the defined strength requirements. Both early strength and 28-day strength were acceptable. However, the concretes without slag reach the compressive strength target only with a certain amount of fly ash or a very low water volume of 125 L/m³. The relative clinker demand, which represents the required clinker mass per 1 N/mm² compressive strength, is shown in Fig. 4.6. It is obvious that the efficiency of the developed mixtures is very high compared with that of conventional concretes. Moreover, clinker demand is relatively constant in all strength categories.

To enable a further water reduction and consequently an additional reduction in clinker content, systematic optimization of the complete particle-size distribution of the granular powder is necessary.

Successive substitution of the ordinary limestone powder with finer limestone raised the 28-day compressive strength from the reference value of 32 N/mm² up to 46 N/mm² (Fig. 4.7). It is assumed that the more-homogeneous microstructure and the improved interface between the cement matrix and aggregates have positive effects. A great benefit of the fine limestone is also seen to be the considerable reduction in concrete viscosity as well as the lower demand in superplasticizer. The minimum of both was reached with a replacement ratio of 30% (see Table 4.3).

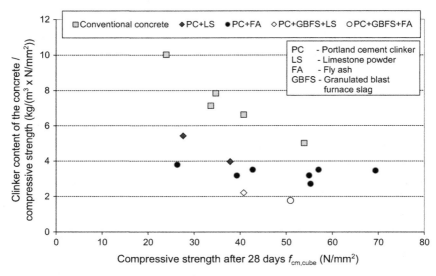

**Figure 4.6** Relative clinker demand of the concretes tested.

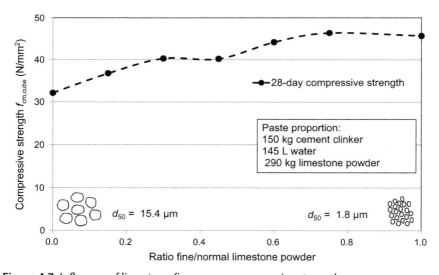

**Figure 4.7** Influence of limestone fineness on compressive strength.

## 4.4.2 Carbonation of the Concrete

Table 4.4 shows the measured carbonation depth as a result of the ACC test method and the long-term $CO_2$ storage. The concrete mix C1-270-FA10-w165 with high-strength cement has a relatively low

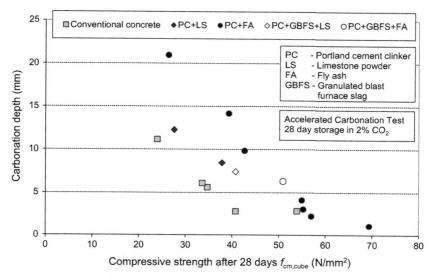

**Figure 4.8** Carbonation depths of the concretes versus compressive strength.

carbonation depth of 3 mm compared with concrete C3–270–FA10–w165 of low cement strength and a depth of 6 mm.

Compared with the carbonation depth of a conventional concrete (C3–270–FA10–w165), approximately the same value was measured for the concretes with only 90 kg/m³ of clinker and 60 kg/m³ of slag (C4–90–GBFS60–FA250–w145). In contrast, the concretes with 150 kg/m³ of Portland cement had a considerably higher carbonation depth than the reference concrete. Requirements for exterior structures could be met by reducing the water content or by a slight increase in cement content. The effect of fly ash, slag, and limestone powder on carbonation was considerable. As a result of higher strength and lower porosity, fly ash reduced the carbonation depth much more than limestone powder, notwithstanding the consumption of calcium hydroxide. However, for a constant strength, the mixtures with limestone powder show a lower carbonation depth than the concretes with fly ash. Hence, the influence of fly ash on the compressive strength was much more remarkable than the contributions to the carbonation resistance.

A reduction in cement clinker content to 125 and 100 kg/m³ tends to produce values that are significantly higher than the carbonation depths of the reference concretes. These concretes are preferable for application in interior members.

Carbonation depth versus compressive strength is presented in Fig. 4.8, which reveals that carbonation depth is not very well correlated with compressive strength based on all mixtures. However, a good correlation exists when the results are categorized by their SCMs and limestone additions. What is obvious is a lower carbonation depth of concretes with a conventional amount of cement clinker. The reduction in cement clinker was connected with a higher carbonation depth at the same compressive strength. For a given compressive strength, concretes with limestone had a better resistance to carbonation than those with fly ash. The used slag seems not to produce negative effects on carbonation resistance.

The long-term carbonation tests under normal $CO_2$ exposure showed a remarkably different trend compared to the results with higher $CO_2$ concentration (Table 4.4). In particular, the correlation between compressive strength and carbonation depth is more significant. This is probably due to the fact that the delayed pozzolanic reaction of fly ash increases the density and compressive strength over time and therefore leads to a higher carbonation resistance compared to the results of the accelerated test.

It can be concluded that, compared with conventional concrete, the same carbonation depth on concretes with reduced cement clinker is only achievable by providing higher compressive strength. However, adequate performance of such concretes with a moderate increase in strength can be verified based on the performance concepts of new design standards and long-term tests. An analytical model for the prediction of carbonation depth based on an existing general model [26] is presented in Ref. [28]. This new model considers the specific mix proportion of concrete with already used additives. It includes the contribution of cement strength class, amount of fly ash and limestone powder, as well as water content on the carbonation resistance of concrete.

### 4.4.3 Environmental Performance Evaluation

The optimization of cement clinker volume in the mix composition leads to a significant reduction in the environmental impact compared with the reference concrete mixtures. This improvement is specifically based on the use of fly ash, slag, and limestone powder. It has to be kept in mind that an allocation of environmental impacts to byproducts such as fly ash and slag can change this conclusion noticeably. In this case, the results depend significantly on the allocation criteria for the SCM. The allocation burdens can be associated with both the relative mass and current economic values of products and byproducts [17]. Every allocation has the effect that the calculated impact of the used waste (SCM) is in some cases and for different impact categories much higher than the replaced material (Portland

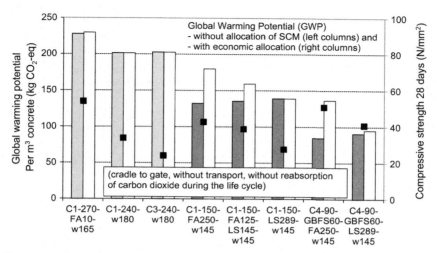

**Figure 4.9** GWP of selected cement-reduced concretes without allocation of SCM and with economic allocation [17] (cradle to gate, CML method, ökobau.dat 2010) and compressive strength.

cement clinker). This leads to problematic results and potentially prevents the use of byproducts that cannot be used for other applications in a reasonable way. However, the allocation procedure supports the efficient use of fly ash and slag in the mix design.

The GWP, which considers the distinctive effect of different greenhouse gases, was calculated using the environmental performance evaluation based on data for the constituents according to Ökobau.dat 2010 and the GaBi database [29], Netzwerk Lebenszyklusdaten [16], and the European Federation of Concrete Admixture Associations [30]. In a first step, no allocations were considered for slag and fly ash except for the secondary process. Also the reabsorption of carbon dioxide was not considered, due to the fact that the degree of carbonation in concrete members over the life cycle and the life cycle itself is uncertain.

Fig. 4.9 shows the GWP of selected concretes without allocation to SCM and with economic allocation according to Ref. [17]. For a comparable concrete strength the GWP without allocation was reduced by approximately 35% by using fly ash and limestone and by approximately 60% when using slag as cement clinker replacement. According to the environmental performance evaluation, other impact factors as well as primary energy consumption are also reduced significantly [8]. If the economic allocation to SCM is considered, the reduction of the GWP is only

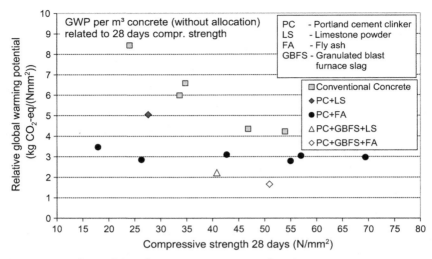

**Figure 4.10** Relative GWP of appropriate cement-reduced concretes versus compressive strength and used SCM without allocation.

15–25% and 35–45% using slag as a replacement. For other environmental categories (e.g., for acidification, photochemical oxidation) the calculated impact of the cement-reduced concretes based on the allocation to SCM is even higher.

To consider the performance of the concrete, the GWP was also related to the compressive strength. The results without allocation to SCM are shown in Fig. 4.10 for appropriate mixtures. The reduction of water and cement clinker leads to a relative GWP of approximately 3 kg $CO_2$-eq/(N/mm$^2$) notwithstanding the compressive strength. It can be further noticed that the reduction of environmental impact in comparison to conventional concrete was more remarkable for low- and medium-strength concrete.

## 4.5 APPLICATION IN PRACTICE

In cooperation with a producer of prefabricated concrete elements, the application of cement-reduced concretes in practice was tested. The mixture development was focused on semifinished concrete slabs and walls. At the construction site, the completion of the structural element is carried out with ready-mix concrete. The processing as well as the field of application specifies the requirements for the fresh and hardened concrete

**Table 4.5** Mix design of cement-reduced concretes for the precast industry

| Selected concretes | | DIN C25/30 | ECO-Concrete 1 | ECO-Concrete 2 | ECO-Concrete 3 |
|---|---|---|---|---|---|
| | | (Reference) | C25/30 | C30/37 | C30/37 |
| **Exposition class** | | XC4/XF1 | XC1 | XC1 | XC4/XF1 |
| Cement | kg/m³ | 275[a] | 150[b] | 150[b] | 180[b] |
| Fly ash | kg/m³ | 30 | 23 | 50 | 90 |
| Limestone powder | kg/m³ | — | 222 | 222 | 119 |
| Sand 0–4 mm | kg/m³ | 693 | 721 | 721 | 721 |
| Gravel 4–16 mm | kg/m³ | 1183 | 1126 | 1126 | 1126 |
| Superplasticizer | kg/m³ | 4.0[c] | 4.0[d] | 5.0[d] | 4.0[d] |
| Total water | kg/m³ | 172 | 146 | 135 | 146 |
| w/c | — | 0.63 | 0.97 | 0.9 | 0.81 |
| $w/c_{eq}$ | — | 0.6 | 0.92 | 0.8 | 0.72 |
| Flow diameter | cm | 53 | 53 | 50 | 55 |
| Air volume | % | 2.2 | 2.4 | 1.4 | 1.5 |

[a]CEM II/A-LL 42.5R.
[b]CEM 152.5R.
[c]SP 1.
[d]SP 2.

properties. Concrete slabs are usually used for interior elements (XC1, C20/30), while concrete walls are used for both interior as well as exterior elements (XC4, XF1, C30/37, and C35/45). To meet the processing requirements a compressive strength of 7 N/mm² was targeted after 5 h, including heat treatment with 50°C.

As a result of the mixture optimization, the cement content (CEM I equivalent) for interior and exterior elements was decreased from 255 to 150 kg/m³ in ECO-Concrete 1 and from 300 to 180 kg/m³ in ECO-Concrete 3, respectively (see Table 4.5 and Fig. 4.11). Hence, the optimization allowed a cement reduction of approximately 40%. At the same time, the powder content was increased by 100 kg/m³. The cement content of the developed mixes is below the minimum value accepted according to DIN 1045–1:2008–08. This requires a building authority approval for the production of reinforced concrete elements.

The performance of the laboratory tests was verified by producing the elements in a precast concrete plant and by full-scale tests. It was shown that an increased mixing time is necessary if the standard technology is used. The workability of the concretes during the processing time was sufficient. However, the quality and quantity of the materials must be

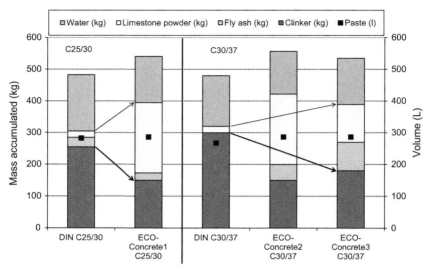

**Figure 4.11** Paste proportion of the cement-reduced concretes and the reference mixes.

**Table 4.6** Hardened concrete properties according to DIN EN 12390

| Hardened concrete properties tests after 28 days according to DIN EN 12390 | | DIN C25/30 reference | ECO-1 C25/30 | ECO-2 C30/37 | ECO-3 C30/37 |
|---|---|---|---|---|---|
| Compressive strength $f_{cm, cube\,150}$ | N/mm² | 30 | 35.9 | 40.8 | 49.1 |
| Compressive strength $f_{cm, cyl\,150}$ | N/mm² | 23.9 | 29.3 | 31.2 | 38.3 |
| Splitting strength $f_{ctm, sp}$ cylinder 150/300 mm | N/mm² | 2.63 | 3.14 | 3.56 | 3.81 |
| Flexure strength $f_{ctm,fl}$ prisms 150/150/700 mm | N/mm² | 4.61 | 4.48 | 5.38 | 4.86 |
| Modulus of elasticity $E_{cm}$ cylinder 150/300 mm $\sigma_{max} = 0.333 f_{cm,cyl}$ | N/mm² | 23.557 | 30.277 | 34.095 | 38,819 |
| Bond concrete/rebar $f_{bm,0,1mm}$ Pull-out test (RILEM), $d_s = 10\,mm, 0.1\,mm$ | N/mm² | 6.4 | 6 | 7.6 | 15.1 |
| Depth of penetration of water | cm | 6.3 | 6.6 | 3.9 | 2.1 |

controlled exactly. The measurement of aggregate moisture is of particular importance. The high early strength allowed the integration of the new concretes in the normal production process.

Several mechanical properties of the cement–reduced concretes were tested (Table 4.6). It must be noted that the concrete specimens

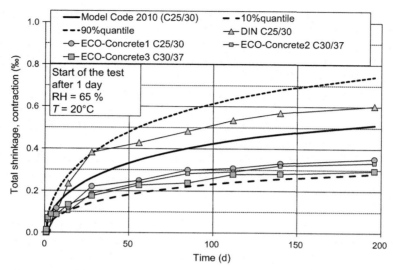

**Figure 4.12** Time-dependent shrinkage of the cement-reduced and reference concretes compared to the approach of Model Code 2010.

were produced with the maximum water content, which is expected in the later fabrication process. The properties of the concretes developed are mostly equivalent to the conventional concretes and the normative standards.

The low heat of hydration and the lower shrinkage (see Fig. 4.12) can be considered as advantages resulting from the decreased water and cement content. The reason for the increased modulus of elasticity is probably the lower porosity of the matrix due to the lower water–powder ratio for a given paste volume. In addition, ECO-3 fulfilled the requirements for exposure class XC4 (carbonation resistance) and XF1 (freeze–thaw resistance).

The environmental performance evaluation highlighted the advantages of the cement reduction for the construction of ordinary concrete structures with prefabricated elements. The used basic data are summarized in Table 4.7. Fig. 4.13 shows the GWP of reinforced concrete slabs produced with conventional concrete as well as eco-friendly concrete.

A decrease of the environmental impact by 50%, including the reinforcement and energy for the processing, is possible. However, cement-reduced eco-friendly mixtures should be used for both the prefabricated and the ready-mix concrete. For the latter case, the cement with low clinker content and moderate early strength (CEM III) is recommended.

**Table 4.7** Basic data for the environmental performance evaluation of concrete and concrete elements according to Ökobau.dat 2010 and GaBi database [29]

| Constituent | Reference unit | GWP (kg $CO_2$-eq.) | Primary energy (MJ) |
|---|---|---|---|
| Cement, CEM I | kg | 0.8198 | 3901 |
| Cement, CEM III A | kg | 0.5021 | 2389 |
| Cement, CEM III B | kg | 0.2887 | 1389 |
| Fly ash | kg | 0.0110 | 0.154 |
| Limestone powder | kg | 0.0278 | 0.444 |
| Superplasticizer | kg | 0.7721 | 16,915 |
| River sand | kg | 0.0023 | 0.037 |
| River gravel | kg | 0.0023 | 0.036 |
| Crushed aggregates | kg | 0.0068 | 0.107 |
| Electricity | kWh | 0.6550 | 12,465 |
| Heating oil | l | 3.3434 | 46,390 |
| Reinforcement | kg | 0.8744 | 13,407 |

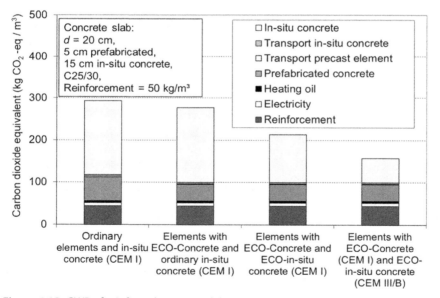

**Figure 4.13** GWP of reinforced concrete slabs.

The properties of these concretes were investigated in an additional joint research project [8].

The material costs for the cement–reduced concretes are either approximately the same compared to conventional ordinary concrete or slightly lower. However, the costs are highly influenced by the price for the additives, especially for the limestone powder and superplasticizer.

## 4.6 CONCLUSIONS

A stepwise approach was implemented in order to develop eco-friendly concretes with reduced cement and water contents. In parallel, several experimental attempts were conducted to evaluate the required concrete performance for practical purposes. According to the results, the following conclusions can be derived:

1. $CO_2$ emissions can be reduced significantly in structural concretes. A significant reduction in Portland cement demand may be achieved by using high-performance superplasticizer, high-strength cement and optimized particle-size distribution.
2. Replacement of Portland cement and water with mineral fillers such as limestone powder provides an optimal paste volume in the low water mixture. It was shown that concretes with cement clinker and slag contents as low as $150 \, kg/m^3$ were able to meet the usual requirements of workability, compressive strength ($\sim 40 \, N/mm^2$), and other mechanical properties.
3. The carbonation depth on eco-friendly concretes with at least $175 \, kg/m^3$ clinker and slag was observed to be lower than that of conventional concretes for exterior structures.
4. A reduction in the GWP of up to 35% compared with conventional concrete can be seen as well as a reduction of more than 60% when using GBFS. The allocation decreases the reduction, which is in the case of economical allocation 15–25% and 35–45% using slag.
5. Practical application was verified in precast and ready-mix concrete plants. Results showed an acceptable capability for eco-friendly concretes to be used in the aforementioned industries in both fresh and hardened states.

## ACKNOWLEDGMENTS

These studies were part of research projects in the field of Sustainable Concrete Construction. The authors are grateful for the financial support from the German Federal Ministry of Economics and Technology and the AiF Projekt GmbH. In addition, we would like to thank Beton Kemmler GmbH, Waibel KG, Mapei Betontechnik GmbH, BASF Construction Polymers GmbH, Dyckerhoff AG, Heidelberg Cement AG, and Lafarge Zement GmbH for their generous support.

# REFERENCES

[1] Graubner C-A, Hock C, Proske T, Schneider C. Leadership through sustainability—new challenges for the concrete industry. Proceedings of the 11th annual international fib symposium. Concrete: 21st century superhero—building a sustainable future London. Fédération Internationale du Béton (fib), Lausanne; 2009.

[2] Nielsen CV, Glavind M. Danish experiences with a decade of green concrete, ACT 5, 2007. p. 3–12.

[3] Fib Bulletin 67, Fib Bulletin 67: Guidelines for green concrete structures, 2012.

[4] U.S. Geological Survey. Mineral commodity summaries, 2011.

[5] Dhir RK, McCarthy MJ, Paine KA. Engineering property and structural design relationships for new and developing concretes. Mat Struct 2005;38:1–9.

[6] Bilodeau A, Malhotra VM. High-volume fly ash system: concrete solution for sustainable development. ACI Struct J 2000;97:41–8.

[7] Härdtl R, Koc I. Evaluation of the performance of multi-component cements. ZKG 2012;4:66–79.

[8] Proske T, Hainer S, Jakob M, Garrecht H, Graubner C-A. Stahlbetonbauteile aus klima- und ressourcenschonendem Ökobeton. Beton- und Stahlbetonbau 2012;107:401–13.

[9] Fennis-Huijben S. Design of ecological concrete by particle packing optimization. S.A.A.M. Fennis-Huijben, [S.l.] 2010.

[10] Fennis-Huijben S, Walraven J, Nijland T. Measuring the packing density to lower the cement content in concrete Walraven JC, Stoelhorst D, editors. Tailor made concrete structures: new solutions for our society. London, UK: Taylor & Francis Group; 2008. p. 19–424.

[11] Wallevik OH, Mueller FV, Hjartarson B, Kubens S. The green alternative of self-compacting concrete, Eco-SCC Proceedings 17. Internationale Baustofftagung (ibausil). Bauhaus-Universität Weimar; 2009;1105–16.

[12] Haist M, Müller HS. Nachhaltiger Beton—Betontechnologie im Spannungsfeld zwischen Ökobilanz und Leistungsfähigkeit Proceedings of the 9th Symposium Baustoffe und Bauwerkserhaltung, Nachhaltiger Beton, Werkstoff, Konstruktion und Nutzung. Karlsruhe: KIT Scientific Publishing; 2012;29–52.

[13] Müller C, Palm S, Graubner C-A, Proske T, Hainer S, Rezvani M, et al. Cements with a high limestone content—durability and practicability. Cement Int 2014;109:78–85.

[14] Müller C, Palm S, Graubner C-A, Proske T, Hainer S, Rezvani M, et al. Zemente mit hohen Kalksteingehalten—Dauerhaftigkeit und praktische Umsetzbarkeit. Beton 2014;01/02:43–50.

[15] Proske T, Hainer S, Rezvani M, Graubner C-A. Eco-friendly concretes with reduced water and cement contents—mix design principles and laboratory tests. Cement Concrete Res 2013;51:38–46.

[16] Nemuth S, Kreißig J. Datenprojekt Zement im Netzwerk Lebenszyklusdaten, Projektbericht im Rahmen des Forschungsvorhabens FKZ 01 RN 0401 im Auftrag des Bundesministeriums für Bildung und Forschung, 2007.

[17] Chen C, Habert G, Bouzidi Y, Jullien A, Ventura A. LCA allocation procedure used as an incitative method for waste recycling: an application to mineral additions in concrete. Resour Conserv Recycl 2010;54:1231–40.

[18] Palm S, Müller C. Zemente mit erhöhten Kalksteingehalten—Festigkeit und Dauerhaftigkeit (Cements with high limestone content—strength and durability) Proceedings of the 18. Internationale Baustofftagung (ibausil). : Bauhaus-Universität Weimar; 2012;0200–5.

[19] De Larrard F. Concrete mixture proportioning—a scientific approach, Modern Concrete Technology Series, 1999.

[20] Ramge P, Lohaus L. Robustness by mix design—a new approach for mixture proportioning of SCC Design, Production and Placement of Self-Consolidating Concrete, Proceedings of SCC 2010. Montreal, Canada: Springer; 2010;37–49.

[21] Schwanda F. Der Hohlraumgehalt von Korngemischen (Open porosity of granular mixtures). Beton 1959;12–17.

[22] Graubner C-A, Proske T. Influence of the coarse aggregates on the concrete properties of SCC Proceedings of the second North American conference on the design and use of self-consolidating concrete and the fourth RILEM, international symposium on self-compacting concrete, vol. 1. Chicago: Hanley Wood, LCC; 2005.

[23] Vorhersagemodell für die Carbonatisierung von zementreduzierten Ökobetonen (Prediction model for the carbonation of eco-friendly concretes with low cement content), Proceedings of the 18. Internationale Baustofftagung (ibausil), Bauhaus-Universität Weimar. 2012. p. 0435–0444.

[24] Mittermayr F, Rezvani M, Baldermann A, Hainer S, Breitenbücher P, Juhart J, et al. Sulfate resistance of cement-reduced eco-friendly concretes. Cement Concrete Comp 2015;55:364–73.

[25] Hainer S, Proske T, Graubner C-A. Modell zur Vorhersage der Karbonatisierungstiefe von klinkerreduzierten Betonen Proceedings 19. Internationale Baustofftagung (ibausil). Bauhaus-Universität Weimar; 2015.

[26] Fib Bulletin 34: Model Code for Service Life Design, fib Task Group 5.6, International Federation for Structural Concrete (fib), 2006.

[27] Proske T, Rezvani M, Hainer S, Graubner C-A. Highly workable eco-friendly concretes—influence of the constituents on the rheological properties. Proceedings of the 7th RILEM conference on self-compacting concrete and 1st international RILEM conference on rheology and processing of construction materials. Paris; 2013.

[28] Proske T, Hainer S, Graubner C-A. Carbonation of cement reduced green concrete Proceedings of the international congress on durability of concrete. Trondheim: ICDC; 2012.

[29] PE Europe GmbH(Hrsg.): Manual GaBi 4, PE EuropeGmbH, Leinfelden-Echterdingen, 2003.

[30] European Federation of Concrete Admixture Associations (EFCA) (Hrsg.): EFCA Environmental, Declaration—Plasticising Admixtures, 2006.

# CHAPTER 5

# Effect of Supplementary Cementitious Materials on Reduction of $CO_2$ Emissions From Concrete

**K.-H. Yang[1], Y.-B. Jung[2,3], M.-S. Cho[2,3] and S.-H. Tae[4]**
[1]Kyonggi University, Suwon, Republic of Korea
[2]Korea Hydro & Nuclear Power Co., Ltd, Seoul, Republic of Korea
[3]Hyundai Engineering & Construction, Seoul, Republic of Korea
[4]Hanyang University, Ansan, Republic of Korea

## 5.1 INTRODUCTION

Concrete is predominantly utilized in buildings and infrastructure worldwide; it is mainly produced by using ordinary Portland cement (OPC) as a binder. In recent years, the annual world cement production has grown from 1.0 billion tons to approximately 1.7 billion tons, which is enough to produce $1 \, m^3$ of concrete per person [1]. As a result, the cement industry is commonly regarded as being in a period of high growth. However, the industry has been confronted since the late 1990s by the need to reduce its environmental load, including carbon dioxide ($CO_2$) emissions. Some estimates [2] suggest that the amount of $CO_2$ emitted from the worldwide production of OPC may be as high as 7% of the total global $CO_2$ emissions. Furthermore, the production of OPC involves serious collateral environmental impacts, such as environmental pollution caused by dust and the enormous energy consumption required from having a plasticity temperature of over 1300°C. For these reasons, the cement industry has been challenged in the past 10 years to effectively reduce and control $CO_2$ emissions.

Four alternative technologies to reduce $CO_2$ in the cement industry have been commonly discussed [1,3]: (1) a change in fuel to one with a lower carbon content, such as from coal to natural gas, during limestone calcination; (2) adding a chemical absorption process that would capture $CO_2$; (3) changing the clinker manufacturing process by using efficient grinding and conversion from a wet to a dry process; and (4) adding high volumes

*Handbook of Low Carbon Concrete.*
DOI: http://dx.doi.org/10.1016/B978-0-12-804524-4.00005-1
89

of supplementary cementitious materials (SCMs), such as ground granulated blast furnace slag (GGBFS), fly ash (FA), and/or silica fume (SF). Of these four technologies, using cement blended with SCMs is the most practical and economical method, and one that can be straightforwardly applied in the ready-mixed concrete field. Moreover, the use of GGBFS or FA can provide additional environmental advantages, including natural resource conservation and recycling of industrial byproducts. However, available data [4] for quantitatively evaluating the effect of SCMs on $CO_2$ emissions from the concrete production process are very rare, although it is essential to design the replacement level of SCMs for OPC according to targeted concrete requirements such as initial slump, 28-day compressive strength, and $CO_2$ reduction in order to determine the sustainable concrete mix proportions.

The present study aims to propose design equations for determining the replacement level of SCMs and unit binder content needed to achieve the targeted compressive strength and $CO_2$ reduction during concrete production. The proposed equations also provide a straightforward means of assessing the $CO_2$ footprints for a given concrete mix condition. The effect of SCMs and unit binder content on the reduction of $CO_2$ in concrete was examined using a comprehensive database including a total of 5294 laboratory concrete mixes and 3915 ready-mixed concrete mixes. All of the mixes were evaluated in terms of binder and $CO_2$ intensities [4,5]. The $CO_2$ emissions of the concrete were calculated in accordance with the life-cycle assessment (LCA) procedure specified in the ISO 14040 series [6] based on the Korean life-cycle inventory (LCI) database [7]. This means that the studied boundary conditions are from cradle to preconstruction system, including various contributions from the constituent steps, transportation to the plant, in-plant production, and transportation from the plant to the construction site.

## 5.2 LIFE-CYCLE CO$_2$ ASSESSMENT PROCEDURE FOR CONCRETE

The $CO_2$ assessment considered in this study closely followed the LCA procedure specified in the ISO 14040 series [5], and is summarized in the following sections.

### 5.2.1 Objective and Scope

The objectives of the current LCA were to evaluate the $CO_2$ footprint for a given concrete mix proportion and to ascertain the effect of SCMs on the reduction in $CO_2$ emissions in OPC concrete production. The functional unit of concrete was selected to be $1\,m^3$. The system boundary that was studied was from cradle to preconstruction, which included the following phases: (1) procurement of all constituents in a materials inventory taken from cradle to gate, (2) transportation of the constituents to a ready-mixed concrete plant, (3) in-plant production of the concrete, and (4) transportation of the concrete to a work site, as presented in Fig. 5.1. Hence, the investigated system boundary satisfies the minimum requirements of the ISO 14040 series, which is defined to be from the cradle to the gate of the concrete plant. The assumed time and regional boundaries for concrete mixes were between 1990 and 2012, and Seoul, South Korea, respectively. The typical manufacturing process conducted in a ready-mixed concrete plant in standard weather (under temperature of 15–25°C and relative humidity of 60–75%) was selected as a process specification of the concrete.

To assess the CO$_2$ footprint of the transportation phase, the transportation distance for each concrete constituent material was estimated as being from the gate of each production facility to the ready-mixed concrete plant. The cementitious materials, aggregates, and chemical admixtures were transported by a 23-ton capacity bulk trailer, 15-ton capacity diesel truck, and 1.5-ton capacity diesel truck, respectively. Fresh concrete produced at the plant was transported to a construction site by a 6-m$^3$ capacity in-transit mixing truck. The distance from the plant to the

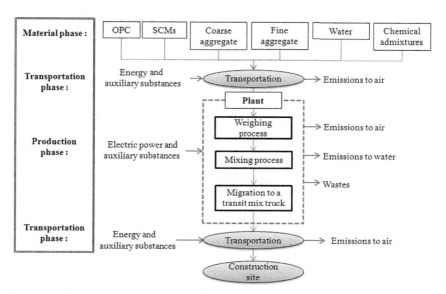

**Figure 5.1** Schematic diagram of concrete production from cradle to preconstruction.

construction site was assumed to be 30 km, considering the allowable casting time of fresh concrete.

## 5.2.2 LCI Database

Considering the regional location, the Korean LCI database [7] was primarily used for $CO_2$ assessment in each phase of concrete, as given in Table 5.1. The LCI for a building material provides a collective data set that covers everything from the cradle to the grave. The Japanese Society of Civil Engineering (JSCE) LCI database [8] was also used for a data set that is not provided in the Korean LCI database, because the climate conditions and energy sources for concrete sources are similar in both countries. The $CO_2$ inventory for the concrete production phase was obtained from the conversion of energy sources consumed in the plant for annual productivity. The energy sources in the plant include electric power and crude oils.

## 5.2.3 $CO_2$ Assessment Procedure

For the studied system, a total $CO_2$ footprint ($C_e$) for 1-$m^3$ concrete can be assessed by the individual integration method [4] using the following equation:

$$C_e = CO_{2-M} + CO_{2-T} + CO_{2-P} \tag{5.1}$$

where $CO_{2-M}$, $CO_{2-T}$, and $CO_{2-P}$ indicate the $CO_2$ emissions in the materials, transportation, and production phases, respectively. Because the materials phase includes the cementitious materials, water, fine aggregate, coarse aggregate, and chemical admixtures, $CO_{2-M}$ can be calculated as follows:

$$CO_{2-M} = \sum_{i=1}^{n}(W_i \times CO_{2(i)-LCI}) \tag{5.2}$$

where $i$ represents a raw material constituent of the concrete, $n$ is the total number of constituents added for concrete production, and $W_i$ and $CO_{2(i)-LCI}$ are the unit volume weight (kg/$m^3$) and $CO_2$ emission inventory ($CO_2$-kg/kg), respectively, of raw material $i$. The amount of $CO_2$ generated during the transportation process can be calculated by summing the amount generated during transportation of each constituent $i$, as well as that generated from transporting the produced concrete. Therefore, $CO_{2-T}$ can be obtained as follows:

**Table 5.1** Examples for $CO_2$ assessment of concrete in the studied system ($f'_c=35$ MPa)
Functional unit (FU): m³

| Unit | Material | | | Transportation | | |
|---|---|---|---|---|---|---|
| | A | B | A·B | D | E | A·D·E |
| Item | kg/FU | $CO_2$-kg/kg | $CO_2$-kg/FU | km | $CO_2$-kg/kg·km | $CO_2$-kg/FU |
| OPC | 336 | 0.931 | 312.8 | 277 | $5.18 \times 10^{-5}$ | 4.82 |
| GGBFS[a] | 60 | 0.0265 | 1.59 | 339 | $5.18 \times 10^{-5}$ | 1.05 |
| FA[a] | 5 | 0.0196 | 0.098 | 322 | $5.18 \times 10^{-5}$ | 0.08 |
| Sand | 855 | 0.0026 | 2.223 | 47 | $6.3 \times 10^{-5}$ | 2.53 |
| Coarse | 893 | 0.0075 | 6.6975 | 37.6 | $6.3 \times 10^{-5}$ | 2.12 |
| Water | 171 | $1.96 \times 10^{-4}$ | 0.034 | — | — | — |
| Admixture | 3.23 | 0.25 | 0.8075 | 70.6 | $2.21 \times 10^{-4}$ | 0.05 |
| | Sum | | 324.3 | — | | 30.88 |
| Production (fresh concrete) | 2323 | 0.00768 | 17.84 | 30 | 0.674 $CO_2$-kg/m³·km | D·E= 20.22 |
| | | | | | Sum | 30.88 |

Total = 372.98 $CO_2$-kg/FU (= 324.3 + 30.88 + 17.84 $CO_2$-kg)

[a]LCI data provided in JSCE are referenced wherever the Korean LCI database is unavailable.

$$CO_{2-T} = \sum_{i=1}^{n}(W_i \times D_i \times CO_{2(i)-LCI(TR)}) + D_{FC} \times CO_{2(FC)-LCI(TR\_con)}$$

$$(5.3)$$

where $D_i$ is the transportation distance of each concrete constituent material $i$ from the gate of the raw material–producing facility to the concrete plant, $CO_{2(i)-LCI(TR)}$ is the $CO_2$ inventory for the vehicles to transport material $i$, $D_{FC}$ is the transportation distance for $1 \, m^3$ of the produced fresh concrete from the ready-mixed concrete plant to the construction site, and $CO_{2(FC)-LCI(TR\_con)}$ is the $CO_2$ inventory of the in-transit mixing truck for fresh concrete. The $CO_2$ inventory for the in-transit mixer is expressed in the units $CO_2$-kg/(m$^3$·km).

An example of the $CO_2$ assessment for a given concrete mix proportion using the above equations is given in Table 5.1. The concrete mix proportion (column A in the table) is sampled from a comprehensive database that is introduced in detail in the following section. In the table, the $CO_2$ inventories for each constituent material and plant for concrete production are listed in column B, and those for vehicles are given in column E. The $CO_2$ emissions per functional unit of concrete were calculated to be 324.3, 17.84, and 30.88 kg for the material, production, and transportation phases, respectively. The $CO_2$ emissions from the OPC material made up 96.5% of the emissions for the material phase, which corresponded to 83.9% of the total $CO_2$ emissions. The $CO_2$ emissions in the transportation phase resulted mostly from the transportation of the mixed concrete because the $CO_2$ emissions from the 6-m$^3$ capacity in-transit mixing truck were considerably higher than those of the bulk trailer and diesel trucks used to transport the concrete materials.

## 5.3 DATABASE OF CONCRETE MIX PROPORTIONS

To assess the $CO_2$ footprint of concrete production under a wide variety of mix conditions and different 28-day compressive strengths, a comprehensive database was established. The database included 5294 concrete mixes tested in the laboratory and 3915 concrete mixes produced in ready-mixed concrete plants. The incidence of the various parameter values in the database is given in Table 5.2. When the data sets were classified according to the addition of SCMs, the laboratory specimens were found to consist of 3037 OPC mixes, 1000 OPC + FA mixes, 341 OPC + GGBFS mixes, 697 OPC + SF mixes, 135 OPC + FA+ GGBFS mixes,

Table 5.2 Incidence of various parameter values for 5294 laboratory and 3915 plant concrete mixes

| Mixing type | Type of binder | Range | | | | | | | | | Total |
|---|---|---|---|---|---|---|---|---|---|---|---|
| B (kg/m³) | | 50–250 | 250–300 | 300–400 | 400–500 | 500–600 | 600–700 | 700–800 | 800–1000 | 1000–1400 | |
| Laboratory mix | OPC | 34 | 187 | 1057 | 882 | 575 | 235 | 44 | 23 | — | 3037 |
| | OPC + FA | 6 | 74 | 375 | 289 | 171 | 58 | 25 | 1 | — | 1000 |
| | OPC + GGBFS | 2 | 10 | 105 | 55 | 32 | 31 | 31 | 55 | 20 | 341 |
| | OPC + SF | — | 4 | 31 | 80 | 193 | 151 | 126 | 91 | 21 | 697 |
| | OPC + FA + GGBFS | 2 | — | 45 | 48 | 18 | 14 | 5 | 3 | — | 135 |
| | OPC + FA + SF | — | — | — | 6 | 2 | 8 | 6 | 2 | — | 24 |
| | OPC + SF + GGBFS | — | — | — | 2 | 5 | 10 | 19 | 23 | 1 | 60 |
| Plant mix | OPC | 29 | 184 | 458 | 42 | — | — | — | — | — | 713 |
| | OPC + FA | 63 | 425 | 733 | 150 | 30 | 1 | — | — | — | 1402 |
| | OPC + GGBFS | 1 | 7 | 342 | 184 | 13 | 2 | — | 1 | — | 550 |
| | OPC + FA + GGBFS | 57 | 414 | 625 | 107 | 21 | 3 | 1 | — | — | 1228 |
| | OPC + FA + SF | — | — | — | 18 | 2 | 2 | — | — | — | 22 |

(Continued)

Table 5.2 (Continued)

|  | Mixing type | Type of binder | Range | | | | | | | | | Total |
|---|---|---|---|---|---|---|---|---|---|---|---|---|
|  |  |  | 7–20 | 20–30 | 30–40 | 40–60 | 60–80 | 80–100 | 100–120 | 120–140 | 140–170 |  |
| $f_c'$ (MPa) | Laboratory mix | OPC | 118 | 488 | 604 | 981 | 507 | 224 | 105 | 9 | 1 | 3037 |
|  |  | OPC+ FA | 138 | 360 | 187 | 208 | 98 | 9 | — | 7 | — | 1000 |
|  |  | OPC+ GGBFS | 13 | 47 | 61 | 73 | 68 | 45 | 27 | 7 | — | 341 |
|  |  | OPC+ SF | 1 | 4 | 21 | 95 | 170 | 176 | 116 | 94 | 13 | 690 |
|  |  | OPC+ FA+ GGBFS | — | 5 | 19 | 92 | 16 | 2 | 1 | — | — | 135 |
|  |  | OPC+ FA+ SF | — | — | 1 | 7 | 11 | 3 | 0 | 2 | — | 24 |
|  |  | OPC+ SF+ GGBFS | — | — | — | 3 | 13 | 14 | 12 | 13 | 5 | 60 |
|  | Plant mix | OPC | 116 | 505 | 86 | 6 | — | — | — | — | — | 713 |
|  |  | OPC+ FA | 311 | 907 | 124 | 57 | 3 | — | — | — | — | 1402 |
|  |  | OPC+ GGBFS | 8 | 302 | 167 | 61 | 12 | — | — | — | — | 550 |
|  |  | OPC+ FA+ GGBFS | 293 | 724 | 163 | 45 | 1 | 1 | 1 | — | — | 1228 |
|  |  | OPC+ FA+ SF | — | — | 20 | — | — | 2 | — | — | — | 22 |

|  | Mixing type | Type of binder | Range | | | | | | | | | Total |
|---|---|---|---|---|---|---|---|---|---|---|---|---|
|  |  |  | 3–10 | 0–20 | 20–30 | 30–40 | 40–50 | 50–60 | 60–70 | 70–80 | 80–100 |  |
| $R_G$ (%) | Laboratory mix | OPC + GGBFS | 22 | 41 | 112 | 62 | 47 | 35 | 14 | 8 | — | 341 |
|  | Plant mix | OPC + GGBFS | 69 | 145 | 210 | 111 | 3 | 7 | 1 | 4 | — | 550 |

| $R_F$ (%) | Mixing type | Type of binder | Range | | | | | | | | | Total |
|---|---|---|---|---|---|---|---|---|---|---|---|---|
| | | | 3–10 | 10–20 | 20–30 | 30–40 | 40–50 | 50–60 | 60–70 | 70–80 | 80–90 | |
| | Laboratory mix | OPC + FA | 284 | 469 | 126 | 85 | 20 | 10 | 6 | — | — | 1000 |
| | Plant mix | OPC + FA | 1402 | — | — | — | — | — | — | — | — | 1402 |
| $R_S$ (%) | Laboratory mix | OPC + SF | 348 | 293 | 56 | 3 | — | — | — | — | — | 700 |

| $R_F + R_G$ (%) | Mixing type | Type of binder | Range | | | | | | | | | Total |
|---|---|---|---|---|---|---|---|---|---|---|---|---|
| | | | 9–10 | 10–20 | 20–30 | 30–40 | 40–50 | 50–60 | 60–70 | 70–80 | 80–90 | |
| | Laboratory mix | OPC + FA + GGBFS | 2 | 50 | 43 | 15 | 5 | 14 | 6 | — | — | 135 |
| | Plant mix | OPC + FA + GGBFS | 1 | 52 | 210 | 482 | 227 | 225 | 30 | — | 1 | 1228 |

| $R_S + R_G$ (%) | Mixing type | Type of binder | Range | | | | | | | | | Total |
|---|---|---|---|---|---|---|---|---|---|---|---|---|
| | | | 0–10 | 10–20 | 21–30 | 30–40 | 40–50 | 50–60 | 60–70 | 70–80 | 80–90 | |
| | Laboratory mix | OPC + SF + GGBFS | — | — | 19 | 26 | 9 | 5 | 1 | — | — | 60 |

| $R_F + R_S$ (%) | Mixing type | Type of binder | Range | | | | | | | | | Total |
|---|---|---|---|---|---|---|---|---|---|---|---|---|
| | | | 0–10 | 16–20 | 21–30 | 30–40 | 40–50 | 50–60 | 60–70 | 70–80 | 80–90 | |
| | Laboratory mix | OPC + FA + SF | — | 1 | 15 | 8 | — | — | — | — | — | 24 |
| | Plant mix | OPC + FA + SF | — | 20 | 2 | — | — | — | — | — | — | 22 |

24 OPC + FA + SF mixes, and 60 OPC + GGBFS + SF mixes, while the plant productions consisted of 713 OPC mixes, 1402 OPC + FA mixes, 550 OPC + GGBFS mixes, 1228 OPC + FA + GGBFS mixes, and 22 OPC + FA + SF mixes.

The replacement level of SCMs in the laboratory specimens ranged between 3% and 70% for FA, between 3% and 80% for GGBFS, and between 3% and 40% for SF, while that in the plant productions ranged between 3% and 10% for FA, and between 3% and 80% for GGBFS. The range of concrete compressive strength in the laboratory specimens was as follows: 7–170 MPa for OPC mixes, 7–100 MPa for OPC + FA mixes, 7–140 MPa for OPC + GGBFS mixes, 7–170 MPa for OPC+ SF mixes, 20–120 MPa for OPC + FA + GGBFS mixes, 30–140 MPa for OPC + FA + SF mixes, and 40–170 MPa for OPC + GGBFS + SF mixes. The range of concrete compressive strength in the plant-produced concrete was as follows: 7–60 MPa for OPC mixes, 7–80 MPa for OPC + FA mixes, 7–80 MPa for OPC + GGBFS mixes, 7–120 MPa for OPC + FA + GGBFS mixes, and 30–100 MPa for OPC + FA + SF mixes. The unit binder content ranged between 50 and 1400 kg/m$^3$ for laboratory specimens, and between 50 and 1000 kg/m$^3$ for the plant mixes.

Table 5.1 clearly demonstrates that the $CO_2$ emission of concrete results primarily from the OPC content in the material phase. Hence, it is essential to determine the unit binder content with the minimum OPC proportion for reducing the $CO_2$ emissions in concrete production. For this reason, several recent studies [4,5,9] have discussed the efficient use of a binder to reduce $CO_2$ emissions. The increase in the compressive strength of concrete commonly accompanies the consumption of a greater amount of binder, because a higher strength requires a lower water-to-binder ratio. On the other hand, a higher concrete strength can minimize the necessary member size, which contributes to a decrease in $CO_2$ emissions by using a smaller amount of concrete. Hence, the $CO_2$ emissions of concrete need to be assessed in terms of the binder content necessary to develop a unit compressive strength (1 MPa). Considering the interrelation of the concrete compressive strength ($f'_c$ in MPa), the unit binder content ($B$ in kg/m$^3$), and the corresponding $CO_2$ emissions ($C_e$ in kg/m$^3$), the following binder intensity ($B_i$) and $CO_2$ intensity ($C_i$) [4,5] were introduced:

$$B_i = B/f'_c \tag{5.4}$$

$$C_i = C_e/f'_c \tag{5.5}$$

## 5.3.1 Effect of SCMs on $B_i$

Fig. 5.2 shows the effect of different SCMs on $B_i$ with the variation of $f_c'$. Best–fit curves determined according to the data of each binder type are also compared in Fig. 5.3. Although the mix design methods for the concrete and the materials used differ widely, the variability of the data

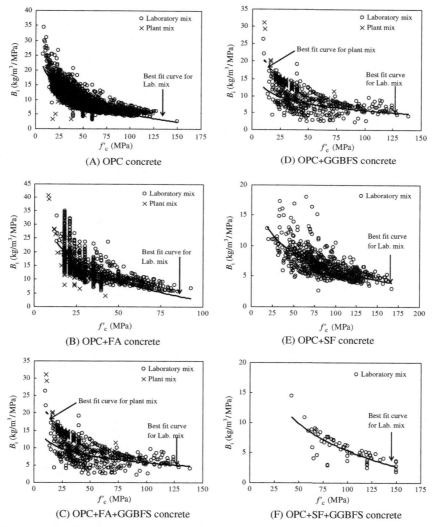

**Figure 5.2** Binder intensity ($B_i$) of different concrete types as a function of compressive strength ($f_c'$). (A) OPC concrete; (B) OPC + FA concrete; (C) OPC + FA + GGBFS concrete; (D) OPC + GGBFS concrete; (E) OPC + SF concrete; and (F) OPC + SF + GGBFS concrete.

is considerably small, especially for high-strength concrete with an $f'_c$ of more than approximately 50 MPa. The differences among $B_i$ values in both laboratory and plant mixes depend on $f'_c$ and the type of binder. When $f'_c$ is 20 MPa, the $B_i$ value determined from the OPC laboratory concrete mixes is 1.25 times higher than that obtained from the corresponding plant mixes, while the $B_i$ value determined from the OPC+ GGBFS laboratory mixes is 1.19 times higher than that obtained from the corresponding plant mixes. On the other hand, when $f'_c$ increases to 80 MPa, the $B_i$ values determined from the OPC and OPC+ GGBFS laboratory mixes are lower than those determined from the corresponding plant mixes by

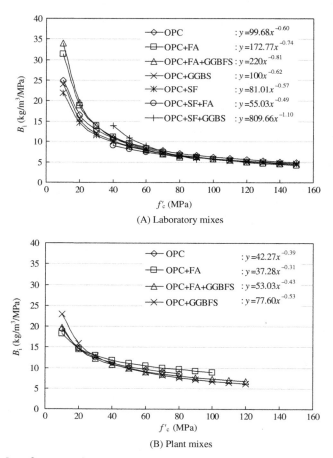

**Figure 5.3** Best-fit curves determined from $B_i$ data of Fig. 5.2. (A) Laboratory mixes and (B) plant mixes.

7% and 10%, respectively. However, both mixes commonly tend to have similar $B_i$ values for a given $f'_c$, although the number and coverage range of data in the plant mixes are considerably smaller than those in the laboratory mixes. For example, when $f'_c$ is 40 MPa, $B_i$ varies between 9.02 and 14.05 kg/m$^3$/MPa for the laboratory mixes and between 10.85 and 12.77 kg/m$^3$/MPa for the plant mixes. Furthermore, a similar relation of the $B_i$ value and $f'_c$ is found in both mixes. The binder intensity commonly tends to decrease with the increase in $f'_c$, regardless of the binder type. This indicates that the amount of binder to develop the unit strength decreases as $f'_c$ increases. The decreasing value of $B_i$ with the increase in $f'_c$ is gradually mitigated beyond a concrete strength of 50 MPa, and the values of $B_i$ converge towards a minimum value. The convergence value of $B_i$ is marginally affected by the type of binder, showing a value of approximately 5 kg/m$^3$/MPa. The value of $B_i$ for OPC+ GGBFS concrete is very similar to that for OPC concrete with the same $f'_c$. The OPC+ SF concrete gives a lower value of $B_i$ than OPC does concrete when $f'_c$ is less than 40 MPa. On the other hand, the OPC+ FA concrete and OPC+ FA+ GGBFS concrete have slightly higher $B_i$ values than does OPC concrete, indicating that the former two require more unit binder content than the latter with the same $f'_c$.

## 5.3.2 Effect of SCMs on $C_i$

Fig. 5.4 shows the effect of different SCMs on $C_i$ with the variation of $f'_c$. Fig. 5.5 also presents comparisons of the best-fit curves determined from the relationship of $f'_c$ and the $C_i$ value, according to the type of binder. As was observed for the $B_i$ value, the laboratory and plant mixes have similar $C_i$ values for a given $f'_c$. The values of $C_i$ range between 11.58 and 18.0 kg/m$^3$/MPa for the laboratory mixes and between 11.3 and 16.36 kg/m$^3$/MPa for the plant mixes when $f'_c$ is 20 MPa. As $f'_c$ increases to 80 MPa, the values of $C_i$ vary between 4.66 and 7.21 kg/m$^3$/MPa for the laboratory mixes and between 5.56 and 8.41 kg/m$^3$/MPa for the plant mixes. Furthermore, both the laboratory and plant mixes reveal the same trends for the effect of SCMs on $C_i$. The variation of the $C_i$ value with $f'_c$ is very similar to the trend observed in $B_i$. At the same $f'_c$, the OPC concrete gives the highest $C_i$ value, while the OPC+ FA+ GGBFS concrete commonly has a lower $C_i$ value than any of the other concrete types. This indicates that the combined substitution of FA and GGBFS is more favorable for the reduction of $CO_2$ emissions in developing the unit strength of concrete than are the other SCMs. Compared with the OPC concrete

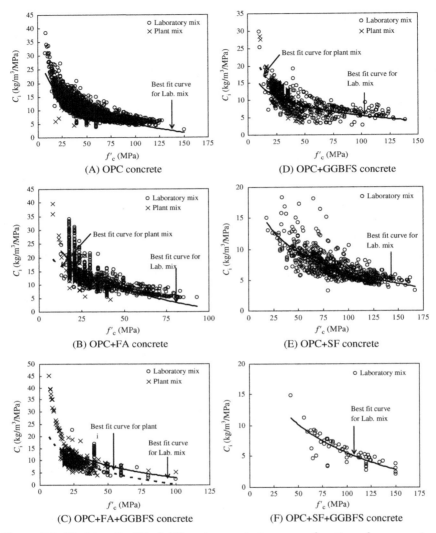

**Figure 5.4** $CO_2$ intensity ($C_i$) of different concrete types as a function of compressive strength ($f'_c$). (A) OPC concrete; (B) OPC + FA concrete; (C) OPC + FA + GGBFS concrete; (D) OPC + GGBFS concrete; (E) OPC + SF concrete; and (F) OPC + SF + GGBFS concrete.

laboratory mix, the $C_i$ value for OPC+ FA+ GGBFS concrete decreases by approximately 12% when $f'_c$ is 30 MPa, and then the decreasing ratio increases by as much as 48% as $f'_c$ increases to 100 MPa. The $C_i$ value for concrete with FA or SF alone is also lower than that for OPC concrete, resulting in a decrease of between 10% and 20%. When $f'_c$ is higher than

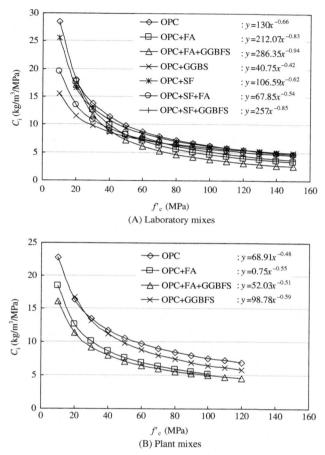

**Figure 5.5** Best-fit curves determined from $C_i$ data of Fig. 5.4. (A) Laboratory mixes and (B) plant mixes.

30 MPa, the OPC+ GGBFS concrete gives a slightly lower $C_i$ value than does the OPC concrete, although both concrete mixes have a similar $B_i$ value. This observation is particularly noticeable for the laboratory concrete mixes.

Fig. 5.6 shows the effect of the substitution level of SCMs on the $C_i$ values of concrete mixes for similar compressive-strength ranges. In general, the $C_i$ value of concrete decreases sharply as the substitution level of SCMs increases up to approximately 15–20%, beyond which the decreasing rate tends to gradually slow, as demonstrated by the best-fit curves of the test data. Hence, the relation of the $C_i$ value and the SCM substitution

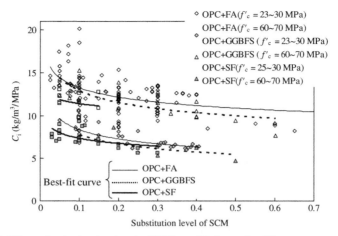

**Figure 5.6** Effect of substitution level of SCM on $CO_2$ intensity ($C_i$).

level is nonlinear. This observation is insignificantly affected by $f_c'$. At the same level of the substitution, the OPC+ GGBFS concrete gives lower $C_i$ values than does the OPC+ FA concrete. Moreover, when the substitution level is below 25%, the $C_i$ values of OPC+ SF concrete are lower than those of the OPC+ FA concrete. Therefore, it can be concluded that the substitution of GGBFS or SF is more favorable than that of FA in reducing the $CO_2$ emissions of concrete.

## 5.3.3 Relation of $B_i$ and $C_i$

The $CO_2$ emissions of concrete significantly depend on $f_c'$, the corresponding unit binder content, and the substitution level of SCMs. The $CO_2$ emissions of concrete for a given $f_c'$ can be simply assessed if the $C_i$ value is known. To propose a simple closed-form equation to determine the $C_i$ value, the important parameters, including $B_i$ and the SCM substitution level, were adjusted by a nonlinear multiple-regression analysis. The boundary conditions for the analysis were as follows: (1) a binder content of 0 results in a $CO_2$ emission of 0; (2) the effect of each SCM on the reduction in the $C_i$ value should be considered individually, because the $CO_2$ emission of concrete is assessed by the individual integration method; and (3) the decrease in $C_i$ with the substitution level of each SCM can be realized by the power function, as shown in Fig. 5.6. Overall, the relation of $B_i$ and $C_i$ can be formulated in the following way:

$$C_i = A_1[1 - (R_F^{B_1} + R_G^{C_1} + R_S^{D_1})]B_i \quad (kg/m^3/MPa^{-1}) \quad (5.6)$$

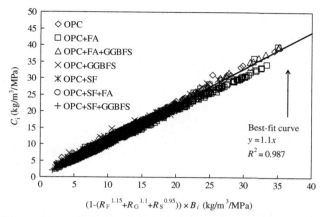

**Figure 5.7** Regression analysis for relationship between $CO_2$ intensity ($C_i$) and binder intensity ($B_i$).

where $R_F$, $R_G$, and $R_S$ are the substitution levels for OPC of FA, GGBFS, and SF, respectively. In Eq. (5.6), the experimental constants, $A_1$, $B_1$, $C_1$, and $D_1$ were determined to be 1.1, 1.15, 1.1, and 0.95, respectively, by the regression analysis using the database (see Fig. 5.7). In the determination of experimental constants, all data sets in the database were used without any separation of laboratory and plant mixes, because the two mixes had substantially similar $B_i$ and $C_i$ values for a given $f_c'$. Ultimately, from the definition of $B_i$ and $C_i$, the $CO_2$ emission ($C_e$) per functional unit of concrete can be formulated as follows:

$$C_e = 1.1[1 - (R_F^{1.15} + R_G^{1.1} + R_S^{0.95})]B \quad (CO_2 - kg/m^3) \quad (5.7)$$

The $CO_2$ emission of OPC concrete without SCMs can be straightforwardly assessed as $1.1B$. Hence, the term $(R_F^{1.15} + R_G^{1.1} + R_S^{0.95})$ in Eq. (5.7) indicates the reduction ratio of $CO_2$ according to the substitution level of SCMs in OPC-based concrete. In OPC concrete, the value of $(R_F^{1.15} + R_G^{1.1} + R_S^{0.95})$ cannot exceed 1.0.

## 5.3.4 Determination of Unit Binder Content

The binder intensity is inversely and nonlinearly proportional to $f_c'$, and affected by the type and amount of SCM, as shown in Fig. 5.3. Compared with the unit binder content ($B$) of OPC concrete with the same $f_c'$, OPC+ FA concrete mixes tend to require 1.0–1.16 times more binder up to an $f_c'$ of 50 MPa, beyond which the required binder content decreases by 3–15%. On the other hand, the OPC+ GGBFS and OPC+

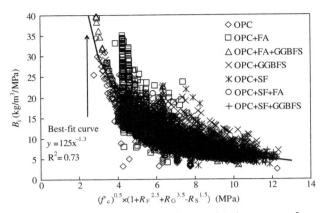

**Figure 5.8** Regression analysis to determine the unit binder content for a given compressive strength.

SF concrete mixes require a slightly lower binder content than does the OPC concrete with the same $f_c'$, and the required ratio varies nonlinearly with $f_c'$. Considering these observations and the individual integration procedure of each SCM, $B_i$ can be formulated from the regression analysis using all data sets as follows (see Fig. 5.8):

$$B_i = 125(f_c')^{-0.65}[1 + (R_F^{2.5} + R_G^{3.5} - R_S^{1.5})]^{-1.3} \quad (\text{kg/m}^3/\text{MPa}^{-1}) \quad (5.8)$$

Because the unit binder content (or unit water content) for a given $f_c'$ somewhat depends on various parameters such as the curing condition of the concrete, designed workability and durability, and empirical ability of the mixing designers, the correlation coefficient ($R^2 = 0.73$) between the test data and the fit curve for $B_i$ is lower than that ($R^2 = 0.987$) for $C_i$. However, Eq. (5.8) is expected to be of practical use in determining the unit binder content and design of SCMs for a given $f_c'$. From the definition of $B_i$, the unit binder content of OPC concrete with different SCMs can be expressed as follows:

$$B = 125(f_c')^{0.35}[1 + (R_F^{2.5} + R_G^{3.5} - R_S^{1.5})]^{-1.3} \quad (\text{kg/m}^3) \quad (5.9)$$

## 5.4 DESIGN OF SCMs TO REDUCE $CO_2$ EMISSIONS DURING CONCRETE PRODUCTION

The OPC constituent predominantly governs the $CO_2$ emissions in concrete production, with a contribution rate roughly equivalent to 80–90%. Hence, the determination of the substitution level of different SCMs is a very critical parameter for a low $CO_2$ emissions–based mixing design of

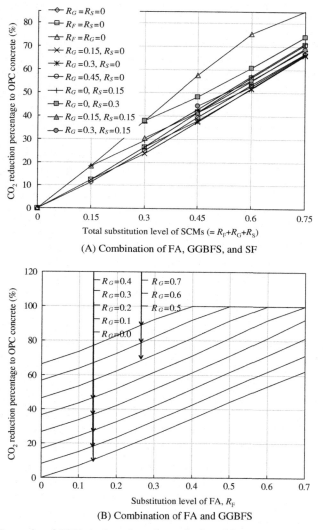

**Figure 5.9** Example of SCM design chart for reducing CO$_2$ emissions from concrete production. (A) Combination of FA, GGBFS, and SF; (B) Combination of FA and GGBFS.

concrete that will achieve a targeted $f'_c$. The unit binder content for a targeted $f'_c$ can be easily obtained using Eq. (5.9), according to the different substitution levels of each SCM, and then, the total CO$_2$ emissions of a particular concrete can be straightforwardly calculated by substituting the determined $B$ into Eq. (5.7). Fig. 5.9 shows examples of an SCM design chart designed to achieve the targeted CO$_2$ reduction rate in OPC–based concrete production. The International Green Construction Code [10] specifies that the LCA shall demonstrate that the building project achieves not less than a 20% improvement in environmental impact as compared to a reference building of similar usable floor area, function, and configuration. The substitution of FA gives a lower $CO_2$ reduction rate than that of either GGBFS or SF. To achieve a $CO_2$ reduction rate of more than 20%, various combinations of each SCM are required (Fig. 5.9A); for example, $R_G$ above 30%, or combination of $R_F$ above 15% and $R_S$ above 15%. For normal-strength concrete, FA and GGBFS rather than SF are primarily used as SCMs because of their economical and workability efficiencies. Fig. 5.9B clearly shows that the combinations of (1) $R_G$ above 20% and $R_F$ above 10%, and/or (2) $R_G$ above 10% and $R_F$ above 25% achieve a $CO_2$ reduction rate of 20% relative to OPC concrete. Hence, using Eqs. (5.7) and (5.9) or the design chart example in Fig. 5.9, the unit binder content and the type and substitution level of SCM can be straightforwardly determined for the targeted $f_c'$ and $CO_2$ reduction rate. Furthermore, Eq. (5.7) is of practical use in predicting the $CO_2$ emissions of concrete for a given mixing proportion.

## 5.5 CONCLUSIONS

The $CO_2$ emissions of concrete production were assessed in accordance with the life-cycle procedure specified in the ISO 14040 series. The effect of SCMs including FA, GGBFS, and SF on reduction in $CO_2$ emissions from OPC-based concrete was analyzed by using binder intensity ($B_i$) and $CO_2$ intensity ($C_i$) concepts in terms of the unit strength (1 MPa) of concrete. Based on the data from 5294 laboratory concrete mixes and 3915 plant mixes, equations were formulated to calculate the $B_i$ and $C_i$ values, and then to straightforwardly determine the unit binder content and the type and substitution of SCM for the targeted $f_c'$ and $CO_2$ reduction rate. The proposed models do not consider the $CO_2$ emissions during the casting of concrete, the service life of the concrete structures, or the disposal process for concrete. However, the $CO_2$ absorption of concrete due to carbonation during the service life of the structures is estimated to be only 5–7% of the $CO_2$ emissions caused by the constituent materials of the concrete [11]. It is also not an easy task to obtain reliable $CO_2$ emissions during the casting and disposal process because they are significantly dependent on the size, function, and configuration of the structure. Moreover, no reliable LCI database is available at this time for various types of concrete construction equipment and waste-crushing equipment. The $CO_2$ emissions calculated in the current transportation phase would vary slightly with the locations of the ready-mixed concrete plant and the original supplier of a raw material constituent of the concrete, but this variation is expected to be less than 2% of the total $CO_2$ emissions, which is very small and can be regarded as negligible. Hence, the proposed equations hold considerable promise as a guideline for low $CO_2$-based concrete mixing design.

From the analysis of a total of 9209 concrete mixes, the following conclusions may be drawn:

1. The binder intensity $(B_i)$ tends to decrease with an increase in the compressive strength $(f_c')$ and converges towards a minimum value of $5\,kg/m^3/MPa$, regardless of the type of binder.

2. The variation of the $CO_2$ intensity $(C_i)$ with $f_c'$ is very similar to the trend observed in $B_i$. At the same $f_c'$, the OPC concrete gives the highest $C_i$ value, while the OPC+ FA+ GGBFS concrete has a lower $C_i$ value than any of the other concrete types.

3. The $C_i$ value of the concrete decreases sharply as the substitution level of the SCMs increases up to approximately 15–20%, beyond which the decreasing rate gradually slows. At the same level of substitution, GGBFS is more favorable than FA in reducing $CO_2$ emissions of concrete.

4. The binder and $CO_2$ intensities can be formulated as a function of the individual substitution level of each SCM. The significance of the proposed models can be summarized as follows: (1) the total $CO_2$ emission of concrete can be straightforwardly calculated for a given concrete mix proportion, (2) the unit binder content of concrete can be reasonably determined for the targeted $f_c'$, and (3) the type and substitution level of SCMs can be easily selected to achieve the design strength and targeted $CO_2$-reduction rate.

## ACKNOWLEDGMENTS

This work was supported by Grant No. 12CCTI-C063722-01 from the Construction Technology Innovation Program (CTIP) funded by the Ministry of Land, Transportation, and Maritime Affairs (MLTM) of the Korea Government and by Grant No. 2011T100200161 from the Nuclear Research and Development Program of the Korea Institute of Energy Technology Evaluation and Planning (KETEP) funded by the Korea Government Ministry of Knowledge Economy.

## REFERENCES

[1] Gartner E. Industrially interesting approaches to 'low-$CO_2$' cement. Cem Concr Res 2004;34(9):1489–98.

[2] Malhotra VM. Introduction: sustainable development and concrete technology. Concr Int 2002;24(7):22.

[3] Ogbeide SO. Developing an optimization model for $CO_2$ reduction in cement production process. J Eng Sci Technol Rev 2012;3(1):85–8.

[4] Yang KH, Song JK, Song KI. Assessment of $CO_2$ reduction of alkali-activated concrete. J Clean Prod 2013;39(1):265–72.

[5] Damineli BL, Kemeid FM, Aguiar PS, John VM. Measuring the eco-efficiency of cement use. Cem Concr Compos 2010;32(8):555–62.

[6] ISO 14040 Environmental management–life cycle assessment—principles and framework. International Standardisation Organisation; 2006.

[7] KEITI (Korea Environmental Industry & Technology Institute). Korea LCI database information network, http://www.edp.or.kr/lcidb.

[8] Sakai K, Kawai K. JSCE guidelines for concrete No. 7: recommendation of environmental performance verification for concrete structures. Japan Society of Civil Engineering; 2006.

[9] Aïtcin PC. Cements of yesterday and today: concrete of tomorrow. Cem Concr Res 2000;30(9):1349–59.

[10] IgCC Public Comment Hearing Committee, USA. International green construction code. : International Code Council, Inc; 2010.

[11] Lee SH, Lee SB, Lee HS. Study on the evaluation CO2 emission-absorption of concrete in the view of carbonation. J Korea Concr Inst 2009;21(1):85–92. (in Korean).

# CHAPTER 6

# Binder and Carbon Dioxide Intensity Indexes as a Useful Tool to Estimate the Ecological Influence of Type and Maximum Aggregate Size on Some High-Strength Concrete Properties

**A.M. Grabiec, D. Zawal and J. Szulc**
Poznan University of Life Sciences, Poznań, Poland

## 6.1 INTRODUCTION

A growing intensity of extreme weather conditions brought about by climate change has become a serious social and economic problem of 21st century. Natural changes of the climate are particularly connected with solar activity, occurrence of Milankovitch cycles (variations in eccentricity, axial tilt, and precession), volcanic activity and ENSO (El Niño–Southern Oscillation) phenomenon. However, the increasingly higher and higher ratio of carbon dioxide concentration in the atmosphere, which has been observed for the last 150 years, undoubtedly, is becoming increasingly significant. In 2013, for the first time, the $CO_2$ content in the atmosphere exceeded 400 ppm (http://climate.nasa.gov/news/916/) [1]. C.D. Keeling has been conducting research on $CO_2$ content in the atmosphere since the 1950s. Although they do not constitute a direct evidence for anthropogenic reasons for the intensity of climate changes, today, in light of other evidence, it is difficult to find a more rational interpretation. According to measurements conducted by R.F. Keeling (C.D. Keeling's son) in nine observatories around the world [2,3], the rise in $CO_2$ content is associated with a simultaneous lowering of oxygen-to-nitrogen proportion, which is presented in the upper part of Fig. 6.1. The only plausible explanation of this phenomenon is the using up of oxygen in the processes of combustion of fossil fuels. It is more difficult to agree with one

*Handbook of Low Carbon Concrete.*
DOI: http://dx.doi.org/10.1016/B978-0-12-804524-4.00006-3

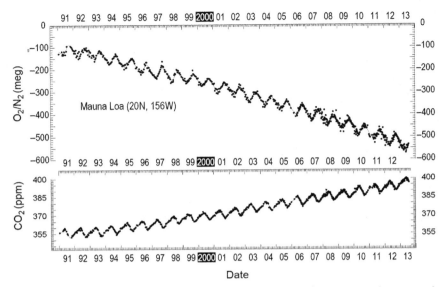

**Figure 6.1** Correlation between carbon dioxide quantity and oxygen-to-nitrogen ratio in the atmosphere based on Mauna Loa observations [2,3].

of most recognized alternative hypotheses, according to which the increase in $CO_2$ content in the atmosphere is a result of a rise in the temperature of the oceans and seas, brought about by increased solar activity. Had this hypothesis been true, the proportion of oxygen to nitrogen would not have changed. It is only the $CO_2$ content in the atmosphere that would have changed. In fact, there is an observed, on the one hand, correlation between the drop in the proportion of oxygen to nitrogen [2,3] with a commonly known C.D. Keeling curve (lower part of Fig. 6.1) illustrating the increase in the $CO_2$ concentration in the atmosphere and, on the other hand, the lower activity of the sun in the last 11-year cycle (http://solarscience.msfc.nasa.gov/predict.shtml) [4].

The notion of sustained development became equivalent to the notion of environment for human life in 21st century. It is expressed in three dimensions: social, economical, and ecological (Fig. 6.2). The three-dimensional concept of world development, in some sense initiated in the report *Our Common Future* issued by the UN Brundtland Commission in 1987 [5], is an obvious fact, accepted by the majority of people, especially when one realizes the changes the world underwent in the last 200 years.

According to Ref. [6] mankind created its own geological age, the Anthropocene, which commenced in the period 1850–1900. There were

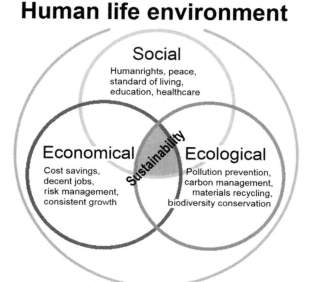

**Figure 6.2** Sustainable development in the context of three-dimensional human life environment.

and are many mutually linked phenomena occurring in the Anthropocene. Demographic and social development intensified as well as that of towns and cities; in 2007 the world rural and urban populations became balanced [7]. The industrial revolution brought spectacular achievements. Development of railways and other means of commonly understood modern transport took place. Electricity was implemented, empiric sciences progressed (biology, chemistry, and experimental physics, where Edison was the pioneer of creative inventions).

In the Anthropocene age, Portland cement, which was invented by Aspdin in 1824, got into mass production. At the end of the 19th century Monier invented reinforced concrete. Thanks to these two inventions the development of industrial and hydrotechnical engineering was stimulated, including construction of dams with the highly interesting project MOSE which aimed to protect Venice against periodic flooding [8], and high–rise building engineering [9] with the world's highest building, Burj Khalifa (in Dubai) reaching 828 m, with forecasts indicating that in 2020 it will be only one of at least 30 buildings exceeding 400 m in height (sky-scraperpage.com/diagrams/?searchID=201&page=1) [10]. The above-mentioned inventions of the 19th century enabled the development of transport infrastructure, road transport (substrates and pavement of roads and motorways), railway transport (railway sleepers), and sea transport (quays and breakwaters), too. In all those branches unceasing progress is observed, and is strongly influenced by the breakthrough invention in concrete technology from the end of the 20th century, namely, self-compacting concrete [11].

Conscious and responsible actions of mankind in concrete technology do not only involve reaching high technical, strength, and durability parameters, but also producing environment-friendly concrete. This regards acquiring of raw materials, the processing and preparation of materials, including cement and aggregates, and carrying out the particular stages of concrete production. Unfortunately, concrete technology is one of the areas of human activity involved in the increase in $CO_2$ emission; in particular production of cement, a key constituent of concrete. In addition, high demand for energy and water, erection of buildings and their demolition are the reasons why concrete is not considered to be particularly environmentally friendly or compatible with the demands of sustainable development [12]. Hence, in the opinion of not only the environmentalists but also the average users of the environment, the concrete construction sector is obliged to undertake immediate action aimed at reducing its environmental impacts, including the reduction of $CO_2$.

The cement industry occupies a significant place in the spectrum of negative influences onto the environment. According to the current assessments [13] it is responsible for a 5–7% emission among all anthropological sources of carbon dioxide. Assuming the yearly cement production rate of over 4.3 Gt [14] and taking into account (because of the increasing content of mineral additives in cement production) the value of unit emission coefficient of about 0.6 t of $CO_2$ per one cement ton, one gets about 2.5 Gt of $CO_2$ worldwide emission due to the cement industry. The $CO_2$ emission problem is very important because of its relation to climate changes. Measurements of $CO_2$ content in the atmosphere based on investigations carried out in the second half of the 20th century in the Mauna Loa Observatory by C.D. Keeling as well as ice-core analyses [15] indicate a significant relation between the "awakening of technological creativity" of mankind due to the Industrial Revolution and temperature changes. Previously, S. Arrhenius had observed the glass-house effect. He concluded that it might have been advantageous for cold regions of Scandinavia (encyclopedia.com/topic/Svante_August_Arrhenius.aspx) [16]. Unfortunately, climate change endangers mankind's environment, especially in the economical dimension, similar to the nonrational processing of natural resources. Severe harm, e.g., in civil engineering, may follow. For instance in Siberian Russia, in the regions with permafrost melting because of global warming, damage to buildings is reported [17,18].

Yet, it cannot be doubted that the most common danger to civil engineering itself on the side of progressing climate change is the increasing $CO_2$ content in the air. According to some research [19,20] it can lead to the necessity to verify some models of carbonation and restrict all the current recommendations related to slowing down and limiting carbonation effects in order to protect reinforcing steel.

## 6.2 ENVIRONMENTAL FRIENDLINESS IN CIVIL ENGINEERING

The regard for the environment has led to the appearance of such notions as green building, green concrete, green cement, biocement, and eco-cement [12,21–23], which define industrial materials and processes that are environmentally friendly, but at the same time are economically viable and thus meet the expectations of concrete technology. Aïtcin's [24] opinion, according to which "the concrete of tomorrow will be green, green, and green," seems to be the best summary of the significance of this aspect.

The principles of sustainable development regarding the construction and concrete industries have also begun to be a concern of not only environmentalists, but also governments on the local, state, and national levels. Slowly, owners and developers have consequently begun to implement "going green" principles, for political reasons as a source of impalpable benefits and promotion. Still, they also find it as a way to improve the quality of the environment where they live. They have realized that such aspects as reduced energy consumption and reduced life-cycle costs [12] are truly worth it.

The aim of present-day technologists is to produce increasingly sophisticated concrete, in terms of technical parameters, that is absolutely environmentally friendly. The potential tools and strategies needed to meet the environmental challenges in the construction and concrete industries could be achieved in different ways. They are as follows:

1. replacement, as much as possible, of Portland cement by supplementary cementitious materials, especially by-products of industrial processes, such as fly ash, ground granulated blast furnace slag and silica fume [12,25–29];

2. using eco-cements [22,30,31];
3. using recycled materials, including recycled concrete aggregate, in place of natural resources [12,32–36];
4. improvement of durability and service life of structures and as a result reducing the amount of materials needed for their replacement [27,32,37,38];
5. improvement of concrete mechanical and other properties, which can also reduce the amount of materials needed [12,39]; and
6. reusing wash water in concrete plants [12].

## 6.2.1 Conception of Binder and Carbon Intensity Indexes

One of the criteria that can be a measure of sustainable development in the cement and concrete industry are those of Damineli et al. [40] propositions on binder intensity index and carbon dioxide intensity index ($b_i$ and $c_i$, respectively). The first one describes the cement mass per $1\,m^3$ of concrete necessary to achieve $1\,MPa$ strength. The second presents the mass of carbon dioxide emitted in the production process of such a volume of cement that make it possible to attain the concrete strength of $1\,MPa$. Hence, the $b_i$ index makes it possible to estimate the efficiency of a given cement binder in the process of obtaining the durable concrete. The $c_i$ index means a unitary contribution of the binder into the $CO_2$ emission. If a complementary assessment of the eco-efficiency of cement is to be achieved, it is essential to use both indexes simultaneously.

As in all processes of concrete production, most of the carbon dioxide emission comes from the production of cement, and the data can be treated as nearly estimating this emission for the needs of concrete production with specified parameters. Obviously, estimations should be carried out individually, since emissivity of $CO_2$ varies from cement plant to cement plant and different raw materials needed to produce clinker are used, and since selected concrete components vary in terms of quality and quantity.

More precise estimations could be achieved by including the volume of $CO_2$ emitted in the cause of transport of raw materials and cement as a final product, and emitted from technological operations connected with execution of concrete structure (production, transport, and compaction of concrete).

In light of the concept of indexes [40], use of cements with mineral additives results in a lower value of the $c_i$ index compared to the value of

$c_i$ for Portland cements without mineral additives. On the basis of local and international data, Damineli et al. [40] estimated that the $CO_2$ emission index in the production of purely clinker cements is approximately 4.3 and 1.5 kg/MPa in the production of cements with mineral additives [40]. The $c_i$ index is related to the binder intensity index $b_i$, which for concretes with compressive strength exceeding 50 MPa is approximately equal to 5 kg/m$^3$/MPa, and for concretes with strength of 20 MPa is 13 kg/m$^3$/MPa. Hence, there are two ways of reducing $CO_2$ emission, i.e., selecting cements with mineral additives and producing concretes of high strength, more durable by nature, which can bring a meaningful effect when a low proportion of Portland clinker cements are used for the production of high-strength concretes (HSCs).

Reports on HSC in a "green" option refer to replacing a part of cement binder in the concrete with mineral additives [41]. There are few data on achieving this goal using eco-cements [22,30,31], but, according to the literature review made by the authors of this chapter, no studies have been done on the use of pozzolana cement CEM IV/B-V 32.5 R in HSC technology. Since the binder is still not widely recognized in this application area, the authors accepted this as a sort of a research challenge. Attempts were taken to determine the extent, in terms of HSC needs, this type of cement works with the simultaneous addition of natural and crushed aggregates to this concrete, concerning the effect of their maximum particle size. Likewise, there are relatively little data in the available literature covering this area.

## 6.2.2 Sustainable Technology for HSC

Properties of HSC, to a larger extent, are determined by the type of aggregate, both on a qualitative selection as well as graining. Both gravel aggregates and crushed aggregates of various rocks are used. However, the use of harder rock aggregates does not always result in higher concrete strength [32]. Jamroży [42] stresses the need for selection of aggregates according to their cement paste demand. He has shown in one of his studies that crushed basalt aggregate has lower cement paste demand than granite. The rationality of using fractionated aggregate, both fine and coarse, has been emphasized [32]. The issue of influence of maximum size of coarse aggregate particles on concrete strength has been raised i.a. by Aïtcin [32], Venkateswara et al. [43], Kurdowski [44], and Neville [45], who agree on the sense behind the use of smaller aggregate-particle size if higher concrete strength is sought. Aïtcin [32] indicates that the increase of maximum size of aggregate may evoke some problems with the quality of the interfacial transition zone (ITZ), which can be larger and more heterogeneous. Furthermore, as smaller aggregate particles are more durable than larger ones, use of the latter ones involves risk of occurrence of uncontrolled microcracks, which could lead to weakening of the concrete structure. However, Aïtcin [32] claims that it is possible to obtain both a good workability of concrete mixes and strength of concretes made of aggregates of a maximum particle size of 25 mm, provided they originate from sufficiently strong and homogenous rocks. In practice, particles of a smaller size are generally used mainly in order to eliminate the abovementioned effect in the face of a lack of suitable procedures of optimization of aggregate testing. According to Chen and Liu [46] aggregate size significantly influences the fracture behavior of high-performance concrete (HPC). Fracture energy of concrete increases with the increase of the maximum aggregate size. The larger the size of the aggregate, the more significant the deflection of propagating crack and the greater a fracture process zone can form [45,46]. It is in agreement with Zhang and Sun's studies [47], which claim that autogenic shrinkage of HPC decreases with the increase of maximum coarse aggregate size.

However, generally there are not enough studies in this field, and opinions are not unambiguous. For instance, there is lack of data on the simultaneous impact of maximum size of aggregate particles and cement content on the concrete strength, which is significantly essential in carbon dioxide emission.

The undertaken research is associated with this point and is largely connected with an ecological aspect and traditional technologies, focused on the role of aggregate in HSC technology. This ecological aspect is increasingly raised in concrete technology, i.e., in this case, HSC.

Sustainable concrete technology should take into account the environment's action on concrete, too. Concrete subjected to an aggressive corrosive environment will lose its original quality, which will mean a violation of the balanced condition. Among the factors destructively influencing concrete are carbonation and chloride diffusion, especially coming from sea water and cyclic freezing and thawing. The action of the latter can be reduced by the application of admixtures aerating a fresh concrete structure. Referring to the concept by Damineli et al. [40], cyclic freezing and thawing negatively influences the value of coefficients $b_i$ and $c_i$. From this point of view, if frost-aggressive action is foreseen, it is vital to design concrete that will preserve its strength the longest, i.e., concrete for which the values of $b_i$ and $c_i$ would increase as slowly as possible. Doubtlessly, it is an indispensable requirement for HPC.

## 6.3 MATERIALS AND METHODS

### 6.3.1 Cement

The pozzolana cement CEM IV/B-V 32.5 R, used for the needs of this thesis, was created in the laboratory by mixing Portland fly-ash cement CEM II/B-V 42.5N (commercially produced) with fly ash from a CHP (combined heat and power) plant complying with requirements and compatibility criteria according to the standard EN 450-1:2010 (*Fly ash for concrete. Definition, specifications and conformity criteria*) in mass percentages 86.5% and 13.5%, respectively. Additionally, in one recipe, CEM I 42.5 R was used. Properties of the binders, both factory-manufactured (the abovementioned CEM I 42.5 R) and laboratory-prepared (CEM IV/B-V 32.5 R) are presented in Table 6.1.

### 6.3.2 Aggregate

The following types of aggregate were used to produce concretes: pit sand (0/2 mm), gravel aggregate divided into two fractions: 2/8 and 8/16 mm, and granite and basalt crushed aggregate (2/8 and 8/16 mm). All the

**Table 6.1** Chemical composition, physical, and mechanical properties of cements

| Characteristic | Result | |
|---|---|---|
| | **CEM IV/B-V 42.5 R** | **CEM I 42.5 R** |
| *Chemical compounds (%)* | | |
| $SO_3$ | 2.65 | 3.23 |
| $SiO_2$ | 30.45 | 19.55 |
| $Al_2O_3$ | 12.08 | 5.32 |
| $Fe_2O_3$ | 4.92 | 3.24 |
| CaO | 40.95 | 62.74 |
| $Na_2O_{eq}$ | 0.65 | 0.56 |
| Insoluble residue (%) | 3.03 | 0.99 |
| Ignition loss (%) | 4.43 | 3.18 |
| Blaine specific surface ($m^2$/kg) | 418 | 335 |
| Density (kg/$m^3$) | 2740 | 3190 |
| *Compressive strength (MPa)* | | |
| 2 days | 17.4 | 26.8 |
| 28 days | 36.4 | 56.1 |

aggregates complied with requirements of the EN 12620:2002 standard (*Aggregates for concrete*).

In the experiment focused on freezing–thawing durability and its connection with $b_i$ and $c_i$ indexes only two fractions of coarse granite aggregate (2/8 and 8/16 mm) were used.

### 6.3.3 Superplasticizer and Air-Entraining Agent

The authors used a highly effective new-generation superplasticizer based on polycarboxylate ether and air-entraining agent (AEA) based on saponified fluid turpentine resin. The physical and chemical parameters of both are given in Tables 6.2 and 6.3, respectively.

### 6.3.4 Microsilica

Amorphous microsilica was used. The characteristics are listed in Table 6.4.

### 6.3.5 Concrete Mix Recipes

Concrete recipes were determined for the following variants: cement in the amount of $600 \, kg/m^3$ along with aggregate mix of dense pile up to

**Table 6.2** Physical and chemical properties of superplasticizer

| Property | Result |
|---|---|
| Chemical base | Polycarboxylate ether |
| Physical state | Water solution |
| Color | From beige to brown |
| Density at 20°C ($kg/m^3$) | 1063 |
| pH at 20°C | 5.5 |
| Chloride content (%) | 0.1 |
| Boiling point (°C) | 100 |
| Absolute viscosity (MPa · s) | 30 |

**Table 6.3** Physical and chemical properties of AEA

| Property | Result |
|---|---|
| Physical state | Liquid |
| Color | Light yellow |
| Density at 20°C ($kg/m^3$) | 1000 |
| pH at 20°C | $10 \pm 1.0$ |
| Chloride content (% of mass) | 0.09 |
| Alkaline content (% of mass) | 0.08 |
| Boiling point (°C) | 100 |
| Water solubility at 20°C | Soluble |

**Table 6.4** Physical and chemical properties of microsilica

| Parameter | Result |
|---|---|
| Physical state | Powder |
| Color | Gray |
| Odor | Odorless |
| Bulk density (kg/m³) | 150–170 |
| pH at 20°C | 5.0–7.0 |
| Chloride content (%) | ≤0.15 |
| Melting point (°C) | 1.550–1.570 |

**Table 6.5** Recipes of HSC concretes

| Constituent (kg/m³) | Recipe designation | | | |
|---|---|---|---|---|
| | Rec-1 (600–0/8) | Rec-2 (600–0/16) | Rec-3 (700–0/8) | Rec-4 (700–0/16) |
| CEM IV/B–V 32.5 R | 600 | 600 | 700 | 700 |
| Sand 0/2 mm | 615 | 525 | 574 | 490 |
| Coarse aggregate 2/8 mm | 1020 | 555 | 952 | 518 |
| Coarse aggregate 8/16 mm | 0 | 555 | 0 | 518 |
| Water | 140 | 140 | 140 | 140 |
| Microsilica | 60 | 60 | 70 | 70 |
| Superplasticizer | 6.0 | 6.0 | 7.7 | 7.7 |
| w/c | 0.23 | 0.23 | 0.20 | 0.20 |

8 mm and up to 16 mm and 700 kg/m³ with aggregate mix of dense pile up to 8 mm and up to 16 mm (Table 6.5). The class of consistency of concrete mixes was beyond S1–S4, corresponding to the recommendation of the EN 206-1:2000 standard (*Concrete–Part 1: Specification, performance, production and conformity*).

In the experiment focused on freeze–thaw durability (Recipes: Rec-5, Rec-6 and Rec-7, Table 6.6) the cement amount was different (350–450 kg/m³); moreover, in Rec-5 only CEM I 42.5 R was used. The class of the concrete mixes' consistency was S4.

## 6.3.6 Testing Procedure

Firstly, ingredients of the concrete mix are as follows: coarse aggregate of the 8/16-mm fraction, coarse aggregate of the 2/8-mm fraction, sand 0/2 mm and microsilica were placed in a laboratory concrete mixer and mixed for 1 min. The next step (in all series, except for the recipe based on CEM I 42.5 R) was the addition of 75% of the total amount of water, cement CEM II B-V 42.5N (86.5%), and fly ash (13.5%). The whole was

**Table 6.6** Recipes of concretes subjected to freeze–thaw cycles

| Constituent (kg/m$^3$) | Recipe designation | | |
|---|---|---|---|
| | Rec-5 | Rec-6 | Rec-7 |
| CEM I 42.5 R | 350 | – | – |
| CEM IV/B-V 32.5 R | – | 400 | 450 |
| Sand 0/2 mm | 715 | 700 | 656 |
| Coarse aggregate 2/8 mm | 601 | 588 | 551 |
| Coarse aggregate 8/16 mm | 613 | 601 | 564 |
| Water | 155 | 155 | 155 |
| Superplasticizer | 3.2 | 3.2 | 18.7 |
| AEA | 8.5 | 10.0 | 11.3 |
| w/c | 0.43 | 0.39 | 0.35 |

mixed for 3 min. The last stage was a gradual addition of 25% of water along with the superplasticizer (the remaining 25% of water was added with an appropriate amount of superplasticizer), with switching the laboratory concrete mixer on for the next 3 min. In the case of air-entrained concrete mixes the AEA was added at the end and all ingredients were mixed to obtain a fully homogeneous concrete mix.

The concrete mix was placed in forms of plastic cubic molds of side 150 mm, and was laid in two layers. After 24 h, the samples were unmolded and placed in a climatic chamber for a period of 28 days, where temperature was kept at 18 ± 2°C and relative humidity at 95% providing standard curing conditions.

Consistency of concrete mixes was marked by slump cone method according to the requirements of the EN-12350-2:2009 standard (*Testing fresh concrete. Slump flow*). Measurements of density of concrete mixes were conducted in compliance with the EN 12350-6:2009 standard (*Testing fresh concrete. Density*). Air contents were measured (immediately after making the mix and after 45 min of its preparation) in accordance with the EN 12350-7:2009 standard (*Testing fresh concrete. Air content. Pressure methods*). Determination of concrete absorption was performed on samples after 28 days of hardening, according to the procedure specified in the Polish standard PN-88/B-06250 (*Ordinary concrete*). The mentioned standard is no longer compulsory, but is still used in engineering practice. Tests on compressive strength, in compliance with requirements of the EN 12390-3:2009 standard (*Testing hardened concrete. Compressive strength of test specimens*), were performed after:

- 1, 7, 28 days (on five samples for each series of concrete)
- 56, 90, and 180 days (on a single sample for given series).

Tests results of 1-, 7-, and 28-day concrete compressive strength were subjected to a statistical analysis using the method of analysis of variance complemented with contrast analysis. The analysis was supplemented with post hoc tests including the most sensitive testing method, which is the least significant difference (LSD) test. Discussing the results, attention was focused on statistically essential differences between compared groups for results obtained after 28 days of hardening. For the remaining cases ($f_{c1}$ and $f_{c7}$) only the registered trends were discussed. Comparisons by the LSD method were performed by considering the following variants:

- a maximum size of aggregate particles (a type of aggregate, with the cement content distinguished),
- a maximum size of aggregate particles (a type of aggregate, without distinguishing the cement content),
- a maximum aggregate size (cement content, without distinguishing the type of aggregate).

For the mean results of the compressive strength obtained after 28 days, the $b_i$ and $c_i$ (binder and carbon dioxide intensity) indexes were calculated. For the results obtained in later periods (after 56, 90, and 180 days of hardening) and for HSC mixes (Rec-1 to Rec-4) extreme values were noted (minimum and maximum). For Rec-5, Rec-6, and Rec-7 only the 28- and 90-day results were taken into consideration, both of which were based on five replications; $b_i$ and $c_i$ were also calculated with and without presence of AEA. In order to calculate a unitary emission for the CEM IV/B-V 32.5 R cement, 2011 data from a Polish cement plant were used, assuming the cement content in the amount of 86.5%. The rest was accepted as an emission-free ash. The calculations included the following cement value, i.e., 740 kg $CO_2$ per 1 t of CEM I 42.5 R cement (used only in Rec-5) and 448 kg $CO_2$ per 1 t of CEM IV/B-V 32.5 R cement. The emission from the combustion of biomass, 5% of the remaining additives according to the EN 197-2: 2000 standard (*Cement—Part 2: Conformity evaluation*), and the participation of the setting regulator were not taken into account.

## 6.4 RESULTS AND DISCUSSION

Test results of the air content in concrete mixes (immediately after making the concrete mix, and after 45 min of its implementation), the density

**Table 6.7** Air content and density of concrete mixes and water absorption of HSCs

| Aggregate type–cement content–graining of aggregate[a] | Air content (%) | | Density (kg/m³) | Water absorption (%) |
|---|---|---|---|---|
| | 0′ | 45′ | | |
| Gravel–700–0/8 | 2.8 | 2.1 | 2300 | 2.79 |
| Gravel–700–0/16 | 0.8 | 0.8 | 2290 | 2.39 |
| Gravel–600–0/8 | 2.6 | 2.6 | 2330 | 2.96 |
| Gravel–600–0/16 | 1.1 | 1.1 | 2330 | 2.80 |
| Granite–700–0/8 | 2.6 | 2.4 | 2300 | 2.66 |
| Granite–700–0/16 | 1.2 | 1.2 | 2280 | 2.93 |
| Granite–600–0/8 | 1.1 | 1.1 | 2340 | 2.54 |
| Granite–600–0/16 | 1.0 | 1.0 | 2310 | 2.83 |
| Basalt–700–0/8 | 1.7 | 1.4 | 2350 | 2.65 |
| Basalt–700–0/16 | 0.7 | 0.7 | 2320 | 2.68 |
| Basalt–600–0/8 | 1.2 | 1.2 | 2390 | 2.64 |
| Basalt–600–0/16 | 0.8 | 0.8 | 2390 | 2.48 |

[a]Cement content and graining of aggregate are given in $kg/m^3$ and mm, respectively.

of concrete mixes, and water absorption of HSC concrete after 28 days are summarized in Table 6.7.

## 6.4.1 Air Content and Density

Low air contents in concrete mixes, ranging from 0.7% to 2.8%, both immediately after making of the concrete mix as well as 45 min later, seem to be particularly noticeable. For a given amount of cement, 600 and 700 $kg/m^3$, respectively, mixes of aggregates of maximum particle size up to 16 mm were characterized by lower air contents. The concrete mixes obtained during research exhibited self-compacting properties, spontaneously flowing and de-aerating. There are no obligatory requirements concerning air content for HSCs or self-compacting concretes. Some literature sources [48,49] indicate the value of 2% but some of them indicate a higher value, even 6% [48]. The results should therefore be considered satisfactory, and the concrete mixes' properties should be considered as improved, especially in a variant of the ready-mix concrete, where in practice application of concrete mixes of a consistency close to S4 class is required. The density of concrete mixes ranged from 2300 to 2390 $kg/m^3$ for HSC concretes and from 2222 to 2328 $kg/m^3$ for ordinary concrete. This is obviously related to the density of the aggregates, the smallest for gravel aggregate and the largest for crushed basalt aggregate (in ordinary concrete only granite was used).

In the case of concrete mixes used for frost-resistance studies, air content ranged from 1.9% to 2.7% for non air-entrained and from 5.3% to 6.3% for air-entrained concrete mixes.

## 6.4.2 Water Absorption

Values of 28-day water absorption of HSC concrete samples did not exceed 3%. There was no significant variation of water absorption of concrete on the type of aggregate used. The range of changes depending on the amount of cement (600 and 700 kg/m³) was also insignificant, amounting to approximately 0.2%. As regards ordinary concretes subjected to the influence of natural atmospheric factors, water absorption rate should not exceed 5% and 9% in case of concretes protected from the impact of absorption (according to PN-88/B-06250). The concretes made for this study belong to the self-compacting and HSCs. Hence, they are special in two ways. Such cement matrix composites should be expected to have better parameters in comparison with ordinary concrete's parameters. The results obtained here prove that pozzolana cement has got very good characteristics in various configurations with aggregate, regarding aggregate type and maximum particle size. Water absorption is a property relatively easy to study and, at the same time, a factor determining the durability index of concrete in some way. Research results appear to portend positively in terms of concrete strength in concrete made of pozzolana cement.

The small values of water absorption obviously have a relation with participation of microsilica in concretes. It is microsilica that, interacting with the binder, contributes to the compaction of the concrete structure due to formation of greater amounts of hydrated calcium silica phases. Obtaining concretes with low water absorption suggests that the pozzolana cement is compatible in cooperation with microsilica, which of course would require a deeper research in this area.

## 6.4.3 Compressive Strength

A tested key property of concrete was compressive strength. It was assumed that application of pozzolana cement would lead to obtaining concretes of high strength parameters, provided that appropriate qualitative and quantitative selection of aggregates and assistance by a highly effective superplasticizer and microsilica has been carried out.

With a minimum value of 60 MPa, taken as a criterion of HSC, all concretes of 28 days and older (56, 90, 180 days) fulfilled this condition

**Table 6.8** Compressive strength of HSC concretes

| Aggregate type–cement content–graining of aggregate[a] | $f_{c1}$ (MPa) | $f_{c7}$ (MPa) | $f_{c28}$ (MPa) | $f_{c56}$ (MPa) | $f_{c90}$ (MPa) | $f_{c180}$ (MPa) |
|---|---|---|---|---|---|---|
| Gravel–700–0/8 | 51.03 | 68.80 | 86.10 | 89.50 | 93.70 | 95.00 |
| Gravel–700–0/16 | 38.57 | 64.33 | 82.57 | 84.30 | 91.40 | 92.90 |
| Gravel–600–0/8 | 38.67 | 58.77 | 76.57 | 78.90 | 88.30 | 91.50 |
| Gravel–600–0/16 | 40.20 | 61.20 | 78.30 | 82.40 | 85.20 | 88.90 |
| Granite–700–0/8 | 49.23 | 75.60 | 94.97 | 97.20 | 100.10 | 101.90 |
| Granite–700–0/16 | 42.93 | 69.97 | 88.07 | 91.60 | 95.40 | 96.50 |
| Granite–600–0/8 | 37.93 | 63.93 | 80.27 | 89.70 | 97.00 | 98.90 |
| Granite–600–0/16 | 35.57 | 66.67 | 87.13 | 93.90 | 96.50 | 98.00 |
| Basalt–700–0/8 | 48.03 | 75.60 | 82.27 | 86.60 | 91.20 | 93.10 |
| Basalt–700–0/16 | 42.13 | 71.60 | 85.77 | 91.50 | 93.50 | 101.80 |
| Basalt–600–0/8 | 42.23 | 67.20 | 73.73 | 74.80 | 89.10 | 94.60 |
| Basalt–600–0/16 | 41.20 | 72.67 | 86.40 | 90.20 | 102.10 | 106.50 |

[a]Cement content and graining of aggregate are given in kg/m$^3$ and mm, respectively.

(Table 6.8). Moreover, 7–day strengths exceeded a value of 60 MPa, except for one type of concrete, i.e., that made of gravel aggregate with a maximum particle size up to 8 mm, combined with cement content of 600 kg/ m$^3$. A maximum 28-day compressive strength was reached by concrete made of cement CEM IV/B-V 32.5N, used in the amount of 700 kg/m$^3$, applying granite aggregate with a maximum particle size of 8 mm (Fig. 6.3), which was also confirmed by a statistical analysis. This also covered the period of 7 and 56 days (Table 6.8). After 90 and 180 days, in terms of concrete strength, basalt aggregate concrete with a maximum particle size up to 16 mm, proved to be slightly better than crushed granite aggregate concrete. However, particular care should be exercised when interpreting this trend, keeping in mind that the 180-day strength was for a single concrete sample for a given series. Concrete compressive strength increased after longer periods of hardening (90- and 180-day periods), which should be attributed to the characteristic properties of pozzolana cement. Undoubtedly, an explanation of the attainment of these high strength values must be sought in the precise selection of dense pile of aggregate, good quality of aggregate, relatively large amount of cement (600 and 700 kg/m$^3$), effectiveness of superplasticizer, and microsilica, compatible with pozzolana cement. With such a large number of factors influencing

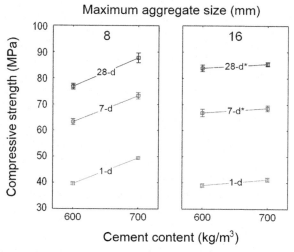

**Figure 6.3** Influence of maximum aggregate size and cement content on 1-, 7-, and 28-day compressive strength of concrete (marked as 1, 7, and 28-days, respectively) depending on aggregate type (Note: if asterisk is added there is lack of statistical difference between compared groups).

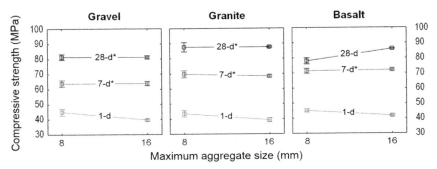

**Figure 6.4** Influence of maximum aggregate size on 1-, 7-, and 28-day compressive strength of concrete (marked as 1, 7, and 28 days, respectively) depending on aggregate type, without differentiating amount of cement used (Note: if asterisk is added there is lack of statistical difference between compared groups).

the high value of strength, and therefore a possibility of a synergistic effect, it is hard to precisely pinpoint which of them had a dominant influence. Although, according to the statistical analysis presented below, the maximum particle size of the aggregate seems to have a relevant meaning in the case of greater cement content, provided that an aggregate of 8 mm is used.

Fig. 6.4 presents the effect of maximum particle size of an aggregate on the increase of concrete strength as a function of the cement content, without taking into account the type of aggregate. The graphs show that it is easier to obtain an additional increase in strength by increasing the cement content by 100 kg/m³ for concrete of aggregate to 8 mm than for concrete of aggregate 0/16 mm. Analyzing Fig. 6.5, it can be concluded that without taking into consideration the impact of diverse cement content (600 and 700 kg/m³), and taking into account only aggregate type and maximum particle size, for the three aggregates used in the test, only in case of the strongest rock, i.e., basalt, did the maximum particle size decide the increase in concrete strength after 28 days of hardening. When 0/16-mm aggregates were used instead of 0/8-mm aggregates an increase of strength by 10% was registered. This can be a confirmation for Aïtcin's hypothesis [32], which states that the size used (0/8 mm) and greater cement content (700 kg/m³) allows for gaining better strength, which makes no difference in the case of aggregates up to 16 mm, and using a greater maximum particles size could lead to a risk of appearance of cracks in larger, originally weaker, grains of aggregate. As it can be noted, in the case of good-quality rock from which aggregate is obtained, no such effect is registered. For other types of aggregate, lack of differences

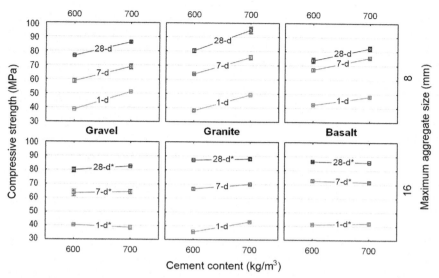

**Figure 6.5** Influence of cement content on 1-, 7-, and 28-day compressive strength of concrete (marked as 1, 7, and 28 days, respectively) depending on maximum aggregate size, without differentiating type of aggregates (Note: if asterisk is added there is lack of statistical difference between compared groups).

was not detected even in the post hoc LSD test, a method considered to be the most liberal of the method of testing, i.e., allowing for detecting of even the smallest of differences between the mean values of the tested characteristics. However, taking into account the maximum particle size up to 8 mm, the use of granite aggregate made it possible to attain better concrete strength than in the case of 0/8-mm basalt aggregate, although basalt rock is considered to be more durable than granite. This could result from the impact of the type of bonding between basalt aggregate and cement paste, which is different when a high amount of fly ash in cement is used. In case of smaller maximum aggregate size (0/8 mm) it can be expected that ITZ occupies more volume because of the larger specific surface area of aggregate. More detailed results on factors affecting concrete compressive strength are specified in the analysis in Fig. 6.3. In the case of increasing cement content by 100 kg/m$^3$, the largest increase resulting from the use of aggregate with the maximum particle size of 8 mm was obtained for granite aggregate concrete and the lowest for basalt concrete. The aggregate with maximum particle size up to 16 mm, only after one day, did a greater amount of cement decide a minimum increase of concrete strength ($f_{c1}$), but it was only true for the granite aggregate.

A similar effect, statistically significant, though less meaningful, was found for the 7-day strength ($f_{c7}$), and also for concrete from the same aggregate. For all types of aggregate with 0/16-mm graining, concerning their effect on 28-day strength, no difference in results was registered, even statistically. Cement content comes into play in case of a lower maximum aggregate size (0/8 mm). Without taking into consideration the aggregate type, better results were obtained for 700 kg/m³. However, when aggregate up to 8 mm is used, granite seems to be advisable for both cement amounts. The reason for the relatively low increase in 28-day strength, compared to 7-day strength, for concretes made with the use of basalt aggregate with 0/8-mm graining is also worth asking. With a lower maximum particle size, a larger area of contact between the aggregate and cement paste (ITZ) can be expected. It is conceivable that in this case an unfavorable tendency to weaken a transition zone between basalt aggregate and cement paste in the presence of a large amount of fly ash in the binder appeared. A similar effect, although to a lesser extent, was noted in the case of the same aggregate but with 0/16-mm graining. Moreover, it was true only for basalt aggregate and 700 kg/m³ cement content that 28-day compressive strength was better for 0/16-mm graining than for 0/8 mm. Generally, without taking into consideration aggregate type, when smaller maximum aggregate size is used (0/8 mm) greater cement content (700 kg/m³) allows for gaining better strength, which makes no difference in case of aggregate up to 16 mm.

## 6.4.4 Binder and Carbon Dioxide Indexes

Interesting observations were provided by an analysis on the results of calculations concerning binder intensity and carbon dioxide intensity indexes, the reduction of which is key to environmental friendliness.

Based on the data shown in Fig. 6.6 it can be concluded that the best indexes were obtained for HSCs made with the use of 600-kg/m³ cement with the maximum particle size of 16 mm. The most beneficial was the use of granite aggregate ($b_i = 6.9$ kg/m³/MPa, $c_i = 3.09$ kg/MPa) and basalt ($b_i = 7.0$ kg/m³/MPa, $c_i = 3.11$ kg/MPa). Slightly less favorable was the use of gravel ($b_i = 7.7$ kg/m³/MPa, $c_i = 3.44$ kg/MPa) aggregate. The least satisfactory results were obtained with a larger amount of cement (700 kg/m³), in case of the use of gravel and granite aggregate with 0/16-mm graining; for basalt the graining was 0/8 mm. It seems that in order to achieve a relatively good-quality HSC concrete there is no need to increase the proposed proportion of the pozzolana cement CEM

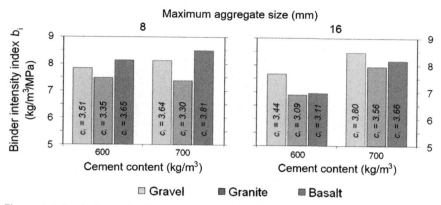

**Figure 6.6** Binder intensity ($b_i$) and carbon dioxide intensity ($c_i$) indexes estimated for 28-day compressive strength of concrete and carbon dioxide emission from cement production at the level of 448 kg $CO_2/t_{CEM}$ (Note: $c_i$ is given in kg/MPa).

IV above $600\,kg/m^3$, provided there are no precise requirements to attain high-class concrete. In general, both $b_i$ and $c_i$ indexes obtained for concretes produced with the use of the CEM IV cement have their values within the range indicated by Damineli et al. [40]. The $c_i$ index is closer to its upper-limit range, but its worst result is more than $1\,kg/m^3/MPa$ lower than the $4.3\,kg/m^3/MPa$ value specified for CEM I. Analysis of indexes of binder intensity, specified for the results of strength tests (Table 6.9), showed a gradual reduction in their value with time and amount of cement used, with preference toward using it in smaller amounts. The analysis on indexes of the binder $CO_2$ emission in the context of concrete compressive strength achieved (Table 6.9) confirmed the reduction of the $c_i$ value with time and the amount of the cement used, with a preference toward using it in a smaller amounts, especially in the case of aggregate up to 16 mm, which was also emphasized in Section 6.3.3.

## 6.4.5 Influence of Freeze–Thaw Cycles

Results of 28- and 90-day compressive strength of ordinary concrete (for series without and with AEA) and compressive strength loss (calculated after 150 freeze–thaw cycles) were presented in Table 6.10. Tables 6.11 and 6.12 summarize calculated values of $b_i$ and $c_i$ referring to the abovementioned results. Based on the data it can be concluded that concrete without AEA will become a bigger burden for the environment because the frost-aggressive action gradually increases the coefficients $b_i$ and $c_i$. Even if

**Table 6.9** Values of binder intensity ($b_i$) and carbon intensity ($c_i$) indexes depending on maximum aggregate size and cement content obtained on the base of 56-, 90-, and 180-day compressive strength-test results

| Maximum aggregate size (mm) | | 8 | | 16 | | Best in group | |
| --- | --- | --- | --- | --- | --- | --- | --- |
| Cement content (kg/m³) | | 600 | 700 | 600 | 700 | 600 | 700 |
| 56-day | Gravel | 7.6 (3.4)[a] | 7.8 (3.5) | 7.3 (3.3) | 8.3 (3.7) | *16mm* | *8mm* |
| | Granite | 6.7 (3.0) | 7.2 (3.2) | **6.4 (2.9)** | 7.6 (3.4) | *16mm* | *8mm* |
| | Basalt | 8.0 (3.6) | 8.1 (3.6) | 6.7 (3.0) | 7.7 (3.4) | *16mm* | *16mm* |
| | Best in group | *Granite* | *Granite* | *Granite* | *Granite* | | |
| 90-day | Gravel | 6.8 (3.0) | 7.5 (3.3) | 7.0 (3.2) | 7.7 (3.4) | *8mm* | *8mm* |
| | Granite | 6.2 (2.8) | 7.0 (3.1) | 6.2 (2.8) | 7.3 (3.3) | *similar* | *8mm* |
| | Basalt | 6.7 (3.0) | 7.7 (3.4) | **5.9 (2.6)** | 7.5 (3.4) | *16mm* | *16mm* |
| | Best in group | *Granite* | *Granite* | *Basalt* | *Granite* | | |
| 180-day | Gravel | 6.6 (2.9) | 6.1 (2.8) | 6.7 (3.0) | 7.5 (3.4) | *8mm* | *16mm* |
| | Granite | 6.1 (2.7) | 6.9 (3.1) | 6.1 (2.7) | 7.3 (3.2) | *similar* | *8mm* |
| | Basalt | 6.3 (2.8) | 7.5 (3.4) | **5.6 (2.5)** | 6.9 (3.1) | *16mm* | *16mm* |
| | Best in group | *Granite* | *Gravel* | *Basalt* | *Basalt* | | |

[a]Carbon intensity ($c_i$, given in kg/MPa) value is presented in parenthesis, after binder intensity ($b_i$, given in kg/m³/MPa) value; the lowest $b_i$ and $c_i$ value for each compressive strength time (56, 90, and 180 days, respectively) are bolded.

**Table 6.10** Results of compressive strength and frost resistance for Rec-5, Rec-6, and Rec-7

| Series description | Compressive strength (MPa) | | Compressive strength loss[a] after 150 freeze–thaw cycles (%) | |
|---|---|---|---|---|
| | 28-day | 90-day | 28-day | 90-day |
| Granite–350 | 57.9 | 58.5 | 35 | 33 |
| Granite–350–AEA | 45.3 | 48.6 | 5 | 9 |
| Granite–400 | 50.8 | 56.2 | 47 | 41 |
| Granite–400–AEA | 49.2 | 53.0 | 32 | 5 |
| Granite–450 | 53.0 | 58.7 | 76 | 100 |
| Granite–450–AEA | 56.7 | 58.8 | 29 | 12 |

[a]Expressed in relation to control samples (unaffected by freeze–thaw cycles).

**Table 6.11** Estimated values of $b_i$ index (kg/m$^3$/MPa) for Rec-5, Rec-6, and Rec-7 series

| Series description | Before freeze–thaw cycles | | After freeze–thaw cycles | |
|---|---|---|---|---|
| | 28-day | 90-day | 28-day | 90-day |
| Granite–350 | 6.0 | 6.0 | 9.4 | 8.9 |
| Granite–350–AEA | 7.7 | 7.2 | 8.1 | 7.9 |
| AEA+[a] | −1.7 | −1.2 | 1.3 | 1.0 |
| Granite–400 | 7.9 | 7.1 | 14.8 | 12.1 |
| Granite–400–AEA | 8.1 | 7.5 | 12.0 | 7.9 |
| AEA+[a] | −0.3 | −0.4 | 2.8 | 4.2 |
| Granite–450 | 8.5 | 7.7 | 35.8 | — |
| Granite–450–AEA | 7.9 | 7.7 | 11.2 | 8.71 |
| AEA+[a] | 0.6 | 0.0 | 24.5 | — |

[a]AEA improvement effect in kg/m$^3$/MPa. AEA improvement effect in %.

**Table 6.12** Estimated values of $c_i$ index (kg/MPa) for Rec-5, Rec-6, and Rec-7 series

| Series description | Before freeze–thaw cycles | | After freeze–thaw cycles | |
|---|---|---|---|---|
| | 28-day | 90-day | 28-day | 90-day |
| Granite–350 | 4.5 | 4.4 | 6.9 | 6.6 |
| Granite–350–AEA | 5.7 | 5.3 | 6.0 | 5.9 |
| AEA+[a] | −1.2 | −0.9 | 0.9 | 0.8 |
| AEA+%[b] | −28% | −20% | 14% | 11% |
| Granite–400 | 3.5 | 3.2 | 6.6 | 5.4 |
| Granite–400–AEA | 3.6 | 3.4 | 5.4 | 3.5 |
| AEA+[a] | −0.1 | −0.2 | 1.2 | 1.9 |
| AEA+%[b] | −3% | −6% | 19% | 35% |
| Granite–450 | 3.8 | 3.4 | 16.0 | — |
| Granite–450–AEA | 3.6 | 3.4 | 5.0 | 3.9 |
| AEA+[a] | 0.2 | 0.0 | 11.0 | — |
| AEA+%[b] | 7% | 0% | 69% | — |

[a]AEA improvement effect in kg/MPa.
[b]AEA improvement effect in %.

it does not undergo the frost destruction, a larger amount of cement will be necessary to keep the appropriate strength level. Simultaneously, it turns out that the cement CEM I 42.5 R has better values of coefficients $b_i$ and $c_i$, if the frost action occurs earlier after producing it, even in the absence of AEA. It is evident in the result for the series of 28-day concrete specimens subjected to the freeze–thaw cycles.

The situation is different if concrete is subjected to freeze–thaw cycles at a later age, which show the results obtained after 90 days. In this case the use of eco-cement (with active mineral additives CEM IV/B-V 32.5 R), i.e., concrete which influences environment to a smaller extent, is more advantageous. The positive effect of use of AEA is nearly the same (compared to the series of concrete made with CEM I 42.5 R), which is indicated by the values of $b_i$, however, the value of $c_i$ decreases significantly. The influence of the delay in the occurrence of frost aggression may have a significant practical meaning. Concrete (reinforced concrete) elements destined for frost aggression should be made with such time reserve before the cold periods in the year, so that the resulting degree of frost-resistance would allow for the use of a more environmentally friendly cement. It is obvious that in extreme cases, e.g., when repair of element exchange is necessary, better effects will be achieved if a cement free of mineral additives is used.

## 6.5 CONCLUSIONS

Based on research carried out, the following conclusions were reached:
- Pozzolana cement CEM IV/B-V 32.5 R with a simultaneous use of a highly effective superplasticizer and microsilica made it possible to obtain HSCs made both of mineral natural (gravel) aggregates as well as from crushed (granite and basalt). In the areas of research conducted, resultant concretes showed self-compacting properties, which places them in a group of "green" composites, fulfilling the principle of sustainable development.
- Increase of the cement content from 600 to 700 kg/m$^3$ resulted in the growth of concrete strength (at least by 8.5%) only when 0/8-mm aggregates were used. For 0/16-mm aggregates, practically no statistically significant difference was noted, for both levels of amount of cement used.
- Taking strength as the only criterion for HSC quality assessment for our studies the use of 700-kg/m$^3$ cement proved to be more favorable in the case of 0/8-mm aggregates. However, when basalt aggregate was

used to produce concrete, 0/16-mm aggregates appeared more effective for this cement content.

- Taking into account both concrete compressive strength as well binder and carbon dioxide intensity indexes ($b_i$ and $c_i$), with a lower cement content ($600 \, kg/m^3$) the use of aggregate with 0/16 mm graining was more rational, regardless of the type of aggregate.
- If the $CO_2$ emission during crushing of aggregates is not factored in, the best solution is to use basalt and granite aggregate. As far as 180-day compressive strength was concerned, granite aggregate of 0/8 mm and basalt aggregate of 0/16 mm were advisable.
- In the case of ordinary concrete durability considerations, related to its frost resistance, the coefficients $b_i$ and $c_i$ should be estimated at the concrete design stage. This would allow for a choice of a cement type appropriate for the season when concrete laying will take place. In an extreme cases, a longer period of time between placing of concrete and initiation of freeze–thaw action will be mandatory.

A selection of a maximum particle size of the aggregates between 8 and 16 mm with a simultaneous use of pozzolana cement CEM IV/B-V 32.5 R, although relevant, is not unequivocal. Each time it requires individual consideration of the following factors: type and local availability of aggregate, concrete strength requirements, and $CO_2$ emission connected with extraction and production of concrete constituents, especially cement. However, the points mentioned above concerning the influence of type and maximum size of aggregate on some HSC properties seem to mark out a new direction of investigations when $CO_2$ emission is taken into consideration. In order to support the conclusions, different types and origins of gravel aggregate as well as other rock sources of crushed aggregate should be tested. Similarly, there is a need to check a wider range of maximum aggregate size. Moreover, all the abovementioned factors require individual consideration of a type and availability of aggregate, strength parameters of the rock, and, finally, real and individual carbon dioxide emission connected with extraction or production of concrete ingredients.

## ACKNOWLEDGMENTS

The authors express sincere appreciation to Daniel Owsiak, MSc, and Paweł Madej, MSc, from Lafarge Poland for providing the necessary materials for research, creative inspiration, and fruitful discussion on the realized research issues.

# REFERENCES

[1] <http://climate.nasa.gov/news/916/> [uploaded 14.10.2013].
[2] <http://scrippso2.ucsd.edu> [uploaded 14.10.2013].
[3] Keeling RF, Manning AC. Studies of recent changes in atmospheric $O_2$ content. Treatise on Geochemistry 2014;4:385–404.
[4] <http://solarscience.msfc.nasa.gov/predict.shtml> [uploaded 14.10.2014].
[5] <www.bne-portal.de/fileadmin/unesco/de/Downloads/Hintergrundmaterial_international/Brundtlandbericht.File.pdf?linklisted=2812> [uploaded 10.10.2015].
[6] Crutzen PJ, Stoermer EF.The anthropocene. Global Change Newsletter 2000;41:17–18.
[7] <http://esa.un.org/unpd/wpp/Publications/Files/Key_Findings_WPP_2015.pdf> [uploaded 20.10.2015].
[8] Nosengo N.Venice floods: save our city!. Nature 2003;424(6949):608–9.
[9] Jasiczak J, Wdowska A, Rudnicki T. *Betony ultrawysokowartościowe: właściwości, technologie, zastosowania. (Ultra high performance concretes – properties, technologies, applications).* Stowarzyszenie Producentów Cementu [in Polish]. 2008.
[10] <http://skyscraperpage.com/diagrams/?searchID=201&page=1> [uploaded 24.03.2014].
[11] Okamura H, Ozawa K. Self-compacting high performance concrete. Struct Eng Int 1996;6(4):269–70.
[12] Meyer C.The greening of the concrete industry. Cem Concr Compos 2009;31(8):601–5.
[13] Chen C, Habert G, Bouzidi Y, Jullien A. Environmental impact of cement production: detail of the different processes and cement plant variability evaluation. J Clean Prod 2010;18(5):478–85.
[14] Schneider M, Romer M, Tschudin M, Bolio H. Sustainable cement production—present and future. Cem Concr Res 2011;41(7):642–50.
[15] <http://www.esf.org/index.php?id=855> [uploaded 20.10.2015].
[16] <http://www.encyclopedia.com/topic/Svante_August_Arrhenius.aspx> [uploaded 20.10.2015].
[17] Anisimov O, Reneva S. Permafrost and changing climate: the Russian perspective. AMBIO 2006;35(4):169–75.
[18] <http://news.bbc.co.uk/2/hi/science/nature/4120755.stm> [uploaded 20.10.2015].
[19] Talukdar S, Banthia N, Grace JR. Carbonation in concrete infrastructure in the context of global climate change–Part 1: experimental results and model development. Cem Concr Compos 2012;34(8):924–30.
[20] Yoon IS, Çopuroğlu O, Park KB. Effect of global climatic change on carbonation progress of concrete. Atmos Environ 2007;41(34):7274–85.
[21] James Hicks PE. Durable "green" concrete from activated pozzolan cement. : Green Streets and Highways, American Society of Civil Engineers; 2010.408–30.
[22] Samer M.Towards the implementation of the Green Building concept in agricultural buildings: a literature review. Agric Eng Int 2013;15(2):25–46.
[23] Hameed MS, Sekar ASS. Self compaction high performance green concrete for sustainable development. J Ind Pollut Control 2010;26(1):49–55.
[24] Aïtcin PC. Cements of yesterday and today: concrete of tomorrow. Cem Concr Res 2000;30(9):1349–59.
[25] Rukzon S, Chindaprasirt P. Utilization of bagasse ash in high-strength concrete. Mater Des 2012;34:45–50.
[26] Smitha M, Sevatham D, Venkatasubramani R. Suitability of utilizing industrial wastes for producing sustainable high performance concrete. Ecol Environ Conserv 2012;18(4):907–10.
[27] Teng S, Lim TYD, Divsholi BS. Durability and mechanical properties of high strength concrete incorporating ultra fine ground granulated blast-furnace slag. Constr Build Mater 2013;40:875–81.

[28]  Volz JS. High-volume fly ash concrete for sustainable construction. In: Advanced materials research, vol. 512. 2012. pp. 2976–81.

[29]  Giergiczny Z. Popiół lotny w składzie cementu i betonu (Fly ash in cement and concrete). Wydawnictwo Politechniki Śląskiej [in Polish]. 2013.

[30]  Huntzinger DN, Eatmon TD. A life-cycle assessment of Portland cement manufacturing: comparing the traditional process with alternative technologies. J Clean Prod 2009;17(7):668–75.

[31]  Paceagiu J, Mohanu I, Draganoaia C. Reuse of metalurgical slags in eco-cement clinker production. In: 10th international multidisciplinary scientific geoconference SGEM2010. vol. 2. 2010. p. 741–8.

[32]  Aïtcin PC. High-performance concrete. London and New York: E & FN Spon; 1998.

[33]  Malhotra VM, Mehta PK. High-performance, high-volume fly ash concrete: materials, mixture proportioning, properties, construction practice, and case histories, 2nd ed. Ottawa: Supplementary Cementing Materials for Sustainable Development Inc.; 2005.

[34]  Corinaldesi V, Moriconi G. Environmentally-friendly self-compacting concrete for rehabilitation of concrete structures. In: International conference on concrete construction. London. 2008. p. 403–7.

[35]  Corinaldesi V, Moriconi G. Recycling of rubble from building demolition for low-shrinkage concretes. Waste Manage 2010;30(4):655–9.

[36]  Silva RV, De Brito J, Dhir RK. Properties and composition of recycled aggregates from construction and demolition waste suitable for concrete production. Constr Build Mater 2014;65:201–17.

[37]  Purushothaman M, Senthamarai R, Mullainathan L. Eco-friendly HPC that helps to agriculture by minimizing the pollution from industrial wastes. Plant Arch 2011;11(2):887–90.

[38]  Yoshida Y, Yamamoto K, Jinnai H. Basic study on environmentally friendly ultra high strength concrete. J Struct Constr Eng 2012;77(672):135–42.

[39]  Bromberek Z. Budownictwo zrównoważone w aspekcie trwałości (Sustainable construction in terms of durability). In: Błaszczyński T, editor, Chapter 2.1. in monograph Trwałość budynków i budowli (Durability of buildings and structures), Dolnośląskie Wydawnictwo Edukacyjne, Wrocław [in Polish]. 2012.

[40]  Damineli BL, Kemeid FM, Aguiar PS, John VM. Measuring the eco-efficiency of cement use. Cem Concr Compos 2010;32(8):555–62.

[41]  Bhikshma V, Florence GA. Studies on effect of maximum size of aggregate in higher grade concrete with high volume fly ash. Asian J Civil Eng 2013;14(1):101–9.

[42]  Jamroży Z. Beton i jego technologie (Concrete and its technology), PWN, Warszawa [in Polish]. 2008.

[43]  Rao SV, Rao MS, Kumar PR. Effect of size of aggregate and fines on standard and high strength self compacting concrete. J Appl Sci Res 2010;6(5):433–42.

[44]  Kurdowski W. Chemia cementu i betonu (Chemistry of cement and concrete). Stowarzyszenie Producentów Cementu [in Polish]. 2010.

[45]  Neville AM. Właściwości betonu (Properties of concrete). Polski Cement, Kraków [in Polish]. 2000.

[46]  Chen B, Liu J. Investigation of effects of aggregate size on the fracture behavior of high performance concrete by acoustic emission. Constr Build Mater 2007;21(8):1696–701.

[47]  Zhang W, Sun W. Effect of coarse aggregate on early age autogenous shrinkage of high-performance concrete. J Chin Ceram Soc 2009;37(4):631–6.

[48]  Szwabowski J, Śliwiński J. Betony samozagęszczalne. Budownictwo, Technologie, Architektura 2003;2:42–5. [in Polish].

[49]  Safiuddin M, West JS, Soudki KA. Air content of self-compacting concrete and its mortar phase including rice husk ash. J Civil Eng Manage 2011;17(3):319–29.

# CHAPTER 7

# $CO_2$ Reduction Assessment of Alkali-Activated Concrete Based on Korean Life-Cycle Inventory Database

**K.-H. Yang[1], J.-K. Song[2] and K.-I. Song[2]**
[1]Kyonggi University, Suwon, Republic of Korea
[2]Chonnam National University, Gwangju, Republic of Korea

## 7.1 INTRODUCTION

While traditional design and evaluation approaches are based on the principle of maximization of economy efficiency and include quality, cost, and time, the new approach of sustainable construction emphasizes the importance of reduction of the environmental impact of building and infrastructure [1]. The concrete industry has faced this transformation process from the traditional approach to the new approach since 2000. As a result, the reduction of $CO_2$ emissions and energy consumption in the cement industry has recently become a contentious issue. It is generally estimated that the amount of $CO_2$ emitted from the worldwide production of ordinary Portland cement (OPC) corresponds to approximately 7% of total greenhouse gases (GHGS) emissions into the Earth's atmosphere [2]. In addition, the cement industry subsector in developing countries consumes about 10% of total energy use [3]. These percentages are gradually increasing because the use of OPC is steady in advanced countries and it is rapidly increasing in developing countries.

The concrete industry has recently introduced different techniques to reduce $CO_2$ emissions in concrete production. The major techniques include capture and storage of $CO_2$ emissions and reducing the amount of clinker by replacing it with supplementary cementitious materials (SCMs) obtained from byproducts such as fly ash (FA) and ground granulated blast furnace slag (GGBFS). Alkali-activated (AA) binder has gradually attracted attention since the late 1990s as another active effort to reduce

*Handbook of Low Carbon Concrete.*
DOI: http://dx.doi.org/10.1016/B978-0-12-804524-4.00007-5

$CO_2$ emissions in concrete production [4–7]. For the AA binder, GGBFS, FA, and/or metakaolin (MK) are commonly used for the source material, while alkali hydroxide (ROH), nonsilicic salts of weak acids ($R_2CO_3$, $R_2S$, RF), and silicic salts of the $R_2O \cdot (n)SiO_2$ type are known to be the most effective activator, where R indicates an alkali metal ion such as Na, K, or Li. Although further investigation and complementary technical efforts are required for the practical application of AA concrete, it is commonly recognized that one of the greatest advantages of such concrete is its ability to reduce environmental impact through the recycling and noncalcination process of byproducts. However, there are very few, if any, available data on the quantitative evaluation of $CO_2$ emissions of AA concrete. Most of the alkali ions used for an activator need a subsequent treatment such as the calcination process. As a result, alkali activators would have a relatively high $CO_2$ emission. In addition, the AA binder based on glassy aluminosilicates such as FA and MK usually requires elevated curing temperatures for good strength development and stable hydrate reaction. Hence, the $CO_2$ reduction of AA concrete is significantly dependent on the type, concentration, and dosage of the alkali activators used and the curing condition of the concrete as well as the mix proportions of ingredients.

The targeted $CO_2$ reduction in concrete mix design would become an essential input together with the targeted compressive strength and workability. The selection of the type and mixing amount of binder is an important factor to meet the $CO_2$ reduction goal in concrete production, because the $CO_2$ emissions of aggregates in the material phase are considerably lower compared with OPC. Damineli et al. [8] proposed performance indicators to determine a benchmark and establish feasible goals in OPC concrete. Using performance indicators would be helpful in determining the type and unit content of binder for reduced $CO_2$ emissions of concrete production. Furthermore, the $CO_2$ reduction efficiency of the AA binder can be examined using performance indicators. Hence, the formulation of performance efficiency indicators for AA concrete can provide one of the major contributable tools for mix design of such concrete.

The present study summarizes examples to assess the $CO_2$ reduction of AA concrete based on the Korean Life-Cycle Inventory (LCI) database and test data mostly compiled from Korean journals. The typical mixing proportions of different concrete samples were compiled from the available literature [8] and ready-mixed concrete companies, according to concrete compressive strength and type of binder. In addition, the $CO_2$

emissions for secondary concrete products using AA GGBFS binder are evaluated with reference to practical examples. The CO$_2$ evaluation procedure of concrete includes the various contributions subdivided into concrete constituents, production, curing, and transportation to the plant and the building site, i.e., the cradle-to-preconstruction system is studied. The performance efficiency indicators, binder, and CO$_2$ intensities, obtained from AA GGBFS concrete test data, are also compared with those calculated from OPC-based concrete using the single OPC or OPC added with SCM as the binder. Overall, the relationship of CO$_2$ and binder intensities in different concretes is formulated by the regression analysis of a comprehensive database [9,10].

## 7.2 ASSESSMENT PROCEDURE OF CO$_2$

### 7.2.1 LCI Database

Korea has built an LCI database [11] that currently includes approximately 400 data sets associated with materials, parts manufacturing, machining process, transportation, and disposal activities. CO$_2$ emissions are most often given as a dimensionless figure, i.e., kg CO$_2$ emissions per kilogram material, while in certain cases it is preferred as an emissions per unit volume ratio. The functional unit for transportation is generally given as the product of the unit weight and unit distance. Hence, the CO$_2$ emission of a material or transportation per functional unit in the LCI database can be defined as a CO$_2$ coefficient. When an inventory for a building material is established, a cradle-to-gate approach is basically used. While evaluating the CO$_2$ emissions of concrete, we refer to the Japan Society of Civil Engineers (JSCE) database [12] for a data set that is not provided in the Korean LCI database, though understandably the LCI data often show somewhat of a difference between countries or regions due to differences in climate, energy sources, and natural resources.

### 7.2.2 CO$_2$ Evaluation Procedure

The CO$_2$ evaluation considered in the present study includes the following phases: all of the concrete constituents of an inventory formed from cradle to gate, transportation of the concrete constituents to a ready-mixed concrete plant, production and curing of the concrete, and transportation of the concrete to a building site. This indicates that the studied system is from cradle to preconstruction of concrete. The extra CO$_2$ emissions due to indirect behavior associated with the concrete production are excluded in such evaluation because these extra emissions are small enough to be ignored. The $CO_2$ evaluation procedure of concrete based on individual integration is summarized below.

The material phase includes source materials and alkali activators for binder, water, fine aggregate, coarse aggregate, and chemical admixtures. The functional unit of concrete is assumed to be $1\,m^3$. Hence, the $CO_2$ emissions $CO_{2-M}$ in the material phase can be calculated using the following equation:

$$CO_{2-M} = \sum_{i=1}^{n} (W_i \times CO_{2(i)-LCI}) \tag{7.1}$$

where $i$ identifies a raw material constituting concrete, $n$ is the number of the raw materials added for concrete production, and $W_i$ and $CO_{2(i)-LCI}$ are unit volume weight and $CO_2$ inventory of a raw material $i$, respectively.

The transportation phase includes the material transportation to the concrete plant and the produced concrete transportation to the building site. The ready-mixed concrete plant is selected to be located in Sadang-Dong, Dongjak-Gu, Seoul, South Korea. In addition, the transportation distance of each concrete constituent is assumed to be the maximum possible from the producing gate to the concrete plant. Fresh concrete produced from the plant is transported to a building site by transit-mix truck. In contrast, the AA concrete using FA or MK as the source material is assumed to be produced into precast concrete products and transported by large diesel truck (of 23-t capacity) to a building site. Overall, the $CO_2$ emissions $CO_{2-T}$ in the transportation phase can be obtained from:

$$CO_{2-T} = \sum_{i=1}^{n} (W_i \times D_i \times CO_{2(i)-LCI(TR)}) + D_B \times CO_{2-LCI(TR\_con)} \tag{7.2}$$

where $D_i$ is the transportation distance of each concrete constituent $i$ from the gate of the raw material-producing area to concrete plant, $CO_{2(i)-LCI(TR)}$ is $CO_2$ inventory related to vehicles, $D_B$ is the transportation distance of the produced concrete from concrete plant to building site, and $CO_{2(i)-LCI(TR\_con)}$ is $CO_2$ inventory of the transit-mix truck. It is noted that the unit of $CO_2$ inventory for the transit-mix truck is given as $CO_2$-kg/$m^3$·km.

The production phase includes mixing $CO_{2-P}$ of each concrete constituent using a concrete mixer and curing $CO_{2-C}$ of the produced concrete. As the OPC concrete and AA GGBFS concrete are commonly cured in air-dried conditions, $CO_2$ emissions due to curing for those concretes are neglected. Alternatively, FA- or MK-based AA concrete is assumed to be cured in a steam room because geopolymerization requires a high temperature [4,7]. The $CO_2$ emission due to steam curing is roughly calculated using the LCI specified by the JSCE [12], because most of the LCI database does not include the necessary data to do so, such as the size of curing room, the kind of fuel used for heating, and temperature profiles according to time.

Overall, the total $CO_2$ emission for the production and transportation to a building site of concrete can be obtained from:

$$CO_2 = CO_{2-M} + CO_{2-T} + CO_{2-P} + CO_{2-C} \qquad (7.3)$$

## 7.2.3 Examples for CO$_2$ Assessment

The typical mix proportions of concrete were compiled from the available literature [9], according to the designed compressive strength $(f_c')$ and type of binder, as given in Table 7.1. The MK-based AA concrete sample gives only a compressive strength of 24 MPa due to the extremely poor available test data. The mix proportions for OPC-based concrete refer to the mixing tables used practically in ready-mixed concrete plants. Examples for the calculation procedure of the $CO_2$ emission of concrete sampled from Table 7.1 are given in Tables 7.2 and 7.3 for a concrete compressive strength of 40 MPa,. As mentioned previously, the $CO_2$ emissions of concrete are classified into four groups, including material, transportation, production, and curing phases. For OPC + SCM concrete, $CO_2$ emissions per 1 m$^3$ can be evaluated to be 352.9, 18.7, and 28.8 kg for material, production, and transportation phases, respectively, as given in Table 7.2. The $CO_2$ emission from OPC material forms 96.5% of those of the material phase, which also corresponds to 85.1% of the total $CO_2$ emissions. The $CO_2$ emission in the transportation phase mostly results from concrete transport. This is why the $CO_2$ coefficients the 6-m$^3$ capacity transit-mix truck are considerably higher than those of the bulk trailer and small diesel trucks. Meanwhile, $CO_2$ emissions in the material and production phases of AA FA concrete are evaluated to be 30% of those of OPC + SCM concrete, showing the highest $CO_2$ footprint due to alkali activators in the material phase. In addition, although the $CO_2$ footprint due to steam curing for stable hydration of

Table 7.1 Samples for mix proportions of different concrete types

| Binder type | $f'_c$ (MPa) | OPC | FA | GGBFS | MK | Fine aggregate[a] | Coarse aggregate[a] | Water | Chemical admixture[b] | W/B | Ca(OH)$_2$ | Na$_2$SiO$_3$ | NaOH | Curing |
|---|---|---|---|---|---|---|---|---|---|---|---|---|---|---|
| OPC | 24 | 280 | — | — | — | 767 | 1214 | 168 | — | 0.6 | — | — | — | Air-drying |
|  | 40 | 479 | — | — | — | 572 | 1106 | 191 | 0.5 | 0.4 |  |  |  |  |
|  | 70 | 540 | — | — | — | 666 | 991 | 172.8 | 5.4 | 0.32 |  |  |  |  |
| OPC + SCM[c] | 24 | 291 | 51 | — | — | 829 | 930 | 178 | 1.7 | 0.52 |  |  |  |  |
|  | 40 | 361 | 23 | 68 | — | 805 | 918 | 163 | 3.1 | 0.36 |  |  |  |  |
|  | 70 | 435 | 56 | 210 | — | 670 | 822 | 163 | 5.25 | 0.23 |  |  |  |  |
| AA GGBFS | 24 | — | — | 540 | — | 700 | 849 | 185 | 5.06 | 0.3 | 46.2 | 6.2 | — |  |
|  | 40 | — | — | 648 | — | 619 | 814 | 185 | 9.62 | 0.25 | 55.5 | 7.4 | — |  |
|  | 70 | — | — | 677 | — | 572 | 752 | 185 | 12.6 | 0.22 | 63.1 | 100 | — |  |
| AA FA | 24 | — | 400 | — | — | 781 | 1035 | 62.4 | — | 0.13 | — | — | 58 | Steam curing 85°C/24h |
|  | 40 | — | 469 | — | — | 623 | 935 | 112 | — | 0.2 | — | — | 75 |  |
|  | 70 | — | 630 | — | — | 551 | 731 | 115 | — | 0.15 | — | — | 106 |  |
| AA MK | 24 | — | — | — | 411 | 759 | 998 | 185 | — | 0.20 | — | — | 62 |  |

Note: $f'_c$ = designed compressive strength of concrete and $W/B$ = water-to-binder ratio by weight.
[a]Sea sand and crushed stone are assumed to be used for the fine and coarse aggregates, respectively, in all of the concrete types.
[b]Polycarboxylate-based superplasticizer is used for the water-reducing agent.
[c]OPC added with SCM such as FA and GGBFS.

**Table 7.2** Examples for CO$_2$ assessment of concrete using OPC + SCM binder ($f_c' = 40$ MPa)

| Functional unit (FU): (m$^3$) | Material and production | | | Transportation | | |
|---|---|---|---|---|---|---|
| | A | B | A·B | D | E | A·D·E |
| Item | $\dfrac{kg}{FU}$ | $\dfrac{CO_2 - kg}{kg}$ | $\dfrac{CO_2 - kg}{FU}$ | km | $\dfrac{CO_2 - kg}{kg \cdot km}$ | $\dfrac{CO_2 - kg}{FU}$ |
| OPC[a] | 361 | 0.944 | 340.8 | 100 | $5.18 \times 10^{-5}$ | 1.869 |
| GGBS[a] | 68 | 0.0265[c] | 1.802 | 300 | $5.18 \times 10^{-5}$ | 1.057 |
| FA[a] | 23 | 0.0196[c] | 0.451 | 200 | $5.18 \times 10^{-5}$ | 0.238 |
| Sand[a] | 805 | 0.0026 | 2.093 | 50 | $6.3 \times 10^{-5}$ | 2.536 |
| Coarse[a] | 918 | 0.0075 | 6.885 | 50 | $6.3 \times 10^{-5}$ | 2.892 |
| Water[b] | 163 | $1.96 \cdot 10^{-4}$ | 0.032 | — | — | — |
| Admixture | 3.1 | 0.25[c] | 0.775 | 50 | $2.21 \times 10^{-4}$ | 0.034 |
| Concrete production[d] | 2341 | 0.008[c] | 18.7 | 30 | 0.674 $\dfrac{CO_2 - kg}{m^3 \cdot km}$ | D·E = 20.22 |
| | Sum | | 371.6 | Sum | | 28.8 |
| Air-dried curing | — | | | — | | |

Total = 400.4 (CO$_2$-kg)/FU

[a]All of the cementitious materials are transported by a bulk trailer, while aggregates are transported by 15 ton capacity dump truck.
[b]Water drawn from Han-river is used for concrete mix.
[c]LCI data given in JSCE are referenced wherever Korean LCI database is unavailable.
[d]The fresh concrete produced from the plant is transported to a building site by the transit-mix truck.

**Table 7.3** Examples for CO$_2$ evaluation of AA FA concrete ($f_c' = 40$ MPa)

| Functional unit (FU): m$^3$ | Material and production | | | Transportation | | |
|---|---|---|---|---|---|---|
| | A | B | A·B | D | E | A·D·E |
| Item | $\dfrac{kg}{FU}$ | $\dfrac{CO_2 - kg}{kg}$ | $\dfrac{CO_2 - kg}{FU}$ | km | $\dfrac{CO_2 - kg}{kg \cdot km}$ | $\dfrac{CO_2 - kg}{FU}$ |
| FA | 469 | 0.0196 | 9.192 | 200 | $5.18 \times 10^{-5}$ | 4.86 |
| Sand | 623 | 0.0026 | 1.62 | 50 | $6.3 \times 10^{-5}$ | 1.96 |
| Coarse | 935 | 0.0075 | 7.01 | 50 | $6.3 \times 10^{-5}$ | 2.95 |
| Water | 112 | $1.96 \times 10^{-4}$ | 0.0219 | — | — | — |
| NaOH | 75 | 1.232 | 92.4 | 30 | $2.21 \cdot 10^{-4}$ | 4.97 |
| Concrete production[a] | 2214 | 0.008 | 17.71 | 30 | $5.18 \times 10^{-5}$ | 0.0015 |
| | Sum | | 127.99 | Sum | | 14.74 |
| Steam curing | 85°C/24 h: 38.5 CO$_2$-kg/FU | | | — | | |

Total = 181.2 CO$_2$-kg/FU

[a]A precast concrete product is transported from the plant to a building site using a 23-t capacity truck.

AA FA is approximately assessed to be 38.5 kg/FU, it is reduced in the transportation phase, because the concrete product is transported not by transit-mix trucks, but by 23-t-capacity diesel trucks with a low $CO_2$ coefficient. Overall, the total $CO_2$ emissions of AA FA concrete can be reduced to as little as 45% of those of OPC + SCM concrete.

## 7.2.4 Comparisons of $CO_2$ Footprints According to Different Concrete Types

Fig. 7.1 shows the comparisons of $CO_2$ footprints of different concrete mixes given in Table 7.1. In general, the $CO_2$ emission of concrete increases with its compressive strength, regardless of the type of binder. In addition, the contribution of binder to the total $CO_2$ emission is more significant in OPC-based concrete than in AA concrete. When concrete compressive strength increases from 24 to 70 MPa, the total $CO_2$ emission per functional unit increases as much as 245, 146, and 146 kg for OPC concrete, OPC + SCM concrete, and AA GGBFS concrete, respectively, which correspond to an increasing rates of 1.76, 1.44, and 2.32 times, respectively. The total $CO_2$ emission in AA FA concrete with a compressive strength of 70 MPa is greater by merely 63 kg compared with that with a compressive strength of 24 MPa.

AA concrete commonly uses different alkali activators according to the type of source materials and the designed compressive strength. For a higher compressive strength of AA concrete, a greater addition of a stronger alkali activators is required [4–7], as given in Table 7.1. According to the Korean LCI database, NaOH, and $Na_2SiO_3$ have a relatively high $CO_2$ inventory of 1.232 $CO_2$-t/t and 1.32 $CO_2$-t/t, respectively, while the $CO_2$ inventory of weak alkaline $Ca(OH)_2$ is 0.517 $CO_2$-t/t. For AA GGBFS concrete, $CO_2$ emission due to alkali activators increases as much as 4.3 times when compressive strength increases from 40 to 70 MPa. The contribution of alkali activators to the total $CO_2$ footprints of AA GGBFS concrete accounts for 31.6% and 64.4% for compressive strengths of 40 and 70 MPa, respectively. This indicates that the $CO_2$ emissions of AA concrete are significantly dependent on the dosage and type of the alkali activators used.

The total $CO_2$ footprints of different concrete mixes normalized by those obtained from the OPC concrete with the same compressive strength are plotted in Fig. 7.2. The $CO_2$ emissions of OPC-based concrete with SCM are approximately 80% of those of OPC concrete when compressive strength is higher than 40 MPa. The reduction rates of $CO_2$

emissions of AA GGBFS concrete relative to those of OPC concrete are approximately 75% for a compressive strength of 24 MPa, and 75% for a compressive strength of 70 MPa. The $CO_2$ emissions of AA FA concrete are reduced by as much as 60% of those of OPC concrete when compressive strength is below 40 MPa. Fig. 7.2 clearly shows that the reduction

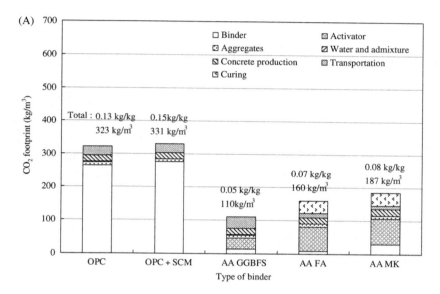

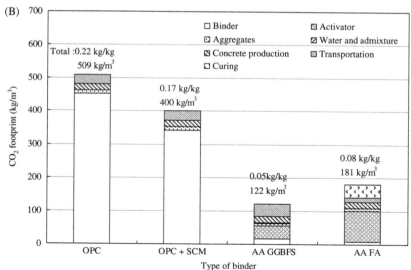

**Figure 7.1** $CO_2$ footprint for different concrete types. (A) $f'_c = 24\,\text{MPa}$, (B) $f'_c = 40\,\text{MPa}$, and (C) $f'_c = 70\,\text{MPa}$.

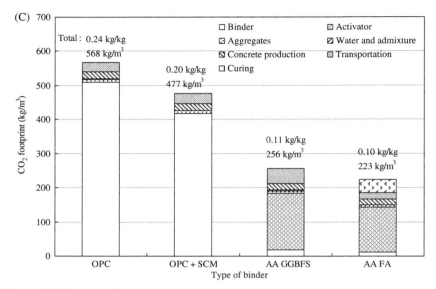

**Figure 7.1** (Continued)

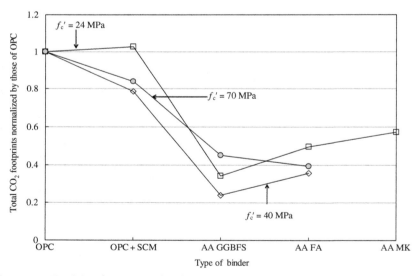

**Figure 7.2** Total $CO_2$ footprints of different concrete mixes normalized by those calculated from OPC concrete with the same compressive strength.

rate of the $CO_2$ emission of AA concrete relative to OPC concrete commonly ranges between 55% and 75%, though the $CO_2$ emissions of AA concrete are somewhat variable according to the type, concentration, and dosage of the added alkali activators.

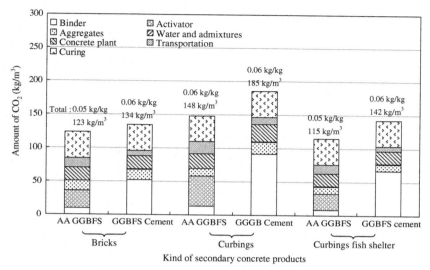

**Figure 7.3** Comparisons of $CO_2$ footprints of secondary precast concrete products.

## 7.2.5 Comparisons of $CO_2$ Footprints in the Secondary Concrete Products

Fig. 7.3 shows the typical comparisons of $CO_2$ emissions for secondary precast concrete products, such as bricks, curbings, and fish shelters according to the type of binder, namely, GGBFS cement (OPC 50% + GGBFS 50%) or GGBFS binder activated by 7.5% $Ca(OH)_2$ and 3% $Na_2SO_3$. Table 7.4 gives also the typical mix details of secondary precast concrete products. The mix proportions of precast concrete products using GGBFS cement refer to the mixing tables practically applied in the plants, while those of AA GGBFS concrete products are determined from mock-up tests [13], considering the performance criteria specified in the Korean Industrial Standard (KS) [14] and economical efficiency. For the concrete bricks, $CO_2$ footprint due to aggregate transportation accounts for 20% total $CO_2$ emission, which matches that obtained from the binder. This is attributed to the fact that the amount of aggregates in the mix proportions is generally as much as 10 times that of the binder. As a result, the reduction of $CO_2$ emissions in the AA GGBFS concrete bricks is minimal compared with the GGBFS cement concrete bricks, indicating a reduction rate of approximately 2.5%. The total aggregate-to-binder ratio by weight in the concrete curbings and concrete fish shelters usually ranges between 3.0 and 4.0, indicating that the portions of binder in

**Table 7.4** Typical mix details of secondary precast concrete products

| Secondary concrete products | Binder type | Unit weight (kg/m³) | | | | | | | | | Curing |
|---|---|---|---|---|---|---|---|---|---|---|---|
| | | GGBFS cement | GGBFS | Fine aggregate | Stone powder (8 mm) | Coarse aggregate (25 mm) | Water | Chemical admixture | Ca(OH)$_2$ | Na$_2$SO$_4$ | |
| Bricks | GGBFS cement | 260 | — | — | 2077 | — | 52 | 2 | — | — | Steam curing 60°C/24 h |
| | AA GGBFS | — | 372 | — | 1937 | — | 107 | 2 | 32.3 | 12.9 | |
| Curbings | GGBFS cement | 478 | — | 597 | 2089 | — | 119 | 2 | — | — | Air-drying |
| | AA GGBFS | — | 500 | 597 | 895 | 445 | 172 | — | 43 | 22 | |
| Fish shelters | GGBFS cement | 350 | — | 776 | — | 1049 | 157 | 2 | — | — | |
| | AA GGBFS | — | 313 | 828 | — | 1119 | 111 | 2 | 26 | 11 | |

the total mixing materials are higher than those in the concrete bricks. The $CO_2$ emissions in the AA GGBFS concrete curbings are lower by 37 kg per functional unit than those in the GGBFS cement concrete curbings, showing a reduction rate of 20%. Further, the $CO_2$ emission of the AA GGBFS concrete fish shelters is reduced by approximately 19% compared with the case of using GGBFS cement as a binder. The $CO_2$ reduction in the AA GGBFS concrete products is significantly dependent on the aggregate-to-binder ratio together with the type and dosage of the added alkali activators.

## 7.3 PERFORMANCE EFFICIENCY INDICATOR OF BINDER

The efficient use of binder to reduce $CO_2$ emissions has been recently discussed [8,15,16]. Damineli et al. [8] concluded that concrete efficiency should be defined in terms of the total binder consumption, the total cost of concrete production, and/or the environmental impact imposed to deliver one unit of functional performance measured by a relevant indicator such as compressive strength and durability factor. The compressive strength of concrete commonly increases with the decrease of water-to-binder ratio [17]. This indicates that the higher the concrete compressive strength the more the binder consumption at the concrete mix. Most $CO_2$ emissions in OPC-based concrete come from the production of binders [8]. In addition, the production of OPC is accompanied by the consumption of natural resources and enormous energy, and various environmental loads. Considering the interrelation of concrete compressive strength, total amount of binder, and environmental impact including $CO_2$ emissions, Damineli et al. [8] proposed the following simplified binder intensity ($B_i$) and $CO_2$ intensity ($C_i$) in order to assess the binder efficient on $CO_2$ emissions.

$$B_i = B/f_c' \tag{7.4}$$

$$C_i = C_d/f_c' \tag{7.5}$$

where $B$ is the total consumption of binder materials (kg/m³), $f_c'$ is the compressive strength of concrete (MPa) at an age of 28 days and $C_d$ is the total $CO_2$ emissions (kg/m³) due to concrete production, which can be approximately obtained from Eq. (7.3).

To examine the efficiency of the AA binder, both intensities in AA GGBFS concrete [10] are calculated and compared with those obtained from the comprehensive database of OPC-based concrete established by Yang [9]. Most of the OPC-based concrete data are compiled from Korean journals issued by the Korea Concrete Institute, Architectural Institute of Korea, and Korea Institute of Building Construction. The intensities of the AA FA concrete are not provided in the following comparisons due to the limited test data. For 34 GGBFS concrete mixes activated by 7.5% $Ca(OH)_2$ and 1% $Na_2SiO_3$ or 7.5% $Ca(OH)_2$ and 2% $Na_2CO_3$, the total amount of binder ranges between 84 and 121 kg/m$^3$, while 28-day compressive strength varies from 8.6 to 42.2 MPa. The OPC-based concrete database includes 2464 OPC concrete mixes, 92 OPC + GGBFS concrete mixes, 481 OPC + FA concrete mixes, and 23 OPC + GGBFS + FA concrete mixes. The total amount of binder and 28-day compressive strength of OPC concrete range between 200 and 1065 kg/m$^3$, and between 7.7 and 147 MPa, respectively, whereas those of OPC + SCM concrete range between 167 and 911 kg/m$^3$, and between 13 and 98 MPa, respectively. For the OPC + SCM concrete, FA and GGBFS are substituted up to 70% and 69%, respectively, for OPC. The total amount of binder at the same $f_c'$ is commonly lower in OPC-based concrete than in $Ca(OH)_2$-based AA GGBFS concrete.

## 7.3.1 Binder Intensity

Fig. 7.4 presents the binder intensity calculated from different concrete mixes according to the type of binder. The best-fit curves determined from the different concrete data are also plotted in the same figure. Although the mix-design method of concrete and materials used are widely different, variability of the data is somewhat small, indicating relatively high coefficients ($R^2$) of correlation between test data and predictions calculated from the best-fit curves. The binder intensity commonly tends to decrease with the increase of the concrete compressive strength. The decreasing rate of the binder intensity calculated for OPC concrete is gradually mitigated beyond a concrete strength of 50 MPa. Above 80 MPa, the binder intensity of OPC concrete reaches the minimum range of approximately 5.5 kg/m$^3$/MPa. Hence, high-strength concrete is more efficient in terms of the binder consumption required to deliver each unit of compressive strength. When concrete strength is lower than approximately 40 MPa, a slightly higher binder intensity is observed in OPC + SCM concrete than in OPC concrete, showing that very similar best-fit curves are deduced in OPC + FA concrete and OPC + GGBFS concrete, though it is difficult to clearly distinguish the fitting curve of OPC

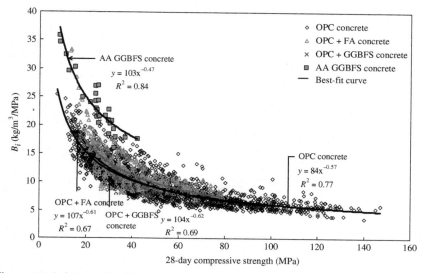

**Figure 7.4** Relationship of binder intensity and concrete compressive strength.

+ GGBFS concrete in Fig. 7.4 due to the overfull data sets. The binder intensity obtained from Ca(OH)$_2$-based AA GGBFS concrete is higher by an average of 1.67 times than that of OPC concrete with the same compressive strength. This indicates that the Ca(OH)$_2$-based AA GGBFS concrete requires a greater binder consumption in order to obtain the same compressive strength as OPC concrete. However, it should be noted that the high binder intensity of AA GGBFS concrete does not indicate the imposition of severe environmental impact, providing that byproducts are recycled.

## 7.3.2 CO$_2$ Intensity

CO$_2$ intensity calculated from different concrete types against 28-day compressive strength is plotted in Fig. 7.5. The coefficients of correlation determined from the OPC + SCM concrete are slightly lower than those of OPC concrete. This may be attributed to the fact that the CO$_2$ intensity is also affected by the replacement level of FA and GGBFS. The data trend is similar to that observed in the binder intensity result, namely, the CO$_2$ intensity tends to decrease as compressive strength increases. Hence, high-strength concrete emits less CO$_2$ emissions to develop a unit of compressive strength. In general, the CO$_2$ intensity calculated from OPC

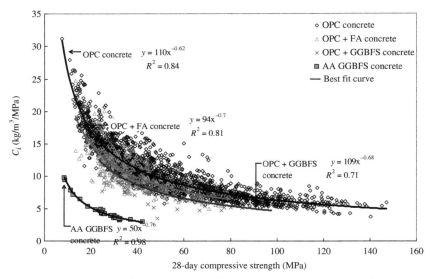

**Figure 7.5** Relationship of $CO_2$ intensity and concrete compressive strength.

+ SCM concrete is slightly lower than that of OPC concrete with the same strength. $Ca(OH)_2$-based AA GGBFS concrete reveals around 3.3 times lower $CO_2$ intensity than OPC concrete with the same compressive strength, though the binder consumption is higher in $Ca(OH)_2$-based AA GGBFS concrete than in OPC concrete. This indicates that replacing OPC with AA GGBFS binder can be regarded as a kind of major tool for achieving sustainable concrete.

The relationship of binder and $CO_2$ intensities of different concrete types is shown in Fig. 7.6. It is noted that the effects of binder intensity on $CO_2$ emissions are significantly affected by the replacement level of SCM. Hence, the OPC-based concrete data in Fig. 7.6 are adjusted by considering the replacement level of FA and GGBFS. Variability of the data is greatly diminished in comparisons with Figs. 7.4 and 7.5. As a result, a relatively high correlation coefficient ($R^2$) between test data and best-fit curves is obtained. Hence, the relation between binder and $CO_2$ intensities is helpful in designing a sustainable concrete. It is possible to formulate the mix proportions of ingredients including the amount of binder for targeted compressive strength and $CO_2$ reduction. Furthermore, the $CO_2$ emission of concrete can be simply predicted from the information of the type and amount of binder.

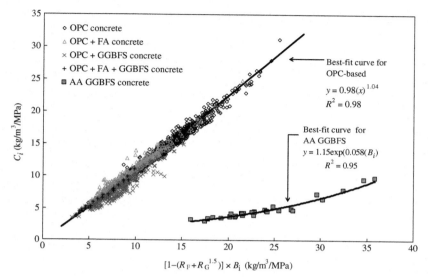

**Figure 7.6** Relationship of binder intensity and $CO_2$ intensity for different concrete types.

The $CO_2$ intensity increases in proportion to the binder intensity, regardless of the type of binder, indicating that $CO_2$ emission in concrete production is significantly dependent on binder consumption. In contrast, the $CO_2$ intensity for OPC + SCM concrete tends to decrease with the increase of the replacement level of FA and GGBFS. To determine the relationship of binder and $CO_2$ intensities, and reflect these trends in a simple closed-form equation, regression analysis was repeatedly conducted using the 3060 OPC-based concrete data and 34 AA GGBFS concrete data until a value of $R^2$ was obtained, as shown in Fig. 7.6. Overall, the $CO_2$ intensity can be formulated as follows:

$$C_i = 0.98[\{1-(R_F + R_G^{1.5})\} \cdot B_i]^{1.04} \quad \text{for OPC-based concrete (7.6a)}$$

$$C_i = 1.15 \cdot \exp(0.058 \cdot B_i) \quad \text{for Ca(OH)}_2\text{-based AA GGBS concrete (7.6b)}$$

where, $R_F$ and $R_G$ are replacement levels of FA and GGBFS, respectively. The slope of the best-fit curve determined from OPC-based concrete is higher than that obtained from Ca(OH)₂-based AA GGBFS concrete. Hence, it can be concluded that the effect of binder consumption on $CO_2$ emission is lower in AA GGBFS concrete than in OPC-based concrete.

## 7.4 FURTHER INVESTIGATIONS

The alkali activators generally have a relative high $CO_2$ inventory because the production process of most oxides involves the calcinations of carbonates. Hence, the $CO_2$ footprint of the AA concrete is somewhat dependent on the type, concentration, and dosage of alkali activators. The source material selected to produce the AA binder also influences the $CO_2$ footprint of concrete, because aluminosilicate materials such as FA and MK need a high-temperature curing condition. The available test data on AA concrete are very rare, though tests on AA pastes or AA mortars are intermittently published. Therefore, it is necessary to collect further experimental data on AA concrete in order to assess $CO_2$ reduction and examine the performance efficiency indicators according to the various mixing proportions.

## 7.5 CONCLUSIONS

The $CO_2$ reduction of AA concrete is examined based on the Korean LCI database. The $CO_2$ evaluation procedure of concrete using individual integration includes material, production, curing, and transportation phases. The performance efficiency indicators of the binder are also ascertained for different concrete types. From the $CO_2$ reduction assessment of AA concrete based on the typical mixing proportion tables and binder efficiency indicators calculated from the extensive database, the following conclusions may be drawn:

1. The $CO_2$ emission of concrete increases with its compressive strength, showing that the contribution of binder to the total $CO_2$ footprint is more significant in OPC-based concrete than in AA concrete.
2. The $CO_2$ emissions of OPC + SCM concrete are approximately 80% of those of OPC concrete when the compressive strength is higher than 40 MPa. On the other hand, the reduction rate of $CO_2$ emission of AA concrete relative to OPC concrete commonly ranges between 55% and 75%, though the $CO_2$ reduction of AA concrete is somewhat dependent on the type, concentration, and dosage of the added alkali activators.
3. The $CO_2$ reduction rate in secondary precast concrete products using AA GGBFS binder instead of GGBFS cement can be evaluated to be approximately 20% when the total aggregate-to-binder ratio ranges between 3.0 and 4.0.
4. The binder and $CO_2$ intensities commonly tend to decrease with the increase of concrete compressive strength. The binder intensity obtained from $Ca(OH)_2$-based AA GGBFS concrete is higher by an average of 1.67 times than that of OPC concrete with the same

compressive strength. In contrast, Ca(OH)$_2$-based AA GGBFS concrete reveals approximately 3.3-times-lower CO$_2$ intensity than OPC concrete, though the binder consumption is higher in Ca(OH)$_2$-based AA GGBFS concrete than in OPC concrete.

5. The CO$_2$ intensity increases in proportion to the binder intensity, showing that the slope of the increasing rate determined from Ca(OH)$_2$-based AA GGBFS concrete data is lower than that obtained from OPC-based concrete.

## ACKNOWLEDGMENTS

This research was supported by a grant (11 Technology Standardization 10-1) from the R&D Policy and Infrastructure Development Program funded by the Ministry of Land, Transport, and Maritime Affairs of the Korean Government.

## REFERENCES

[1] Task Group 3.3. Environmental design. International Federation for Structural Concrete (fib) 2004. Switzerland.
[2] Malhotra VM. Introduction: sustainable development and concrete technology. Concr Int 2002;24:22.
[3] Ali MB, Saidur R, Hossain MS. A review on emission analysis in cement industries. Renewable Sustainable Energy Rev 2001;15(5):2252–61.
[4] Duxson P, Fernández-Jiménez A, Provis JL, Lukey GC, Palomo A, van Deventer JSJ. Geopolymer technology: the current state of the art. J Mater Sci 2007;42:2917–33.
[5] Pacheco-Torgal F, Castro-Gomes J, Jalali S. Alkali-activated binders: a review. Constr Build Mater 2008;22(7):1305–22.
[6] Shi C, Krivenko PV, Roy D. Alkali-activated cements and concretes. Taylor & Francis; 2006.
[7] Davidovits J. Geopolymer: chemistry & applications. Institut Géopolymère; 2008.
[8] Damineli BL, Kemeid FM, Aguiar PS, John VM. Measuring the eco-efficiency of cement use. Cem Concr Compos 2010;32(8):555–62.
[9] Yang KH. Development of green concrete based on alkali-activated binder. Technical report 2010-0027558. Department of Architectural Engineering, Kyonggi University; 2011, (in Korea).
[10] Yang KH, Song JK. Empirical equations for the mechanical properties of Ca(OH)$_2$-based alkali-activated slag concrete. ACI Mater J 2012;109(4):431–40.
[11] Korea LCI database information network, http://www.edp.or.kr/lcidb.
[12] Sakai K, Kawai K. JSCE guidelines for concrete No. 7; recommendation of environmental performance verification for concrete structures. Japan Society of Civil Engineering; 2006.
[13] Yang KH, Song JK. Engineering properties and application of slag-based geopolymer concrete with no cement. Comput Struct Eng (in Korea) 2011;24(1):26–30.
[14] Korean Standards Information Center. Korean industrial standard. South Korea; 2006.
[15] Aïtcin PC. Cements of yesterday and today: concrete of tomorrow. Cem Concr Res 2000;30(9):1349–59.
[16] Gartner E. Industrially interesting approaches to "low-CO2" cements. Cem Concr Res 2004;34(9):1489–98.
[17] Neville AM. Properties of concrete. England: Longman; 1995.

# CHAPTER 8

# Introducing Bayer Liquor–Derived Geopolymers

E. Jamieson[1,2], A. van Riessen[2], B. McLellan[3], B. Penna[1,2], C. Kealley[2] and H. Nikraz[2]

[1]Alcoa of Australia, Kwinana, WA, Australia
[2]Curtin University, Perth, WA, Australia
[3]Kyoto University, Kyoto, Japan

This compilation of work introduces Bayer liquor–derived geopolymers. These alkali–activated materials (AAMs) utilize concentrated Bayer liquors of caustic alumina as the primary activating solution for a reactive silica rich solid, in this case, a class F fly ash.

Formulations with a reactive Si/Al ratio of 2.3 and a Na/Al ratio of 0.8 have achieved compressive strength in excess of 40 MPa, with an elevated-temperature curing process. Production formulations allowing for ambient cure have also achieved minimum compressive strengths in excess of 20 MPa.

The potential to utilize Bayer process liquor and fly ash to manufacture geopolymers leads to significant opportunities for industrial-scale synergies. It is known that geopolymers have the ability to bind a range of contaminants, and Bayer-derived geopolymers have proven similarly effective with cationic species. Consumption of the Bayer process liquor for geopolymer production could achieve significant Bayer process impurity removal by replacement of that liquor. The metallic cations become immobilized by being incorporated into the geopolymer structure. This impurity-removal process provides incentive for the synergistic provision of a concentrated caustic aluminate solution to the geopolymer industry.

Significantly, utilizing this combination of industrial byproducts can dramatically lower the embodied energy of Bayer-derived geopolymers. Because the Bayer liquor is part of an impurity-removal process, from a formal life-cycle assessment (LCA) perspective there is minimal allocation for embodied energy and emissions. Hence, the embodied energy of the Bayer-derived geopolymer binder falls to as little as 0.27 GJ/t. For the first time, Bayer-derived geopolymer binders could be produced with embodied energy intensity at levels comparable to manufactured or recycled sand, gravel, and stone.

*Handbook of Low Carbon Concrete.*
DOI: http://dx.doi.org/10.1016/B978-0-12-804524-4.00008-7

159

To establish a Bayer-derived geopolymer industry of large-enough volume to allow economies of scale, several product markets will be required to generate consistent demand.

One such market is the manufacture of Bayer-derived geopolymer mortar for manufacture of artificial aggregate. Aggregate has a known value within a distribution envelope and can be stored indefinitely. The embodied energy of these Bayer-derived geopolymer mortar aggregates has been calculated to be 0.22 GJ/t, similar to other manufactured or recycled aggregates. The embodied energy being just one of the potential environmental benefits of such a process, with the reduction in land clearing for new aggregate having important local benefit.

## 8.1 INTRODUCTION

Aluminosilicate polymers, inorganic polymers, AAMs, or geopolymers are X-ray amorphous aluminosilicate materials that have the potential to be an alternative concrete binder to ordinary Portland cement (OPC here or CEM1 [1]). While there is ongoing debate over the correct terminology for these materials, for the purpose of consistency we will persist with the terms OPC, geopolymers and Bayer-derived geopolymer.

Geopolymers can be produced from a range of aluminate and silicate materials including metakaolin, fly ash, blast furnace slags, and mineral processing wastes. As such, most industrial precincts would produce a range of suitable feedstock to enable geopolymer production [2].

Other materials required for concrete production that can also be sourced from industrial precincts, such as residue sand from alumina production, have been evaluated for construction purposes [3–6]. An industrial ecology approach, utilizing such residues more effectively, would maximize the community benefit from the consumption of limited resources (improving resource efficiency) and reduce the requirement for production of virgin resources, with its associated environmental impacts. This is something that can only be accomplished if byproducts are produced to a specification [7].

The use of multiple and varied feedstock in geopolymer production has led to a focus on understanding the chemistry of the amorphous reactive components. This in turn allows geopolymer products to be formulated with predictable performance properties in a similar way to OPC, without the requirement for trial and error on the basis of new or variable feedstock flows [2,8–12].

The critical feedstock for geopolymers include:
- concentrated caustic solution with dissolved silica or alumina

- a source of caustic soluble silica
- a source of caustic soluble alumina

All three materials must be available in large quantities and at appropriate cost for commercial application.

A source of heat is often required when casting geopolymer products at elevated temperatures (60–90°C). Thermal curing may be suitable for many product applications, but places limitations on many high-volume markets such as in situ poured paths, roads, and curbing. It is possible to design formulations that can be cured at ambient temperatures by the addition of calcium-containing compounds such a lime or blast furnace slag [13–18].

Experimentally, geopolymers have been produced with alternative reagents such as sodium aluminate solution, especially where precursors have a high Si/Al [19,20]. Moreover, the production of Bayer-derived geopolymers has recently met potential application milestones [21,22].

For geopolymers to be economically viable, large-volume production of products are required along with product acceptance [23–25].

Silicate-derived geopolymer binders have been utilized in applications such as pathways, pavers, mine backfill, railway sleepers, sewerage pipes, and earth retaining walls [8,26–28]. For Bayer-derived geopolymers to penetrate similar markets, in situ and ambient temperature curing are required. This has been achieved in laboratory trials while targeting a compressive strength above 20 MPa (a typical specification for pathways, driveways, etc. [29]).

Another potentially large market is the production of artificial aggregates made from Bayer-derived geopolymer mortar. Additional byproducts such as sand can also be utilized in manufacture. Application of Bayer-derived geopolymer mortar aggregates (Bayer aggregates) as coarse aggregate in the manufacture of concrete utilizing OPC has also been investigated. Concrete contains 70–85 wt% aggregate, hence would provide a larger volume of application for Bayer liquor export and consumption. In effect, the production of Bayer-derived geopolymer would be the limiting factor, not the ability to find a market.

## 8.1.1 The Geopolymer Industry

The primary function of a geopolymer is to act as a cementitious binder and replace OPC in concrete manufacture or provide complementary products. Some of the advantages that geopolymers have over OPC is their high compressive and flexural strength, their very high temperature resistance, a high resistance to acid, and the ability to utilize multiple waste or byproduct streams [13,30–32].

Geopolymers binders have the potential to incorporate cations, anions, and organic species within their three-dimensional structure [33,34]. OPC can also trap impurities, though the acid resistance of geopolymer binders allows for a much wider range of safe receiving environments.

A significant ecological and marketing benefit reported for geopolymer binder over OPC is the reduction in $CO_2$ emissions. One tonne of OPC releases 0.55 t of $CO_2$ from the calcination of limestone and the combustion of carbon-based fuel for heat and power generation produces an average additional 0.40 t of $CO_2$. Comparatively, geopolymer production creates only between 0.2 and 0.5 t of $CO_2$ per tonne of product, depending on inclusion of lifecycle and transport factors [35–37].

Geopolymers have been formed from a range of aluminate and silicate materials including metakaolin, fly ash, blast furnace slags, and mineral processing wastes. It is apparent that with so many feedstocks available and most having variable reactive components, a fundamental evaluation of reagent chemistry is essential to assure product quality suitable for the construction industry [2,8–12]. Geopolymers are commonly cured at slightly elevated temperatures (70–90°C) but the addition of calcium-containing compounds such as lime or blast furnace slags can promote ambient temperature setting [17].

Geopolymer concrete has now been utilized in a wide range of applications traditionally reserved for OPC concrete [8,26–28].

## 8.1.2 The Alumina Industry

The Bayer process is the name for the hydrometallurgical extraction and refinement of alumina from bauxite. Bauxite ore is ground and then digested in highly caustic solutions at elevated temperatures. Gangue solids, usually iron oxides, quartz, and other resistant minerals, are separated from the hot sodium aluminate slurry by physical means such as settling and filtration. The solids (red mud and red sand) are countercurrent washed to recover the caustic solutions, then pumped to specially designed impoundment beds. The mud and sand can be intercepted, neutralized, washed, and stored for reuse [3–8,38–40].

The valuable caustic liquor is recovered via an underdrain system and returned to the Bayer process circuit. The thickener overflow solution concentrated in aluminate is cooled, and then aluminum hydroxide (hydrate seed) is added to induce precipitate growth. The precipitated aluminum hydroxide is then separated from the spent caustic aluminate liquor, which is recycled in the process circuit. During the constant recirculation of caustic liquor, organic and inorganic impurities can build up within the circuit and suppress alumina yield. Some impurities such as sulfate can be removed through the predesilication process in which reactive silica, such as kaolin clay, reacts with caustic and alumina to form an aluminosilicate desilication product, predominantly either sodalite or cancrinite [41]. In some refineries, a significant amount of organic material is also dissolved in the aluminate solution reducing productivity in many different ways. In addition the organics are oxidized to oxalate, which may precipitate with hydrate, thus causing quality issues. There are many impurities for which there are no commercial removal processes. The continued buildup of impurities limits production capacity and the removal processes are often energy intensive and expensive. Thus, a bleed of the recirculating stream could act to keep impurities at required levels.

### 8.1.3 Industrial Synergy

Economic evaluations of the manufacture of geopolymer binder recognize that the supply of concentrated caustic has the highest cost factor. The economic benefit of utilizing concentrated sodium aluminate liquor from the Bayer process varies between locations, and is not normal Bayer process practice [19]. The Bayer process is contingent upon the capture and return of caustic for further processing and is often called the Bayer cycle. Exporting Bayer process liquor as a geopolymer feedstock, then replacing the volume with fresh caustic, is thus counterintuitive, although it is an effective impurity-removal system.

The use of Bayer process liquor for geopolymer manufacture would provide a mechanism for removing soluble impurities from the Bayer process circuit. The Bayer process achieves a substantial "all" impurity-removal bleed and the geopolymer industry gains a potentially economic source of caustic aluminate [21–23].

Further advantages are that the alumina industry is well established with significant infrastructure such as road, rail, and port hubs. There are also highly trained teams of skilled scientists, engineers and operators experienced in working with caustic aluminate solutions. The industry has traditionally been a major consumer of caustic and has dedicated supply and transport systems in place. Being a thermal hydrometallurgical process, significant quantities of "low-value" steam (i.e., saturated temperature <100°C) may be available, something that could be readily utilized by the geopolymer industry for curing. Finally, the development of other secondary-commodity products from the alumina industry is potentially synergistic with geopolymer production (e.g., process sand).

Beyond a number of performance benefits, the environmental impact reduction accorded to Bayer-derived geopolymers could become an important element of alumina industry and cement industry emissions reductions strategies.

### 8.1.4 Carbon and Embodied Energy

A widely reported benefit of geopolymer with large market appeal is the significant improvement in environmental impact compared to OPC. This is largely based on the $CO_2$ emissions from clinker calcination and the comparatively high embodied energy of OPC.

There are many different ways to calculate embodied energy and each method is reliant on assumptions, circumstances, and inclusion of specific processing stages [42]. For example, it can be claimed that the embodied energy of concrete is minimal as it only contains around 15% OPC by dry weight and most of the content is low-energy aggregate [43]. While this is true, the fact remains that carbon dioxide ($CO_2$) release from cement production is a significant proportion of that derived from human activity globally [44]. There are also claims that OPC concretes have become less energy intensive, by utilizing higher levels of pozzolanic materials such as fly ash [45,46], higher-efficiency kilns [47], and by wider use of renewable energy sources [48]. However, it is rare to capture the full energy cost of OPC.

Some assessments of geopolymers have been particularly favorable, ranging from between 10% and 20% of the carbon footprint of OPC [49], to 30% [50,51]. One industrial geopolymer manufacturer claims geopolymers have about 20% of the $CO_2$ footprint of cement [52] while another claims their geopolymer binder is as little as 10–20% of the carbon footprint of OPC [53]. Similar results are noted based on a binder-to-binder comparison, whereas a concrete-to-concrete comparison results in a value closer to 40% of the $CO_2$ footprint of the OPC [54]. Another comparison of OPC, OPC with supplementary cementitious materials (OPC+SCM) and a variety of alkali-activated concretes (geopolymer binders) was made, ensuring products had similar compressive strength [55]. Geopolymer concretes were between 25% and 45% and OPC+SCM were 80% of the OPC carbon emissions.

Not all assessments are as favorable, owing to the system boundaries, geographical limitations, specific mix ratios, and the functional unit of comparison (e.g., kg of binder, $m^3$ of concrete). Some of the key drivers of the embodied energy and emissions from geopolymer production are the impacts associated with the production of feedstocks, particularly activating compounds (e.g., NaOH and sodium silicate). A detailed life-cycle impact assessment for standard geopolymer production demonstrated that the use of sodium silicate as an alkali activator has a large environmental impact [56]. Furthermore, fly ash– and blast furnace slag–based geopolymers have lower environmental impacts than metakaolin-based geopolymer, as both are considered to be waste materials and thus assigned very low embodied energies. Some geopolymer formulations can achieve a 60% reduction in $CO_2$ emissions compared with those of OPC [56].

A comparison has been made of the carbon impacts from OPC concrete and geopolymer concrete in an Australian context, with particular focus on the transportation and grid emissions that can exacerbate or alleviate the energy-related impacts of geopolymer feedstocks [35]. This placed a heavy emphasis upon transport as a contributor to the life-cycle analysis. They reported a wide range of environmental costs and benefits based upon the source material and mode of transport. Following a series of case studies they were able to demonstrate that using typical Australian feedstocks there was a potential 40–60% reduction in greenhouse gas emissions compared to OPC. It was noted that the production of sodium hydroxide and sodium silicate was a major contributing factor to the carbon footprint as well as to cost. Because of the inherent variability in reported comparisons with OPC carbon footprint, it is important to quote the actual embodied energy of the products being compared.

The general conclusion of most recent studies on geopolymers and other cement alternatives has been that the inclusion of materials from waste streams (such as fly ash or slag) tends to provide multiple environmental and cost benefits. The minerals industry offers a number of such waste streams that may be used as activating components or as aggregate in geopolymers [57–59]. For instance, based on global minerals production figures:

at an estimated rate of 0.3 t slag/t steel, the global steel industry (1420 Mtpa raw steel [57]) could contribute approximately 420 Mtpa of slag;

with an ash content of 15%, the global coal industry (7273.3 Mtpa of coal [59]) could contribute approximately 1090 Mtpa of ash.

If these were considered to be entirely economically and technically viable sources of SCM, the combined waste materials from these two industries alone would account for approximately 45% of global cement production (3310 Mtpa in 2010 [57]). Thus, there is currently an ample supply of useful material being largely wasted, which could be transformed into geopolymer or used to supplement cement clinker.

The geopolymer story's Achilles heel appears to be the large embodied energy associated with the production of the activation solution of concentrated sodium silicate or aluminate [35]. It should be apparent that the most efficient production of sodium aluminate solution would be as part of the economically viable alumina production process (Bayer process). This is also likely to be the source of the lowest embodied energy solution. This claim needs to be investigated further as the utilization of industrially produced Bayer liquor for the activation of geopolymers could substantially reduce the embodied energy of the final product, as well as significantly reduce cost structures.

## 8.2 PROCESS AND MATERIALS

In order to assess the properties and embodied impacts of Bayer-derived geopolymers, physical testing and process modeling were undertaken. This section briefly describes the applied methods.

### 8.2.1 Characterization of Materials

Understanding the chemical composition of feedstocks is vitally important for new processes.

The quantification of the reactive components in fly ash and silica fume was determined using a method where the crystalline component (measured by quantitative X-ray powder diffraction, or XRD) is subtracted from the bulk chemical composition (determined by X-ray fluorescence, or XRF) [9].

The XRF samples were prepared by grinding to a particle size of less than 75 μm followed by fusion in a lithium borate flux (Norrish 12:22). Analysis was carried out by a commercial Western Australian analytical laboratory.

The crystalline composition was obtained on an absolute scale using Rietveld refinement with XRD data. The samples were prepared by milling to ~5 μm particle size using a McCrone mill with calcium fluorite (Mesh −325, 99.5+%, Sigma Aldrich) as an internal standard. XRD patterns were obtained using CuKα radiation with a Bruker D8

Advance diffractometer equipped with a LynxEye detector (Bruker-AXS, Karlsruhe, Germany). The patterns were collected from 10° to 100° $2\theta$, with a nominal step size of 0.01° $2\theta$ and a collection time of 0.8 s per step, using a 0.3° divergence slit and 2.5° secondary Soller slit. A knife-edge collimator and tight detector discriminator settings were applied to reduce air scatter and iron fluorescence signals. Crystalline phases were identified by using the Search/Match algorithm, EVA 15.0 (Bruker-AXS, Germany) to search the Powder Diffraction File (PDF4+2009 edition). Relevant crystal structures used to carry out the Rietveld quantitative phase analysis using TOPAS 4.2 (Bruker-AXS, Germany) were obtained from the Inorganic Crystal Structure Database (ICSD 2009/2). Four replicates of each sample were analyzed to allow an estimate of the uncertainty.

Compressive strength analyses were conducted as per ASTM C39 [60] for concrete specimens. For small cylindrical specimens (25-mm diameter and standardized height-to-diameter ratio of 2:1), the method was approximated by utilizing a Lloyd Instruments 6000 R compressive/tensile strength machine fitted with a 50-kN load cell. Six samples from each formulation were tested and the mean and standard deviations reported.

Scanning electron microscopy (SEM) images of samples were collected using a Zeiss Neon 40 EsB scanning electron microscope. Samples were mounted on an aluminum stub and coated with carbon or platinum prior to viewing.

Wet concrete slump was determined by AS 1379 [61]. Concrete evaluations were conducted according to AS 1012 [62] by Boral Technical Services (Maddington, Western Australia).

## 8.2.2 Bayer-Derived Geopolymer Synthesis

For the work reported here, a geopolymer composition with a Si/Al = 2.3, and Na/Al =0.8 was targeted [9,11–13].

Fly ash and silica fume were mixed as dry powders in a sealed environment to prevent dusting. This was followed by the addition of Bayer liquor and water as required. This slurry was mixed for a period of 10 min and placed into 25-mm diameter vials and sealed. Oven-cured samples were cured at 70°C for 24 h, then left at room temperature for a period of 28 days prior to compressive strength testing. Ambient-cured samples were also sealed and allowed to cure at ambient temperature for 7, 28, or 56 days as reported.

## 8.3 COMPARISON OF EMBODIED ENERGY OF OPC WITH BAYER-DERIVED GEOPOLYMER

### 8.3.1 Base Assumptions

The process used to compare the embodied energy of a Bayer-derived geopolymer with that of published silicate-derived geopolymers and OPC is described in this section. The Bayer-derived geopolymer concrete formulation utilized here is targeted to have a compressive strength of 25 MPa, and is shown in Table 8.1 alongside the compared OPC concrete formulation [63]. As this is a new material, this section describes the material characteristics and process of production for clarity.

In this analysis, the approach is based on an "ex-gate" process. That is, the reagents for production of geopolymer and OPC concretes will be assumed to be on location, or in other words, the assumed distance of transport for each of the feedstocks is zero kilometers. This is not true in practice, but both concrete manufacturing processes will require delivery of sand, aggregate, etc., from various locations. Without a specific or representative case study, the preferable method is to assume that the distances will be similar for both processes and assign them to zero for the comparison. For this investigation we have started with the assumption that the binder for both forms of concrete is approximately 17 wt% but acknowledge specific cases will differ depending on feedstock and compressive strength requirements [35]. Both concrete formulations will usually include admixtures (additives designed to enhance performance), but it is noted that the $CO_2$ footprint of admixtures is slightly lower than OPC and the actual amount of superplasticizer use is almost negligible [47]. Most importantly a literature review showed that the environmental impact of admixtures is negligible [56], hence the impact of admixtures is ignored for these calculations. What is remarkable for the Bayer-derived geopolymer formulation is that Bayer liquor, fly ash, silica fume, and possibly the sand are all industrial byproducts. Such waste products provide a triple benefit when used in geopolymers. They reduce the requirement

**Table 8.1** Bayer-derived geopolymer and OPC concrete formulation (dry mass wt%)

| Concrete type | Bayer liquor | Fly ash (Class F) | Lime | Silica fume | OPC | Sand | Coarse aggregate |
|---|---|---|---|---|---|---|---|
| Bayer-derived geopolymer | 3.9 | 9.3 | 0.2 | 3.6 | 0 | 33 | 50 |
| OPC–CEM1 | 0 | 0 | 0 | 0 | 17 | 33 | 50 |

for waste storage, decrease the demand for extraction of raw materials and simultaneously provide potential reductions in greenhouse gas emissions.

### 8.3.1.1 Embodied Energy Calculation

The embodied energies of the OPC and Bayer-derived geopolymer were calculated using data gathered primarily from academic literature supplemented by industrial sources or estimations where necessary. There is a wide range of alternative embodied energy figures for many of the feedstocks – with a factor of two being a typical difference between the lower and upper estimates. The values used here were judged to be most reasonable and reliable out of a wide-range of identified figures. Table 8.2 presents the selected values, as well as the range of values reported for comparison. The following paragraphs offer a brief description of the processes that contribute to the embodied energy of the feedstocks.

**Table 8.2** Embodied energy of key feedstocks (GPC = geopolymer concrete)

| Material | Included stages | Embodied energy (GJ/t) | | Reference |
|---|---|---|---|---|
| | | Assigned value | Range | Underlined is assigned value |
| *Primary materials* | | | | |
| Clay | Mining | 0.1 | 0.1 | [64] |
| | | | 0.07 | [67] |
| Limestone | Mining | 0.1 | 0.1 | [64] |
| | | | 0.03–0.06 | [67] |
| | | | 0.124 | [68] |
| Sand | Mining, separation | 0.1 | 0 | [69] |
| | | | 0.1 | [67] |
| | | | 0.022–0.5 | [66] |
| Coarse aggregate | Mining, crushing, separation | 0.1 | 0.03–0.06 | [67] |
| | | | 0.022–0.5 | [66] |
| | | | 0.0124 | [69] Derived |
| | | | 0.124 | [68] |
| | | | 0.2 | [70] |
| | | | 0.1 Aggregate | [71] |
| | | | 0.25 Recycled aggregate | [71] |
| | | | 1.0 Stone | [71] |
| | | | 0.4 Recycled stone | [71] |
| | | | 0.3 Gravel | [72] |

*(Continued)*

**Table 8.2** (Continued)

| Material | Included stages | Embodied energy (GJ/t) | | Reference |
|---|---|---|---|---|
| | | Assigned value | Range | Underlined is assigned value |
| Fly ash | Capture, separation | 0.05 | 0.05 | [35] |
| | | | 0.1 (Mined) | [64] |
| | | | 0.09 | [71] |
| Silica fume | Capture, separation | 0.05 | 0.05 | [35] |
| | | | 0.1 (Mined) | [64] |
| NaOH | Electrolysis of brine | 5.7 | 4.7–5.7 and 5.3–6.4 For different processes | [73] Allocated 50/50 on mass |
| Bauxite | Mining | 0.04 | 0.0549 | [74] |
| | | | 0.07 | [75] |
| | | | 0.1 | [76] |
| | | | 0.04 | [77] |
| **Secondary (derived) materials** | | | | |
| Bayer liquor Lime (CaO) | Calcining, milling | Not reported 4.5 (fuel) +0.4 (elec) | 4.9 5.63 | [65] [69] |
| OPC | Blending, calcining, milling | 4.5 (fuel) +0.4 (elec) | 4.9 | [65] |
| | | | 3.1–3.7 (Clinker) | [47] |
| | | | 6.0–8.2 | [54] |
| | | | 5.6 | [78] |
| | | | 4.2–7.5 | [69] |
| | | | 6.15 | [68] |
| | | | 5.8–6.5 | [67] |
| | | | 4.6 + −0.2 | [79] |
| | | | 5.5 (94% clinker) | [71] |
| Silicate-activated geopolymer | | 2.6 GJ/m³ GPC | 1.1–5.8 (GJ/m³ GPC) | [55] |
| | | 3.9 GJ/t binder | 1.7–8.9 (GJ/t binder) Utilized figures represent 30 wt% binder produced, assuming emissions equivalent to a 50% coal, 50% natural gas energy | [54] |
| **Onsite operations** | | | | |
| Concrete processing | | 0.1 | | (Estimate) |

The production of OPC involves blending ground clay and limestone, calcination (clinker production), milling, and the addition of gypsum. This cement is then often blended with SCMs such as fly ash to give a final product. Here we consider the mining of clay and limestone and the clinker calcining and milling, and have ignored the intermediate transport of feedstocks. The overall embodied energy of OPC is therefore the sum of the embodied energy for the production of clay and limestone and the processing of this feedstock into OPC. For the current study, the global average mining energy usage of around 0.1 GJ/t is assumed for both limestone and clay [64]. The embodied energy for production of OPC depends significantly on the technology associated with the kiln operation. The weighted average energy intensity from the International Energy Agency's global examination of energy efficiency in the industry is utilized here because of its global coverage [65]. It is likely that this will give a conservative (lower embodied energy) result than most other figures, making the comparison with Bayer-derived geopolymer also conservative.

The embodied energy of the aggregate material varies significantly depending on the source and type of aggregate. A range of values has been reported from 0.02 to 0.5 GJ/t [66]. If minimal processing is assumed, this figure is likely to be lower, whilst crushing, separation, and extraction from deeper deposits will increase energy requirements. In the current study we use 0.1 GJ/t as representative, but the quantity of aggregate is equal in both products; thus, the impact is on the absolute overall figure and should not affect the comparison.

Low-grade silica fume and fly ash are both waste products that require particularly low amounts of energy for capture, and in fact may arguably be given a zero-energy allocation if we consider their capture as a waste-treatment process. However, if these mineral wastes need to be reclaimed from storage dams, or high levels of separation are required to ensure specific feedstock properties are achieved, a higher level of energy is required (estimated at 0.1 GJ/t for remining) [35,64]. This study uses the mean of these extremes (0.05 GJ/t), as the process would be expected to utilize newly produced material rather than stockpiled material. The process energy required for mixing, pumping, and laying both concrete products is estimated to be 0.1 GJ/t.

The most difficult calculation for Bayer-derived geopolymer is the determination of the embodied energy of the Bayer liquor. Several methods were used to determine the validity of the numbers described below.

Fundamentally an LCA process is used to allocate impacts to the liquor, using mass-based and value-based allocation methods.

### 8.3.1.2 Embodied Energy of Bayer Liquor Feedstock

The Bayer process is reliant upon circulating a strongly caustic aluminate stream while managing temperature and manipulating the solubility of alumina (hydroxide). Bayer geopolymer feedstock is an export bleed of spent Bayer liquor subsequently activated for geopolymer production through evaporation as shown in Fig. 8.1. The export of spent Bayer liquor would typically be less than 1% of the circulating liquor flow. As the feedstock stream is an integral part of the Bayer process, which is heavily reliant on the recovery of caustic, calculating the embodied energy of the Bayer liquor feedstock is a nontrivial task.

The starting point in the calculation is to identify the active components within the Bayer liquor feedstock. The processed Bayer feedstock contains in the order of 230-g/L dissolved aluminum (reported as $Al_2O_3$ though in the form of $Al(OH)_4^-$). The Bayer feedstock also contains sodium hydroxide, though this can be separated into 220-g/L freely dissociated caustic soda (NaOH in the form of $Na^+ OH^-$) and 180-g/L bound caustic soda (as NaOH, but in the form of $Al(OH)_4^-$). In addition, the evaporation and preparation of Bayer feedstock from spent Bayer liquor has a significant direct energy requirement.

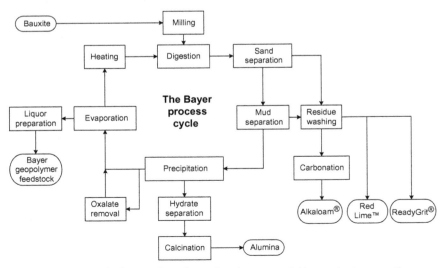

**Figure 8.1** A simplified Bayer flow sheet showing potential byproduct generation.

In the first case, we calculate the embodied energy if the Bayer feed-stock were to be produced from pure raw materials. Currently Bayer liquor is made utilizing a 50% solution of sodium hydroxide derived from the chloralkali process [73]. For this study we have utilized a value of 5.7 GJ/t for the embodied energy of NaOH solution based on an alloca-tion to the two products of NaOH and $Cl_2$ by mass. The resulting contri-bution of sodium hydroxide to the embodied energy of the Bayer liquor feedstock is 2.3 GJ/m$^3$. As the density of the feedstock is 1.6 t/m$^3$, the mass-based contribution from sodium hydroxide is 1.4 GJ/t.

Calculating the embodied energy of dissolved aluminum from the Bayer process is more challenging, as it is typically an intermediate stream. Some estimates are given below.

A Commonwealth Government of Australia report lists alumina ($Al_2O_3$) having an embodied energy of 11 GJ/t [80]. However, alu-mina is a precipitated, filtered, washed, dried, and calcined version of the $Al(OH)_4^-$ species reported in Bayer liquor. Deducting an energy figure for calcination of 3.9$_3$ GJ/t [77], then the embodied energy for dissolved aluminum (reported as $Al_2O_3$) in Bayer liquor is 7.1 GJ/t.

The embodied energy for calcined alumina has been reported else-where as 12.3 GJ/t [77]. Subtracting the reported energy for calcina-tion as above (3.9 GJ/t), the embodied energy for dissolved aluminum (reported as $Al_2O_3$) in Bayer liquor would be 8.4 GJ/t.

The International Aluminum Institute (IAI) represents 57% of global production and has provided a figure for the embodied energy of hydrate ($Al(OH)_3$; but reported as $Al_2O_3$) at 8.9 GJ/t [81]. This figure could be taken as a very conservative estimate for dissolved aluminum given that it includes the subsequent steps of precipitation, filtration, washing, and drying to product hydrate.

It is assumed that the IAI estimate is most reliable given the coverage of data collection. However, it does not include a bauxite-mining compo-nent, reported by Smith et al. [77] to be 0.04 GJ/t bauxite. It takes 2–3 t of bauxite to produce a tonne of alumina, so a figure of 0.1 GJ/t will be added to the IAI number, giving the embodied energy for dissolved alu-minum (reported as $Al_2O_3$) in Bayer liquor to be 9.0 GJ/t.

Utilizing this figure, the contribution of dissolved aluminum to the Bayer liquor feedstock is 2.1 GJ/m$^3$, or 1.3 GJ/t.

The active components of Bayer liquor feedstock combine to 2.7 GJ/t of indirect embodied energy.

When the direct energy cost of 0.8 GJ/t is included, the final embod-ied energy for Bayer liquor feedstock becomes 3.5 GJ/t.

## 8.3.2 Results and Discussion

### 8.3.2.1 Embodied Energy of Concrete Formulations

The feedstock embodied energy and the mix formulation are combined to give the concrete embodied energy (Fig. 8.2).

This analysis shows that the Bayer-derived geopolymer concrete is only 33% of the embodied energy of the OPC concrete. The Bayer liquor is the most important component of the embodied energy in the Bayer-derived geopolymer binder, of which approximately half of the impact is attributable to NaOH and the rest to the alumina content.

Variations in the specific mix ratios and addition of transportation may have significant impacts on the relative performance of the two concretes. If we add a significant transport cost to the sand and aggregate components, the comparative advantage is reduced. For example, in the Australian context, many geopolymer components are transported long distances adding significantly to the feedstock embodied energy, whereas the majority of OPC is locally produced. Adding the transport component in this context would reduce the geopolymer embodied energy as a fraction of OPC to 38% [35].

### 8.3.2.2 Embodied Energy of Binding Agent

Various methods of presenting the comparison are possible, which is one factor in the variability of literature values. It may be useful to present

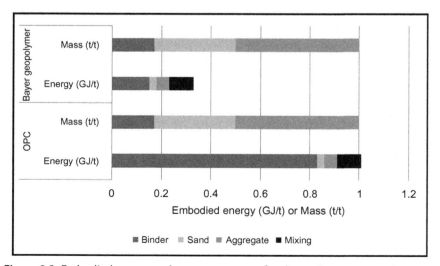

**Figure 8.2** Embodied energy and mass component for alternative concrete mixes.

the figures on the basis of the embodied energy of the binder alone. This allows the concrete producer to readily calculate the relative change in embodied energy based on alternative binder ratios. Bayer-derived geopolymer binder has embodied energy of 0.9 GJ/t compared to dry OPC binder with an embodied energy of 4.9 GJ/t. In this comparison, the embodied energy of Bayer-derived geopolymer is only 18% of that of OPC. However, if cement is assumed to be mixed with water to constitute a binder then the OPC binder embodied energy effectively drops to 2.5–3.3 GJ/t (at 1:2–1:1 ratio water : OPC). The embodied energy of the Bayer-derived geopolymer binder is then 27–36% of that of wet OPC.

### 8.3.3 Bayer Liquor as a Waste Product

The removal of Bayer liquor from the Bayer process circuit and its replacement with fresh caustic is effectively an impurity-removal process [19]. The question is, what to do with this heavily impurity laden sodium aluminate solution? In some cases, the liquor is burnt to remove the organic species with the remaining solids returned to the process. However, inorganic impurities must be dealt with by other processes. It is argued here that Bayer liquor should be treated as a waste material and the energy utilized for its manufacture be attributed to the Bayer process production of alumina (following standard LCA protocols). In this case, the embodied energy of the Bayer liquor bleed stream could reasonably be attributed a value of 0 GJ/t, leaving just the direct energy required to process the liquor to make it a suitable geopolymer feedstock. This would give an overall embodied energy of Bayer liquor feedstock of 0.78 GJ/t.

Applying this approach, the calculated embodied energy of Bayer-derived geopolymer binder becomes 0.27 GJ/t and represents only 6% of the embodied energy of dry OPC. For the first time, we can discuss binding agents within the same embodied energy range as manufactured sand [66], gravel [72], and recycled stone [71].

### 8.3.4 Embodied Energy Implications

Evaluation of the global impact by introduction of Bayer geopolymer into the market place is contingent upon a series of variables. These include the rate of uptake within global plants, their impurity load, the geographical location to markets, the cost of caustic, and value of alumina. Making a series of conservative estimates results in an average available Bayer liquor feedstock volume of 0.012 kL/t, or 0.02 t/t alumina produced.

The estimated alumina production in 2011 was 92 million tonnes; hence the mass of Bayer geopolymer feedstock liquor is 1.8 million tonnes. Given that Bayer feedstock equates to about 25% of binder paste, there is the potential to make 7.3 million tonnes of Bayer-derived geopolymer binder.

Given that cement production in 2010 was 3310 million tonnes, uptake of Bayer-derived geopolymer would equate to less than 1% of the global market. This may seem small, but it is an appropriate entry level into the market. With market forces and a carbon-rated economy, this value could be significantly higher, especially as other byproduct sources of caustic could come to market.

## 8.3.5 Embodied Energy Conclusions

The determination of the embodied energy of a product depends on many assumptions. Reporting these assumptions and how they are derived is essential.

In this case, two methods of calculating the embodied energy of hydrate in Bayer liquor were used. They compared favorably with industry-collected data that was reported for the first time from the IAI. Once the energy of mining is included, that figure is put at a value of less than $9.0\,GJ/t$ of $Al(OH)_3$.

The embodied energy to produce hydrate, caustic, and the energy used to process Bayer liquor into a geopolymer feedstock are combined to provide an embodied energy for Bayer liquor feedstock of $3.5\,GJ/t$.

A new class of construction material, Bayer-derived geopolymer, was then assessed and the embodied energy calculated for binder paste was $0.9\,GJ/t$, while the embodied energy for a subsequent concrete product was $0.33\,GJ/t$.

It is further argued that the Bayer liquor feedstock can be considered as an impurity-removal waste product and should be assigned an embodied energy of zero at the point of it leaving the alumina production process, as is fly ash and other waste streams. In this case, the embodied energy of geopolymer paste made utilizing Bayer liquor feedstock would be $0.27\,GJ/t$.

These conclusions indicate that there is a significant embodied energy advantage to the production of geopolymers from Bayer process liquors. This advantage may reflect a potential commercial advantage allowing large-scale utilization of this low-carbon technology. Further investigation is warranted as production opportunities arise. Moreover, there may be a specific argument for adjusting alumina production from a cross-industry or industrial symbiosis perspective. Further research is recommended to identify the system-wide impacts of undertaking such a process alteration, expanding the scope to include both the production of alumina and the production of geopolymer.

There are many barriers present that are slowing down the adoption of geopolymer products including little or no information about the embodied energy of this relatively new product. The figures presented here reveal that a geopolymer based on Bayer liquor has significantly lower embodied energy than OPC, thus removing one of the barriers to adoption. In addition, the embodied energy figures for Bayer-derived geopolymer are so low that it should act as a catalyst to industry considering moving to the production of this product.

## 8.4 DEVELOPMENT OF BAYER-DERIVED GEOPOLYMERS

There are many steps in achieving production of a new engineering product. Some of these steps require technical problems to be resolved and others require product development, market planning, cost control, and potential production designs. These steps allow for a considered business gate review utilizing cost–benefit analysis. Some of these steps are described below.

### 8.4.1 Ambient Curing: The Impact of Calcium and Fly Ash Sources

High-volume concrete applications typically require ambient in situ curing. The incorporation of $Ca(OH)_2$ or blast furnace slag into a silicate-activated geopolymer mixture facilitates ambient curing and now a similar impact for a Bayer-derived aluminate-activated geopolymer has been confirmed [82]. Blast furnace slag is significantly less expensive than $Ca(OH)_2$; however, larger quantities are required to achieve a compressive strength of 20 MPa (a strength specification typical for pathways, driveways, etc. [29]).

Samples of Collie (Western Australia) fly ash/Bayer liquor geopolymer were produced with different levels of Hylime ($Ca(OH)_2$). Fig. 8.3 shows the 7-day compressive strength for samples cured at ambient temperature (approximately 22°C). The strength increases with increasing $Ca(OH)_2$ content, with mixtures of 4 wt% or higher achieving the targeted strength criteria.

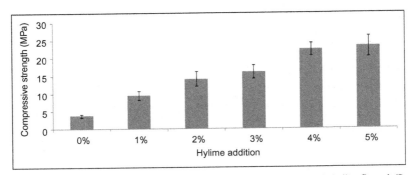

**Figure 8.3** 7-day compressive strength results for ambient-cured Collie fly ash/Bayer liquor geopolymer samples with increasing $Ca(OH)_2$ (*Hylime*) content (wt%).

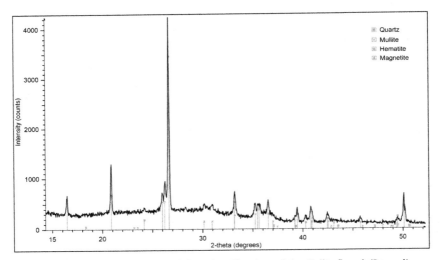

**Figure 8.4** XRD pattern showing phase identification of the Collie fly ash/Bayer liquor–based geopolymer paste with 5 wt% $Ca(OH)_2$ (Hylime).

Fig. 8.4 is an XRD pattern for geopolymer with 5 wt% of $Ca(OH)_2$. The absence of Portlandite indicates that the calcium is incorporated in the geopolymer structure. Phases identified in the geopolymer paste are quartz (PDF# 01-070-7344), mullite (PDF# 01-074-4146), hematite (PDF# 00-033-0664), and magnetite (PDF# 04-009-2285) all originating from the fly ash precursor.

Ground blast furnace slag (builder's slag) is an industrial residue that can be used as a source of calcium (42.9 wt% CaO). Bayer-derived

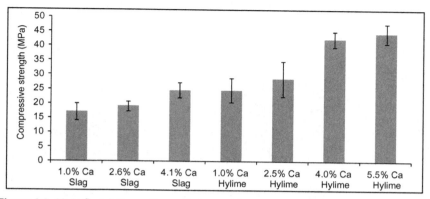

**Figure 8.5** Muja fly ash/Bayer liquor. 7-day ambient cure compressive strength for calcium from different sources.

geopolymers were made utilizing Muja fly ash (Western Australia) and incorporated either blast furnace slag or Hylime for direct comparison of ambient curing. Fig. 8.5 shows the comparative compressive strengths after 7 days of ambient cure.

For similar calcium content, it is evident that Hylime $(Ca(OH)_2)$ is more effective at strength generation during ambient cure. The authors believe that the particle size of the blast furnace slag will dictate how readily the calcium is provided to the geopolymerization process. As the calcium availability appears to have a large impact upon curing and strength development, it is therefore essential that a strict quality control/quality assurance process be put in place for reliable production.

Comparison of Figs. 8.3 and 8.5 demonstrate that Muja fly ash produces a geopolymer of higher compressive strength compared to geopolymers made utilizing Collie fly ash. The Muja fly ash–based geopolymer also requires less calcium to achieve the desired compressive strength. The authors believe that the higher surface area of Muja ash (Muja ash $2.2\,m^2/cc$, Collie at $1\,m^2/cc$) contributed to improved alkali reactivity leading to greater geopolymer formation. This would also suggest that quality assurance and control is an essential element for production of reliable geopolymer products.

Fig. 8.6 shows the XRD patterns collected from $1.0\,wt\%$ and $5.5\,wt\%$ calcium addition (from $Ca(OH)_2$) to Muja fly ash–based geopolymers. There is a marginal increase in the amorphous geopolymer component, recognizable from the broad elevation in the pattern centered at $\sim28°\ 2\theta$.

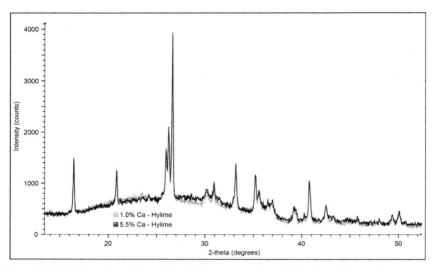

**Figure 8.6** A selected portion of the XRD patterns from Muja fly ash/Bayer liquor geopolymer with 1.0 and 5 wt% addition of $Ca(OH)_2$ (Hylime). Crystalline phases present are the same as shown in Fig. 8.4.

It is apparent that addition of calcium to the formulation of Bayer-derived fly ash geopolymer results in ambient temperature curing with increasing calcium resulting in higher compressive strength. The target compressive strength of 20 MPa is achieved with there being no discernable changes to the formation of geopolymer.

## 8.4.2 Aggregate Production: A Low-Risk, High-Volume Strategic Market

For a Bayer refinery to implement the production of geopolymer feedstock, volumes of liquor exported would need to be between 1 and 20 kL/h to reflect refinery impurity-removal processes. Consumption of this liquor would equate to between 0.1 and 2 million tonnes of Bayer-derived geopolymer concrete per year. This represents up to 50% of the concrete market in Western Australia, which would be disruptive to local industry and difficult to achieve.

Hence, many different applications are required to allow products to enter markets without significant displacement or disruption.

To achieve industrial synergy, large volume markets are required, but also a product that can perform as a base load to allow continuous production. One option for a base-load product is aggregate. This can be summarized as follows:

- The consumption of aggregate is a very large volume market.
- Application of aggregate is wide ranging.
- The value of aggregate can be measured and predicted within a transport envelope.
- Aggregate can be produced in a continuous plant process.
- Aggregate has an indefinite storage life.
- Aggregates in Australia are governed by performance standards, not by composition [83,84].

Other advantages arise due to the geography and location of the Bayer refineries in Western Australia.

The source of all reagents is local or they can be transported by rail or ship. Demand for aggregate is expected to rise significantly and for a prolonged period due to population expansion in the area [85].

Traditional sources of aggregates are disadvantaged by significantly longer transport distance.

Finally, the largest advantage of using Bayer-derived geopolymer to manufacture aggregate is that liquor volume consumed would be much higher. For instance, approximately 4 wt% of Bayer liquor feedstock is used to make geopolymer concrete whereas over 25 wt% would be used if included in aggregate. This dramatic increase in utilization would be less threatening to the conventional cement and concrete industries and could consume a large amount of Bayer liquor.

## 8.4.3 Aggregate Production: Possible Production Design

There are several methods for artificial aggregate production utilizing Bayer-derived geopolymers [23,24]. However, economics, material characteristics, and supply availability are fundamental to plant design.

The Western Australian consumption of aggregates and rock products has been estimated to be up to 7 million tonnes annually (derived from Refs. [86,87]). Some of these products have a value as little as $A14/t while others products such as seawall armor can command $A40/t. These are high-volume, low-value products that are heavily influenced by transport costs. With limited local competition, a market value–transport envelope can be developed.

Bench-scale production has produced mechanically extruded products as well as blocks of material that have been crushed in a jaw crusher and screened (Fig. 8.7).

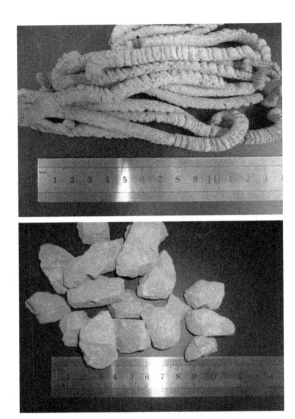

**Figure 8.7** Mechanically extruded product on the top and crushed block on the bottom. Scale in cm.

The following observations were made during the laboratory-based processing:
- Dry reagents can be gravity-fed utilizing hoppers. Vibration is recommended due to a tendency for the powders to arch and bridge over a feed well.
- Components should be made of materials suitable for highly caustic solutions and abrasive sand.
- Standard pug-mixing equipment should suffice for appropriate shear to mix the reagents.
- Extrusion of the mixture was demonstrated by hand and by mechanical screw feed machinery. A screw-fed extruder is believed to be a viable option.

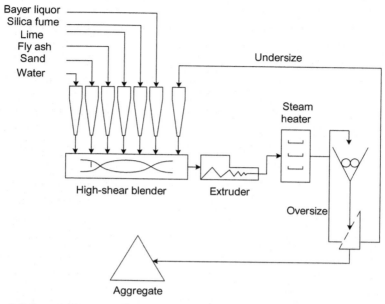

**Figure 8.8** Proposed process flow diagram for aggregate production.

- Mechanical extruders should have vacuum applied to prevent air entrapment.
- Other factors governing design included: (1) important to minimize handling, (2) keep power consumption low, and (3) utilize standard equipment to deliver a suitably priced product. Due to the inherent sticky nature of the geopolymer, a formulation containing up to 50 wt% sand was utilized to reduce machine fouling.

Using these observations and the mix design, potential process diagrams were developed. The availability of low-grade steam for elevated temperature curing is a preferred option but the addition of lime will enable ambient setting. Formation of the aggregate can be applied immediately postextrusion (wet), or by standard jaw crushers postcure. The second option is shown in Fig. 8.8.

## 8.4.4 Aggregate Production: Embodied Energy

Starting with an embodied energy of Bayer-derived geopolymer binder of 0.27 GJ/t and mixing 50/50 with sand having an embodied energy of 0.1 GJ/t [67], the resulting mortar would have an embodied energy of 0.19 GJ/t. An estimate can be made for the production processing energy at 0.05 GJ/t, hence the artificial aggregate would have an embodied energy of 0.24 GJ/t. This result is similar to mined gravel (0.3 GJ/t [72]) and recycled aggregate (0.25 GJ/t [71]). Clearly this is an environmentally significant result allowing for the utilization of byproducts from an industrial precinct while minimizing consumption of virgin materials.

## 8.4.5 Aggregate Consumption: Bayer-Derived Geopolymer Aggregates Utilized in OPC Concrete

Utilizing geopolymer aggregates within OPC concrete might seem like a backward step for the low-carbon construction materials industry; however, it may be initially necessary. Demonstrating the reliability of Bayer-derived geopolymer aggregates has the potential to allow implementation and familiarization while achieving volume production.

Five batches of aggregates were made having different sand, lime, and total water contents. The base geopolymer mortars ranged from 30 to 42 MPa [23], which is significantly lower than the normal aggregate (granite) having a compressive strength of up to 130 MPa.

The aggregate particle size distribution (PSD) is shown in Table 8.3. Wet strength was 99 kN and dry strength was 150 kN as determined in accordance with AS1141 [88]. The dry density was 2020 kg/m$^3$ (granite is 2700 kg/m$^3$) and the saturated surface-dry density was 2170 kg/m$^3$. Water absorption was determined to be 7.7 wt%.

Sand, Bayer aggregate, water, and OPC (450 kg/m$^3$) were added to a high-shear pan mixer to make an 84-kg batch of concrete. The design specification compressive strength was 45 MPa. The aggregates were added

**Table 8.3** PSD of Bayer-derived geopolymer aggregate

| Sieve size | % Passing |
|---|---|
| 19.00 mm | 100 |
| 13.2 mm | 77 |
| 9.5 mm | 52 |
| 6.7 mm | 35 |
| 4.75 mm | 24 |
| 2.36 mm | 14 |
| 1.18 mm | 9 |
| 600 μm | 6 |
| 300 μm | 4 |
| 150 μm | 2 |
| 75 μm | 1 |

in the saturated-surface dry state to prevent water absorption impacting upon the mix.

The batch slump test was shown to be 65 mm, just within the criteria of 80 ±20 mm. It was noted that the slump is not only dependent on rheology, but also on the density of the aggregate. As the density of the geopolymer aggregate was significantly lower than granite, the concrete slump would naturally be smaller.

The wet (plastic) density was recorded at 2080 kg/m³ while the 7–day hardened density was 2125 kg/m³. This was lower than the expected mix design 2400 kg/m³ and is a result of the lower aggregate density.

It is noted that normal concrete has a minimum density of 2100 kg/m³ [61] suggesting a minor formulation modification could result in achieving lightweight concrete standards. The Bayer-derived geopolymer aggregate contained an average 33 wt% sand and the OPC concrete used sand for fine aggregate, which has a density of 2650 kg/m³. As Bayer aggregate is only 2020 kg/m³, replacing the virgin sand with geopolymer sand would lower the density of the resulting concrete further.

The artificial aggregate OPC concrete achieved a compressive strength of 50 MPa after 7–day ambient cure, 57 MPa at day 27 and 69 MPa after day 56. These results are well above the design specification of 45 MPa (Fig. 8.9).

It is concluded that the Bayer geopolymer aggregate has increased the concrete compressive strength while decreasing the product density. Fig. 8.10 gives the first clues as to why the artificial aggregate in OPC concrete was so much stronger than expected. It is apparent that the

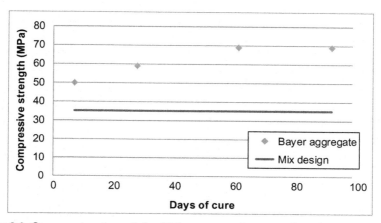

**Figure 8.9** Compressive strength for OPC concrete using geopolymer aggregate.

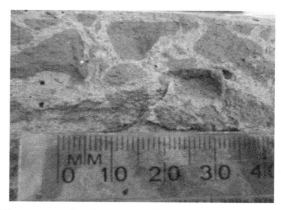

**Figure 8.10** OPC concrete using Bayer geopolymer aggregate.

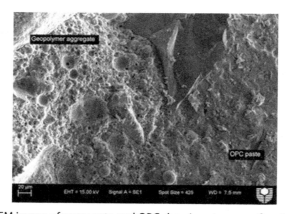

**Figure 8.11** SEM image of aggregate and OPC showing strong surface interaction.

aggregate was sheared during the compression test, in preference to either OPC or surface adhesion sites.

The weakest region in concrete is the interfacial zone between Portland cement paste and aggregate, where aggregate is typically granite and sand [89]. It has also been demonstrated that cracks initiated at the interface boundary before extending into the mortar layer [90]. The interface bond is described as mostly mechanical and dependent upon the aggregate surface roughness and cleanness [91]. For OPC with geopolymer aggregate there appears to be a chemical bond between the binder and aggregate (Fig. 8.11). This shifts the weakest point of the concrete from the surface boundary to the aggregate.

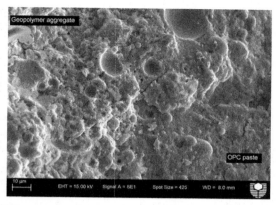

**Figure 8.12** SEM image showing rough aggregate surface providing bind sites for OPC paste.

In addition to the interfacial zone, the surface-area-to-volume ratio of the aggregate controls the degree of mechanical bonding with the OPC binder. For Bayer-derived geopolymer aggregate the microscopically rough surface (Fig. 8.12) facilitates extensive mechanical bonding. This has resulted in a significant improvement in the composite compressive strength.

Similar effects were found using manufactured sand in OPC concrete [33] and by using recycled OPC aggregate in geopolymer [92].

It has also been reported that a large difference in mechanical properties between binder and aggregate would accelerate the development and spread of microcracks [93]. With OPC and Bayer-derived geopolymer aggregate being similar in mechanical properties, the composite concrete is actually stronger in performance.

These results have enabled some important implications to be identified. The lower concrete density will reduce transport cost, vehicle wear, and road deformation. In Western Australia, concrete products can be transported many hundreds of kilometers. The higher than anticipated compressive strength may enable the OPC content to be lowered, reducing the cost and the embodied energy of the concrete. Alternatively, the thickness of the concrete can be reduced thereby reducing the weight and cost.

The successful demonstration of Bayer-derived geopolymer aggregate adds support to the proposal that OPC recycled aggregate be utilized within geopolymer concrete once geopolymers have become the dominant binding agent [94,95]. This is in line with the European Union decree that 70% of demolition waste be recycled by 2020 [96], providing a significant potential future recycling opportunity.

## 8.4.6 Product Application Conclusions

The development of Bayer-derived geopolymers is reliant on large-scale production of products. The supply of caustic Bayer aluminate solution needs to be of large volume and relatively constant demand to facilitate the export being a legitimate impurity-removal process for the alumina refinery. This requirement and the conservative nature of the construction industry places constraints on geopolymer production and application development.

The use of Bayer process liquor as the activating solution for the manufacture of geopolymer has been demonstrated utilizing two sources of fly ash precursors. Two sources of calcium have also been demonstrated to successfully achieve ambient curing, a requirement for in situ concrete market development. The chemical availability of calcium to the geopolymer gel during cure appears critical to the final product compressive strength. This would indicate that well-defined calcium reagents are required for quality control and quality assurance. Smaller but still significant improvements in geopolymer performance are attributed by the authors to changes in fly ash composition and particle size, again highlighting the need for quality control processes. It has been stated that "waste is a product made to poor specifications" [97], so to prevent the manufacture of waste, the starting reagents must be supplied to specification.

It has been demonstrated that Bayer-derived geopolymer mortar is suitable for the manufacture of artificial aggregate. Based on our laboratory test work and an analysis of the economics of a possible geopolymer production process we believe that the production of Bayer-derived geopolymer aggregate is potentially viable and technically feasible. The embodied energy of the Bayer-derived geopolymer mortar aggregate has been calculated to be similar to recycled aggregate and mined gravel.

Initial trials of the artificial aggregate have demonstrated that it can be utilized in the manufacture of OPC-based concrete. The Bayer-derived geopolymer mortar aggregate when added to OPC concrete mixes achieved a 30% lower density, while increasing compressive strength by 50%.

The use of multiple industrial byproducts to manufacture ambient-curing geopolymer products and subsequent demonstration is a major step forward toward the potential commercialization of Bayer-derived geopolymer.

## ACKNOWLEDGMENTS

The authors wish to thank the International Aluminum Institute (IAI) for provision of hydrate production energy data not previously published. In addition, our thanks go to Kirk Moore of Alcoa of Australia for determination of the Bayer evaporation energy consumption.

The authors also wish to thank Gary Jones, Sarvesh Mali, and Jeremy Robertson of Boral Technical Services, Maddington, Western Australian, for assistance with production and assessment of Bayer-derived geopolymer aggregate within OPC concrete.

## REFERENCES

[1] British Standard BS EN 197 Cement. Composition, specifications and conformity criteria for common cements, 2011.

[2] Hart RD, Lowe JL, Southam DC, Perera DS, Walls P, Vance ER, Gourley T, Wright K. Aluminosilicate inorganic polymers from waste materials. In Proceedings of the third international conference on sustainable processing of minerals and metals, green processing (vol. 2006). 2006.

[3] Wahyuni A, Nikraz H, Jamieson E, Cooling D. Sustainable use of bauxite residue sand (red sand) in concrete. In Proceedings of the third international conference on sustainable processing of minerals and metals, green processing (vol. 2006). 2006.

[4] Jitsangiam P, Nikraz H, Jamieson E. Sustainable Use of a Bauxite Residue (red sand) in terms of roadway materials. In 2nd international conference on sustainability engineering and science, Auckland, New Zealand. 2007.

[5] Jitsangiam P, Nikraz H, Jamieson E, Kitanovich R, Siripun K. Sustainable use of a bauxite residue (red sand) as highway embankment materials. In 3rd international conference on sustainability engineering and science, Auckland, New Zealand (vol. 2008). 2008.

[6] Attiwell S, Jamieson E, Jones A, Cooling D, Myers L, Lee F, et al. Demonstration trials for the use of Readygrit®. In Proceedings of the 10th international alumina quality workshop, Perth, Australia. 2015.

[7] Jamieson EJ. By-product utilisation, the bauxite residue program. In First CSRP conference, Melbourne, Australia. 2007.

[8] Avraamides J, van Riessen A, Jamieson E. Transforming mining residues into viable by-products. Energy Gener 2010:1–3.

[9] Williams RP, van Riessen A. Determination of the reactive component of fly ashes for geopolymer production using XRF and XRD. Fuel 2010;89(12):3683–92.

[10] Williams RP, Hart RD, van Riessen A. Quantification of the extent of reaction of metakaolin-based geopolymers using X-ray diffraction, scanning electron microscopy, and energy-dispersive spectroscopy. J Am Ceram Soc 2011;94(8):2663–70.

[11] Chen-Tan NW, van Riessen A, Ly CV, Southam DC. Determining the reactivity of a fly ash for production of geopolymer. J Am Ceram Soc 2009;92(4):881–7.

[12] Rickard WD, Williams R, Temuujin J, van Riessen A. Assessing the suitability of three Australian fly ashes as an aluminosilicate source for geopolymers in high temperature applications. Mater Sci Eng A 2011;528(9):3390–7.

[13] Rangan BV. Studies on fly ash-based geopolymer concrete. Malaysian Constr Res J 2008;3(2):1–20.

[14] Rangan BV. Low-calcium, fly-ash-based geopolymer concrete. In: Nawy Edward G, editor. Concrete construction engineering handbook. Boca Raton, FL: CRC Press; 2008. 26-1–26-19.

[15] Hardjito D, Wallah SE, Sumajouw DM, Rangan BV. On the development of fly ash-based geopolymer concrete. ACI Mater J 2004;101(6).

[16] Davidovits J. Geopolymer chemistry & applications, 2nd edition Saint-Quentin, France: Institut Géopolymère; 2008:526–7 [Chapter 25].

[17] Xu H, Van Deventer JS. Geopolymerisation of multiple minerals. Miner Eng 2002;15(12):1131–9.

[18] Khater HM. Effect of calcium on geopolymerization of aluminosilicate wastes. J Mater Civil Eng 2011;24(1):92–101.

[19] Jamieson E. Method for management of contaminants in alkaline process liquors. International patent WO, 17109, A1. 2008.

[20] Phair JW, Van Deventer JSJ. Characterization of fly-ash-based geopolymeric binders activated with sodium aluminate. Ind Eng Chem Res 2002;41(17):4242–51.

[21] Jamieson EJ, van Riessen A, Kealley C, Hart, RD. Development of Bayer geopolymer paste and use as concrete. In Proceedings of the ninth international alumina quality workshop, Perth, Australia; 2012.

[22] van Riessen A, Jamieson E, Kealley CS, Hart RD, Williams RP. Bayer-geopolymers: an exploration of synergy between the alumina and geopolymer industries. Cem Concr Compos 2013;41:29–33.

[23] Jamieson EJ. Development and utilization of Bayer process by-products. Curtin University, School of Civil and Mechanical Engineering, Department of Civil Engineering; 2014. Available from espace.library.curtin.edu.au.

[24] Jamieson E, Penna B, van Riessen A, Nikraz H. The development of Bayer derived geopolymers as artificial aggregates. Hydrometallurgy. Online 3 May 2016. http://dx.doi.org/10.1016/j.hydromet.2016.05.001.

[25] Van Deventer JS, Provis JL, Duxson P, Brice DG. Chemical research and climate change as drivers in the commercial adoption of alkali activated materials. Waste Biomass Valorization 2010;1(1):145–55.

[26] Gourley JT. Geopolymers; opportunities for environmentally friendly construction materials Materials 2003 conference: adaptive materials for a modern society. Sydney: Institute of Materials Engineering Australia; 2003.

[27] Gourley JT, Johnson GB. Developments in geopolymer precast concrete. In World congress ceopolymer. 2005. p. 139–143.

[28] Southam DC, Brent GF, Felipe F, Carr C, Hart RD, Wright K. Towards more sustainable minefills—replacement of ordinary Portland cement with geopolymer cements. Publications of the Australasian Institute of Mining and Metallurgy 2007;9:157–64.

[29] Australian Standard 3727. Guide to residential pavements. Standards Australia; 1993.

[30] Rickard WD, van Riessen A, Walls P. Thermal character of geopolymers synthesized from class f fly ash containing high concentrations of iron and α-quartz. Int J Appl Ceram Technol 2010;7(1):81–8.

[31] Temuujin J, Minjigmaa A, Lee M, Chen-Tan N, van Riessen A. Characterization of class F fly ash geopolymer pastes immersed in acid and alkaline solutions. Cem Concr Compos 2011;33(10):1086–91.

[32] Rickard WD, Temuujin J, van Riessen A. Thermal analysis of geopolymer pastes synthesised from five fly ashes of variable composition. J Non-Cryst Solids 2012;358(15):1830–9.

[33] Tavor D, Wolfson A, Shamaev A, Shvarzman A. Recycling of industrial wastewater by its immobilization in geopolymer cement. Ind Eng Chem Res 2007;46(21):6801–5.

[34] Van Deventer JSJ, Provis JL, Duxson P, Lukey GC. Reaction mechanisms in the geopolymeric conversion of inorganic waste to useful products. J Hazard Mater 2007;139(3):506–13.

[35] McLellan BC, Williams RP, Lay J, van Riessen A, Corder GD. Costs and carbon emissions for geopolymer pastes in comparison to ordinary Portland cement. J Clean Prod 2011;19(9):1080–90.

[36] Davidovits J. Geopolymer cements to minimise carbon-dioxide greenhouse-warming. Ceram Trans 1993;37:165–82.

[37] Davidovits J. Environmentally driven geopolymer cement applications. In Proceedings of 2002 geopolymer conference, Melbourne, Australia; 2002.

[38] Smith PG, Pennifold RM, Davies MG, Jamieson EJ. Reactions of carbon dioxide with tri-calcium aluminate. Electrometall Environ Hydrometall 2003;2:1705–15.

[39] Jamieson E, Cooling DJ, Fu J. High volume resources from bauxite residue. In Proceedings of the 7th international alumina quality workshop, Perth, Australia. 2005.

[40] Jamieson E, Jones A, Cooling D, Stockton N. Magnetic separation of Red Sand to produce value. Miner Eng 2006;19(15):1603–5.

[41] Smith PG, Xu BG, Wingate CJ. Transformation of sodalite to cancrinite under high temperature Bayer digestion conditions Light Metals. San Francisco, CA: TMS; 2009.51.6

[42] Dixit MK, Fernández-Solís JL, Lavy S, Culp CH. Identification of parameters for embodied energy measurement: a literature review. Energy Build 2010;42(8): 1238–47.

[43] Prusinski JR, Marceau ML, Van Geem MG. Life cycle inventory of slag cement concrete. In Eighth CANMET/ACI international conference on fly ash, silica fume, slag and natural pozzolans in concrete; 2004. p. 26.

[44] Allwood JM, Cullen JM, Milford RL. Options for achieving a 50% cut in industrial carbon emissions by 2050. Environ Sci Technol 2010;44(6):1888–94.

[45] O'Brien KR, Ménaché J, O'Moore LM. Impact of fly ash content and fly ash transportation distance on embodied greenhouse gas emissions and water consumption in concrete. Int J Life Cycle Assess 2009;14(7):621–9.

[46] Huntzinger DN, Eatmon TD. A life-cycle assessment of Portland cement manufacturing: comparing the traditional process with alternative technologies. J Clean Prod 2009;17(7):668–75.

[47] Van den Heede P, De Belie N. Environmental impact and life cycle assessment (LCA) of traditional and 'green' concretes: literature review and theoretical calculations. Cem Concr Compos 2012;34(4):431–42.

[48] CCAA. Sustainable concrete materials, briefing note. In Cement concrete & aggregates, Australia. 2010.

[49] Davidovits J. Properties of geopolymer cements. In First international conference on alkaline cements and concretes (vol. 1). 1994. p. 131–149.

[50] Wimpenny D. Low carbon concrete-options for the next generation of infrastructure. Concr Solutions 2009;9:41-1.

[51] Weil M, Dombrowski K, Buchawald A. Life-cycle analysis of geopolymers Geopolymers, structure, processing, properties and applications. Cambridge: Woodhead Publishing Limited Abington Hall; 2009;194–210.

[52] The Zeobond Group. Life cycle analysis, http://www.zeobond.com/life-cycle-analysis.html; 2012 [retrieved 23.07.15].

[53] Wagners Concrete Pty Ltd. Earth friendly concrete, http://www.wagnerscft.com.au/files/2613/4731/0397/Wagners-Earth-Friendly-Concrete.pdf; 2012 [retrieved 03.01.13].

[54] Van Deventer JS, Provis JL, Duxson P. Technical and commercial progress in the adoption of geopolymer cement. Miner Eng 2012;29:89–104.

[55] Yang KH, Song JK, Song KI. Assessment of $CO_2$ reduction of alkali-activated concrete. J Clean Prod 2013;39:265–72.

[56] Habert G, D'Espinose de Lacaillerie JB, Lanta E, Roussel N. Environmental evaluation for cement substitution with geopolymers. In Proceedings 2nd Int Conf sustainable construction materials and technologies, Ancona, Italy. 2010. p. 1–9.

[57] USGS. Mineral commodity summaries. Washington: United States Geological Survey; 2011.

[58] IEA. Energy balances of non-OECD countries. Paris: International Energy Agency; 2010.

[59] BP. BP statistical review of world energy 2011, http://www.bp.com/statisticalreview.html; 2011 [retrieved 06.06.11].

[60] ASTM C39/C39M. Standard test method for compressive strength of cylindrical concrete specimens. *ASTM* International; 2012.

[61] Australian Standard AS1379. Specifications for the supply of concrete. Standards Australia; 2007.

[62] Australian Standard AS 1012. Methods of testing concrete. Standards Australia; 1993.

[63] Jamieson E, McLellan B, van Riessen A, Nikraz H. Comparison of embodied energies of ordinary Portland cement with Bayer-derived geopolymer products. J Clean Prod 2015;99:112–8.

[64] McLellan BC, Corder GD, Giurco DP, Ishihara KN. Renewable energy in the minerals industry: a review of global potential. J Clean Prod 2012;32:32–44.

[65] IEA. Tracking industrial energy efficiency and CO₂ emissions. Int Energy Agency 2007;34(2):1–12.

[66] Langer W. Sustainability of aggregates in construction. Boca Raton, FL: CRC Press; 2009;1–30.

[67] Alcorn A. Embodied energy and CO2 coefficients for NZ building materials. Centre for Building Performance Research. Victoria University of Wellington; 2001.

[68] Goggins J, Keane T, Kelly A. The assessment of embodied energy in typical reinforced concrete building structures in Ireland. Energy Build 2010;42(5):735–44.

[69] Reddy BV, Jagadish KS. Embodied energy of common and alternative building materials and technologies. Energy Build 2003;35(2):129–37.

[70] Jiao Y, Lloyd CR, Wakes SJ. The relationship between total embodied energy and cost of commercial buildings. Energy Build 2012;52:20–7.

[71] Hammond G, Jones C. Inventory of carbon and energy (ICE). Version 2.0. Sustainable Energy Research Team, Dept. of Mechanical Engineering, University of Bath. 2011. www.bath.ac.uk/mech-eng/sert/embodied.

[72] Hammond G, Jones C. Inventory of carbon & energy: ICE. Bath, UK: Sustainable Energy Research Team, Department of Mechanical Engineering, University of Bath; 2008.

[73] European Commission. Integrated pollution prevention and control (IPPC). Reference document on best available techniques in the Chlor-Alkali manufacturing industry, http://eippcb.jrc.es/reference/BREF/cak_bref_1201.pdf; 2001 [retrieved 03.01.13].

[74] Norgate T, Haque N. Energy and greenhouse gas impacts of mining and mineral processing operations. J Clean Prod 2010;18(3):266–74.

[75] IAI. Life cycle assessment of aluminum: inventory data for the primary aluminum industry. International Aluminum Institute; 2007.

[76] McLellan B. Using transportation to assess optimal value chain configuration for minimal environmental impact. In 2nd world sustainability forum, online, 1–30 November, 2012, Rosen, MA. 2012.

[77] Smith M, Hargroves K, Stasinopoulos P, Stephens R, Desha C, Hargroves S. Engineering sustainable solutions program: sustainable energy solutions portfolio. The Natural Edge Project; 2007.

[78] Milne G, Reardon C. 5.2 Embodied Energy. Australia's guide to environmentally sustainable homes. A joint initiative of the Australian Government and industry, www.yourhome.gov.au/technical/fs52.html; 2010 [retrieved 09.08.12].

[79] Hammond GP, Jones CI. Embodied energy and carbon in construction materials. Proceedings of the Institution of Civil Engineers-Energy; 2008;161(2):87–98.

[80] CGA. Energy efficiency best practice in the Australian aluminum industry: summary report Industry, science and resources; energy efficiency best practice program. A Commonwealth Government Initiative; 2000.

[81] IAI. Personal communications with International Aluminum Institute; 2012.

[82] Jamieson E, Kealley CS, van Riessen A, Hart RD. Optimising Ambient Setting Bayer Derived Fly Ash Geopolymers. Materials 2016;9(5):392.

[83] AS 1141. Methods for sampling and testing aggregates. Standards Australia; 1999.

[84] AS 2758. Aggregates and rock for engineering purposes. Standards Australia; 2009.

[85] CCI. Basic raw materials access and availability 1996–2008. Chamber of Commerce and Industry of Western Australia; 2007.

[86] ABS. 0919 other construction material mining. Australian Bureau of Statistics; 2008.

[87] ABS. 8301 production of selected construction materials. Premixed concrete. Australian Bureau of Statistics; 2013.

[88] Australian Standard AS 1141. Methods for sampling and testing aggregates. Standards Australia; 1974.

[89] Shi C, Roy D, Krivenko P. Alkali-activated cements and concretes. CRC press; 2006.

[90] Zeng F, Lin FB, Subramaniam KV. Experimental investigation on granite masonry behavior under compression Structures congress 2009@ don't mess with structural engineers: expanding our role. ASCE; 2009;1–9.

[91] Brandt AM. Cement-based composites: materials, mechanical properties and performance. CRC Press; 2009.

[92] Shi XS, Collins FG, Zhao XL, Wang QY. Mechanical properties and microstructure analysis of fly ash geopolymeric recycled concrete. J Hazard Mater 2012;237:20–9.

[93] Husem M. The effects of bond strengths between lightweight and ordinary aggregate-mortar, aggregate-cement paste on the mechanical properties of concrete. Mater Sci Eng A 2003;363(1):152–8.

[94] Galvin B, Lloyd N. Fly ash based geopolymer concrete with recycled concrete aggregate. In Concrete 2011. 25th biennial conference of concrete Institute of Australia. 2011. p. 13–4.

[95] Pacheco-Torgal F, Ding Y, Miraldo S, Abdollahnejad Z, Labrincha JA. Are geopolymers more suitable than Portland cement to produce high volume recycled aggregates HPC? Constr Build Mater 2012;36:1048–52.

[96] European Waste Framework Directive 2008/98/EC. Waste Framework Directive, http://ec.europa.eu/environment/waste/framework/; 2008 [retrieved 03.04.13].

[97] Jamieson E. Bauxite Residue Program. Centre for Sustainable Resource Processing, http://asdi.curtin.edu.au/csrp/projects/bauxiteresidue.html; 2008 [retrieved 23.07.15].

# CHAPTER 9

# Alkali-Activated Cement-Based Binders (AACBs) as Durable and Cost-Competitive Low-CO$_2$ Binder Materials: Some Shortcomings That Need to be Addressed

**F. Pacheco-Torgal[1,2], Z. Abdollahnejad[1], S. Miraldo[3] and M. Kheradmand[1]**
[1]University of Minho, Guimarães, Portugal
[2]University of Sungkyunkwan, Suwon, Republic of Korea
[3]University of Coimbra, Coimbra, Portugal

## 9.1 INTRODUCTION

With an annual production of almost 3 Gt ordinary Portland cement (OPC) is the dominant binder of the construction industry [1]. The production of 1 t of OPC generates 0.55 t of chemical CO$_2$ and requires an additional 0.39 t of CO$_2$ in fuel emissions for baking and grinding, accounting for a total of 0.94 t of CO$_2$ [2]. Other authors [3] reported that the cement industry emitted in 2000, on average, 0.87 kg of CO$_2$ for every kilogram of cement produced. As a result the cement industry contributes about 7% of the total worldwide CO$_2$ emissions [4]. The projections for the global demand of Portland cement show that by 2056 it will double, reaching 6 Gt/year [5]. The urge to reduce carbon dioxide emissions and the fact that OPC structures that have been built a few decades ago are still facing disintegration problems points out the handicaps of OPC. Portland cement–based concrete presents a higher permeability that allows water and other aggressive media to enter, leading to carbonation and corrosion problems. The early deterioration of OPC reinforced concrete structures is a current phenomenon with significant consequences both in terms of the cost for the rehabilitation of these structures, and in terms of environmental impacts associated with these operations. Research works carried out so far in the development of alkali-activated cement-based binders (AACBs)

*Handbook of Low Carbon Concrete.*
DOI: http://dx.doi.org/10.1016/B978-0-12-804524-4.00009-9

showed that much has already been investigated and also that an environmentally friendly alternative to Portland cement is becoming more popular [6–8]. However, AACBs still show some shortcomings that need to be addressed so that they can effectively compete against Portland cement. For instance, Zheng et al. [9] mention some AACB problems, namely the difficulty of handling of caustic solutions, poor workability, quality control, and most important, the problem of efflorescences. Heidrich et al. [10] conducted an industry survey in Australia to identify the barriers to the adoption of AACB concrete, and conclude that the fact that this material is not covered by existent Australian standards or any other constitutes the main barrier. Strangely, only 30% of the respondents mention that the cost is a relevant barrier. However, it is important to mention that only 23.1% of the respondents had a detailed knowledge about AACB. The survey also pointed to the need for more research regarding AACB durability. This chapter thus reviews some AACB shortcomings, including its costs and carbon dioxide emissions, and also some durability issues like efflorescences, alkali silica reaction (ASR), and corrosion of steel reinforcement.

## 9.2 AACB COST EFFICIENCY

Currently the cost of AACB concretes is located midway between OPC concretes and high performance concretes [11,12]. These materials only start to become economically competitive compared to OPC concretes with a strength class above C50/60 [13]. Also the average ERMCO concrete class production lies between C25/30 and C30/37 and only around 13% of the concrete ready-mixed production is above the strength class C35/45 [14], which means that currently geopolymer binders are targeting a very small market share. For instance Pacheco-Torgal et al. [15–18] showed that tungsten waste–based AACB mortars can be more cost efficient than current commercial repair mortars. Therefore, in the short term the above-cited disadvantage means that the study of AACB applications should focus only on high-cost construction materials. The authors of Ref. [19] confirm that the high cost of AACB is one of the major factors that still remain a severe disadvantage over Portland cement. These authors suggest that waste-based activators could be used to overcome that gap. McLellan et al. [20] also suggests that the use of less expensive waste feedstocks may reduce AACB costs. Recently, some authors studied [21]. However, these authors did not provide any information regarding the costs of the new waste-based activator. Abdollahnejad et al. [22] recently studied foam fly ash–based two-part (NaOH, NaSiO$_3$) AACBs and reported that the mixtures cost more than 300 euro/m$^3$ (Fig. 9.1).

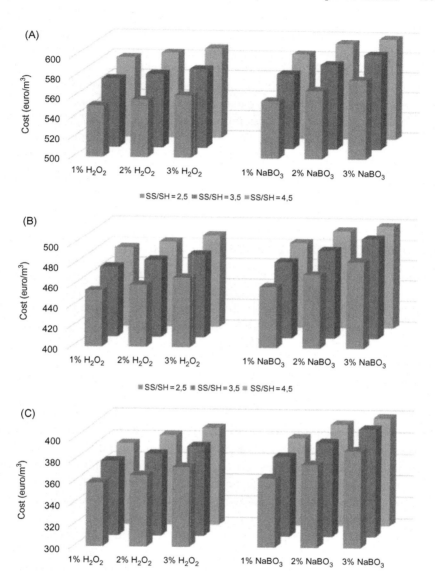

**Figure 9.1** Cost according to activator/binder ratio and sodium silicate/sodium hydroxide mass ratio: (A) activator/binder ratio =1, (B) activator/binder ratio =0.8, and (C) activator/binder ratio =0.6. *Reprinted from Abdollahnejad Z, Pacheco-Torgal F, Félix T, Tahri W, Aguiar A. Mix design, properties and cost analysis of fly ash-based geopolymer foam. Constr Build Mater 2015;80:18–30. Copyright © 2015, with permission from Elsevier.*

Cristelo et al. [23] compared the costs of $3880\,m^3$ of Portland cement and AACB mixtures for jet mix columns and mentioned that the former has an average cost of almost 90% of the latter. However, it is important to emphasize that comparisons should have been made for identical service life assessed by durability parameters. Also these authors made their comparisons against a high-cost Portland cement (type I 42,5 R) that is rarely used for this application. There is no doubt that if they used the less expensive Portland cement type IV/A (V) 32,5 R [24] the cost performance of AACB mixtures would be much less cost competitive. That is why Provis et al. [25] recognized that new activators that allow for cost-efficient AACBs constitute a key aspect that should be further investigated.

## 9.3 CARBON DIOXIDE EMISSIONS OF AACB

Davidovits et al. [26] was the first author to address the carbon dioxide emissions of AACB stating that they generate just $0.184\,t$ of $CO_2$ per ton of binder. Duxson et al. [27] do not confirm these numbers; they stated that although the $CO_2$ emissions generated during the production of $Na_2O$ are very high, still the production of alkali–activated binders is associated to a level of carbon dioxide emissions lower than the emissions generated in the production of OPC. According to those authors the reductions can go from 50% to 100%. Duxson and Van Deventer [28] mention a commercial life-cycle assessment (LCA) conducted by NetBalance Foundation on Zeobond's E–Crete geopolymer, which was compared to standard OPC blends available in Australia in 2007. The binder-to-binder comparison shows an 80% reduction of $CO_2$ emissions while the concrete-to-concrete comparisons show around 60% savings. Such conclusions allow the presentation of E–Crete as a very impressive performer against OPC concretes (Fig. 9.2).

A recent E–Crete geopolymer LCA study [29] used a 100% OPC concrete as the reference concrete although the construction industry uses concrete mixtures with partial replacement of Portland cement by pozzolanic additions. ERMCO [14] reports that the ready-mixed concrete industry in the United States and United Kingdom used 22% of cement additions while some countries like Israel and Portugal used respectively 26% and 28%. Also important is the fact that the study mentioned that a 40-MPa reference OPC concrete requires $440\,kg/m^3$ of Portland cement. However, a similar 40-MPa 28-day compressive strength could easily be achieved with a mixture of just 200-$kg/m^3$ Portland cement type II 42,5

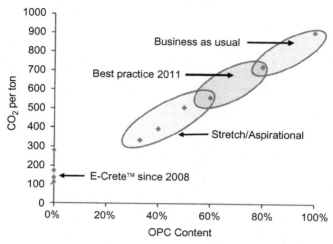

**Figure 9.2** CO$_2$ emissions of various cement binders as a function of OPC content. *Reprinted from Duxson P, Van Deventer JSJ. Geopolymers, structure, processing, properties and applications. In: Provis J, Van Deventer J, editors. Cambridge, UK: Woodhead Publishing Limited Abington Hall; 2009. Copyright © 2009, with permission from Elsevier.*

plus 300-kg/m$^3$ fly ash [30]. The LCA used an OPC with an emissions factor of 0.904 t CO$_2$e/t, which is very far from being the best OPC environmental performance. It also used an alkali activator with a 1.070 t CO$_2$e/t, which does not allow the assessment of which part is from the sodium hydroxide and which part is related to the sodium silicate. Weil et al. [31] confirm that the sodium hydroxide and the sodium silicate are responsible for the majority of CO$_2$ emissions in alkali–activated binders. These authors compared OPC concrete and AACB concrete with similar durability reporting that the latter has 70% lower CO$_2$ emissions. However, these authors' study used 100% OPC concrete and as it was previously mentioned this is not a mix solution used by the construction industry. Habert et al. [19] carry out a detailed environmental evaluation of alkali-activated binders using the LCA methodology, confirming that AACBs have a lower impact on global warming than OPC, but on the other hand, they have a higher environmental impact regarding other impact categories. McLellan et al. [20] reported a 44–64% reduction in greenhouse gas emissions of AACB when compared to OPC. Strangely, other authors [32] who also used Australian-based materials presented very different numbers. They showed that the CO$_2$ footprint of a 40-MPa AACB concrete was approximately just 9% less

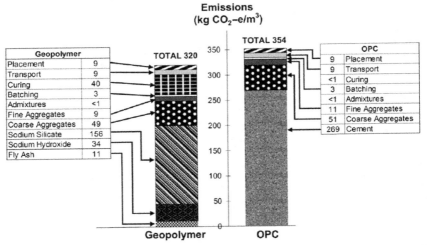

**Figure 9.3** Summary of $CO_2$-e for Grade 40 concrete mixtures with OPC and AACB. *Reprinted from Turner L, Collins F. Carbon dioxide equivalent (CO2-e) emissions: a comparison between geopolymer and OPC cement concrete. Constr Build Mater 2013;43:125–130. Copyright © 2013, with permission from Elsevier.*

than comparable concrete containing 100% OPC binder $(328 \text{kg/m}^3)$. This is much less than the $440 \text{kg/m}^3$ used in the Aurora Construction Materials (ACM) [29] E-Crete LCA. The study of Turner and Collins [32] shows that the major part of geopolymer carbon dioxide footprints is due to sodium silicate (Fig. 9.3). However, it is important to remember (once more) that the construction industry does not use plain 100% OPC concretes; therefore, these mixtures should not be used as a reference comparison. The OPC concrete mixture used in this study could even have a much lower carbon footprint (below the AACB concrete carbon footprint) if fly ash had been used as partial replacement of OPC. A similar 40-MPa 28-day compressive strength could easily be achieved with a mixture of just $200 \text{kg/m}^3$ Portland cement [30]. These results confirm the fact that in some situations AACB can show "an emissions profile worse that of Portland cement-based concretes" as was already recognized by Ref. [7]. More recently, Provis et al. [25] emphasize that AACBs "are not intrinsically or fundamentally 'low-$CO_2$' unless designed effectively to achieve such performance."

Ouellet-Plamondon and Habert [33] confirmed that AACB only has better carbon dioxide emissions when comparisons are made against

100% OPC concrete–based mixtures. These authors mention that only one-part geopolymers show carbon footprint levels much lower than Portland cement, 10–30% when compared to 100% OPC mixtures. One-part geopolymers are considered an important phenomenon in the evolution of low-carbon AACB technology in the "just add water" concept. However, they were associated with low compressive strength [34]. The 2014 investigations of Peng et al. [35] confirm that one-part geopolymers show low mechanical strength. These authors noticed that one-part geopolymer mixtures show an increased reduced compressive strength after being immersed in water. This reduction is dependent on the kaolin thermal treatment. Higher calcination temperatures are responsible for higher compressive losses. Other authors [36] even reported a compressive strength decrease for one-part geopolymers based on calcined red mud and sodium hydroxide blends just after the first week of curing. Abdollahnejad et al. [37,38] investigated one-part geopolymers having obtained relevant compressive strength by using fly ash and minor amounts of OPC. Cristelo et al. [23] compared the carbon dioxide emissions of $3880\,m^3$ of Portland cement and geopolymer-based mixtures for jet-mix columns and mentioned that the AACB solution is responsible for just 77% of the Portland cement–based emissions. These results are only possible because these authors made their comparisons against a high-clinker Portland cement (type I 42,5 R). Also the emission factors that they used for sodium hydroxide and silicate (Table 9.1) are considerably lower than the ones used by Turner and Collins [32], respectively 1915 and $1514\,kg$ $CO_2eq/t$. If they did use those emissions factors they would have to conclude that the AACB-based mixtures had a lower carbon dioxide footprint than the Portland cement–based ones. Strange as it may seem, Poowancum and Horpibulsuk [39] mentioned that AACB is a low-energy-consuming process and does not emit carbon dioxide. This shows the level of misunderstanding about these materials and that is related to the fact that, as it was previously mentioned, AACBs have been advertised as low-carbon footprint materials. However, since Davidovits [40] just mentioned that the carbon footprint calculations of sodium silicate in the paper of Habert et al. (2011) and in the paper of Turner and Collins [32] are wrong because these authors allegedly used the carbon emissions for 100% solid lumps, in place of the actual value of the diluted silicate solution (45% solid), which means that further studies are needed to confirm the real carbon footprint of AACB.

**Table 9.1** Characterization of the activities involved in the production of 3880 m$^3$ of jet mixing columns

| Activity | SC5A1 | | | SA3a | | |
|---|---|---|---|---|---|---|
| | Quant. (kg/m$^3$) | Emission factor (kg CO$_2$-eq/t) (database) | CO$_2$ (eq) (ton) | Quant. (kg/m$^3$) | Emission factor (kg CO$_2$-eq/t) (database) | CO$_2$ (eq) (ton) |
| Materials (prim.)[a] | — | — | 836 | — | — | 630 |
| Cement | 200 | 930 (Sust. conc.) | 720 | — | — | — |
| Steel rebars | 22 | 1351.47[b] | 110 | 22 | 1351.47[b] | 110 |
| Fly ash | — | — | — | 186 | 4 (Sust. conc.) | 3 |
| Water | 100 | 0.3 (AEA, 2012) | 0.1 | 100 | 0.3 (AEA, 2012) | 0.1 |
| Sodium hydroxide | — | — | — | 50 | 999 (Ecoinvent) | 194 |
| Sodium silicate | — | — | — | 75 | 1096 (Ecoinvent) | 319 |
| Energy (prim.) | — | — | 53 | — | — | 53 |
| Diesel | — | 3.6028[c] (AEA, 2012) | 49 | — | 3.6028[c] (AEA, 2012) | 49 |
| Network electricity | — | 0.379285[d] (IEA—CO$_2$ emissions from fuel combustion) | 4 | — | 0.379285[d] (IEA—CO$_2$ emissions from fuel combustion) | 4 |
| Mob/demob (second.) | — | — | 17 | — | — | 13 |
| Freight (second.) | — | — | 54 | — | — | 44 |
| People transp. (second.) | — | — | 4 | — | — | 4 |
| Assets (second.) | — | — | 3 | — | — | 3 |
| Waste (second.) | — | — | 1 | — | — | 1 |

*Source:* Reprinted from Cristelo N, Miranda T, Oliveira D, Rosa I, Soares E, Coelho P, et al. Assessing the production of jet mix columns using alkali activated waste based on mechanical and financial performance and CO$_2$ (eq) emissions. J Clean Prod 2015;102:447–460. Copyright © 2015, with permission from Elsevier.

[a]Only the materials actively contributing to the CO$_2$ (eq) emissions are listed.

[b]The EF value used is a weighted average of the Ecoinvent v2.2 emission factor of the steel rebars (1857 kg CO$_2$-eq/t) and the recycled steel rebars (624 kg CO$_2$-eq/t), considering the 59% and 41% respective shares used.

[c]In this case the EF unit is kg CO$_2$-eq/L.

[d]In this case the EF unit is kg CO$_2$-eq/kWh.

## 9.4 SOME IMPORTANT DURABILITY ISSUES OF AACBs

Duxon et al. [27] state that AACB durability is the most important issue in determining the success of these new materials. Other authors [41] mention that the fact that samples from the former Soviet Union that have been exposed to service conditions for in excess of 30 years show little degradation means that AACBs do therefore appear to stand the test of time. But since those samples were of the (Si +Ca) type that conclusion cannot be extended to geopolymers defined as "alkali aluminosilicate gel, with aluminium and silicon linked in a tetrahedral gel framework" [28]. Juenger et al. [1] argue that "[t]he key unsolved question in the development and application of alkali activation technology is the issue of durability." Also, Van Deventer et al. [8] recognized that "whether geopolymer concretes are durable remains the major obstacle to recognition in standards for structural concrete." Reed et al. [42] stated that the construction industry has not yet fully embraced AACB concrete mainly because the information pertaining to the service life and the durability of AACB concrete applications or infrastructure has yet to be quantified. Scrivener [43] also mentioned that the durability of AACB is not well known. The present section thus reviews three durability issues, namely, efflorescences, ASR, and corrosion of steel reinforcement.

### 9.4.1 Efflorescences

Very few authors have investigated this serious limitation of AACB. Also a search on Scopus journal papers show that the first paper where this problem is mentioned was only published in 2007. Efflorescence is originated by the fact that "alkaline and/or soluble silicates that are added during processing cannot be totally consumed during geopolymerisation" [9]. It is the presence of water that weakens the bond of sodium in the aluminosilicate polymers, a behavior that is confirmed by the Rowles structure model (Fig. 9.4). In the crystalline zeolites the leaching of sodium is negligible, contrary to what happens in the aluminosilicate polymers [45,46]. Recently Skvara et al. [47] showed that Na, K is bounded only weakly in the nanostructure of the AACB (N, K)–A–S–H gel and is therefore almost completely leachable. This confirms that efflorescences are a worrying limitation of AACB when exposed to water or environments with RH above 30%.

Temuujin et al. [48] state that although ambient-cured fly ash AACB exhibited efflorescences, that phenomenon does not occur when the same AACB are cured at elevated temperature, which means the leachate

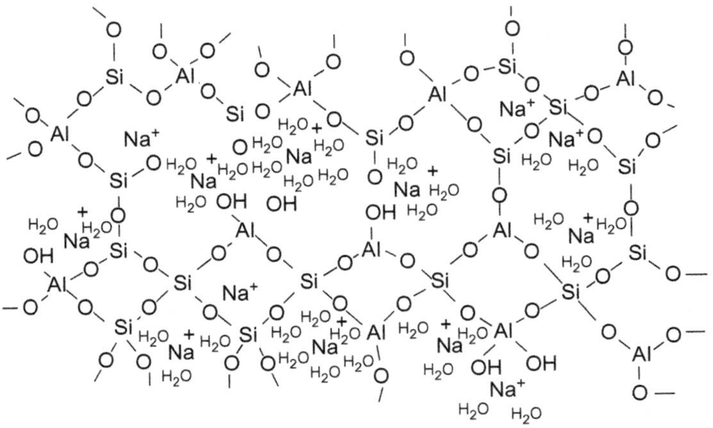

**Figure 9.4** Rowles structure model. *Reprinted from Rowles MR, Hanna JV, Pike KJ, Smith ME, O'Connor BH. 29Si, 27Al, 1H and 23Na MAS NMR study of the bonding character in aluminosilicate inorganic polymers. Appl Magn Reson 2007;32:663–89. Copyright © 2007, with permission from Springer.*

sodium could be a sign of insufficient reaction. Pacheco-Torgal and Jalali [49] found that sodium efflorescences are higher in AACB based on aluminosilicate prime materials calcined at a temperature range below the dehydroxylation temperature with the addition of sodium carbonate as a source of sodium cations (Fig. 9.5). Kani et al. [50] showed that efflorescences can be reduced either by the addition of alumina-rich admixtures or by hydrothermal curing at temperatures of 65°C or higher. These authors found that the use of 8% of calcium aluminate cement greatly reduces the mobility of alkalis, leading to minimal efflorescences. Zhang et al. [51] confirmed that hydrothermal curing can reduce efflorescence. They mentioned that NaOH-activated AACBs possess slower efflorescence than the sodium silicate solution–activated specimens.

According to Fig. 9.6 the lower Na-leaching rate is observed for the NaOH-based mixture (CL1H) while the higher one is related to the sodium silicate AACB (CL2H). Both were cured at 80°C for 90 days. A rather lower leaching behavior is associated with the AACB mixture CL1L made with NaOH and cured at 23°C having just 4.0 MPa at 90 days curing (Table 9.2).

A rather lower leaching behavior is associated to the mixture CL1L made with NaOH and cured at 23°C having just 4.0 MPa at 90 days curing (Table 9.2). A higher leaching behavior is noticed in the mixture CL2H that has a much higher compressive strength (58.4 MPa). These results are

**Figure 9.5** AACB mine-mortar specimens after water immersion. Above mortars based on plain mine-waste mud calcined at 950°C for 2 h. Below mortars based on mine-waste mud calcined at different temperatures with sodium carbonate. *Reprinted from Pacheco Torgal F, Jalali S. Influence of sodium carbonate addition on the thermal reactivity of tungsten mine waste mud based binders. Constr Build Mater 2010;24:56–60. Copyright © 2010, with permission from Elsevier.*

not in line with those of Allahverdi et al. [53] who mentioned that a highest compressive strength is associated with the least tendency for efflorescence formation. These authors also mentioned that slag-containing specimens showed much less and slower efflorescence. Still the role of calcium remains unclear and requires further study. The previous results seem to constitute a step back in the development of AACB. For one, AACBs based only on NaOH solutions without sodium silicate show moderate mechanical strength. Also, the use of hydrothermal curing has serious limitations for onsite concrete placement operations. On the other hand, the use of calcium-based mixtures reduces the acid resistance and raises the chances for the occurrence of the deleterious ASRs. Besides, the use of calcium reduces the global warming emissions advantage over Portland cement.

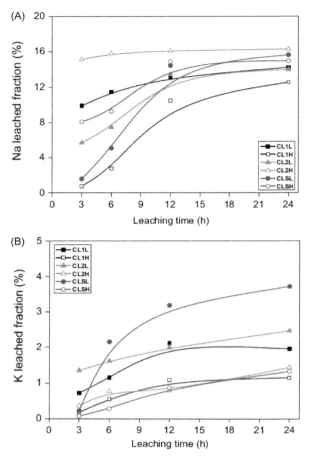

**Figure 9.6** The concentration of Na (A) and K (B) leached from the Callide fly ash mixtures. *Reprinted from Zhang Z, Provis J, Reid A, Wang H. Fly ash-based geopolymers: the relationship between composition, pore structure and efflorescence. Cem Concr Res 2014;64:30–41. Copyright © 2014, with permission from Elsevier.*

## 9.4.2 ASR of AACBs

The chance that ASR may take place in AACBs is still a little-studied subject. For OPC binders, however, the knowledge of ASR has been intensively studied; therefore, some explanations could be also applied to understand the possibility of ASR when AACBs are used. ASR was reported for the first time by Stanton [54] and needs the simultaneous action of three elements in order to occur: (1) enough amorphous silica,

**Table 9.2** Mix proportions and curing conditions of AACBs, and their compressive strengths at 90 days

| Mixtures | Fly ash (g) | Slag (g) | Activator solutions (g) | | Foam (g) | Curing scheme | Compressive strength (MPa) |
|---|---|---|---|---|---|---|---|
| | | | 12 M NaOH | $Na_2O \cdot 1.5SiO_2$ | | | |
| | *Callide* | | | | | | |
| CL1L | 100 | 0 | 23.1 | 0 | 0 | 23°C × 90 days | 4.0 ± 0.3 |
| CL1H | 100 | 0 | 23.1 | 0 | 0 | 80°C × 90 days | 26.2 ± 2.1 |
| CL2L | 100 | 0 | 0 | 35 | 0 | 23°C × 90 days | 53.2 ± 0.9 |
| CL2H | 100 | 0 | 0 | 35 | 0 | 80°C × 90 days | 58.4 ± 12.1 |
| CLSL | 80 | 20 | 0 | 35 | 0 | 23°C × 90 days | 77.4 ± 7.0 |
| CLSH | 80 | 20 | 0 | 35 | 0 | 80°C × 90 days | 58.2 ± 11.2 |

*Source:* Reprinted from Zhang Z, Provis J, Reid A, Wang H. Fly ash–based geopolymers: the relationship between composition, pore structure and efflorescence. Cem Concr Res 2014;64:30–41. Copyright © 2014, with permission from Elsevier.

(2) alkaline ions, and (3) water [55]. The ASR begins when the reactive silica from the aggregates is attacked by the alkaline ions from cement, forming an alkali–silica gel, which attracts water and starts to expand. The gel expansion leads to internal cracking, which has been confirmed by others [56] reporting 4 MPa pressures. Those internal tensions are higher than OPC concrete tensile strength, thus leading to cracking. However, some authors believe that ASR is not just a reaction between alkaline ions and amorphous silica but also requires the presence of $Ca^{2+}$ ions [57]. Davidovits [58] compared AACB and OPC when submitted to the ASTM C227 mortar-bar test, reporting a shrinkage behavior in the first case and an expansion for the OPC binder. Other authors [11,12] reported some expansion behavior for AACB although it was smaller than for OPC binders. However, Puertas [59] believed ASR could occur in slag-based AACB containing reactive opala aggregates. Bakharev et al. [60] compared the expansion of OPC and AACB, reporting that the former had higher expansion. This is clear from the microstructure analysis (Fig. 9.7).

Garcia-Lodeiro et al. [61] showed that fly ash–based AACB is less likely to generate expansion by ASR than OPC. They also showed that the calcium plays an essential role in the expansive nature of the gels. Investigations by Puertas and Palacios [62] show that siliceous aggregates are more prone to ASR than calcareous aggregates in AACB mixtures.

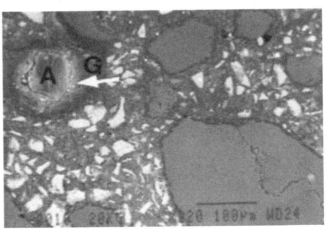

**Figure 9.7** AACB concrete after 10 months curing. Reactive aggregate; G = alkali–silica gel. *Reprinted from Bakharev T, Sanjayan JG, Cheng YB. Resistance of alkali-activated slag to alkali-aggregate reaction. Cem Concr Res 2001;31:331-4. Copyright © 2001, with permission from Elsevier.*

Cyr and Pouhet [63] reviewed the work of several authors concerning the expansion due to ASR (Fig. 9.8) noticing that some mixtures show an expansion above the limit proposed in the standard used for ASR tests. Therefore the study of ASR in AACB is not a closed subject, at least for the AACBs containing calcium.

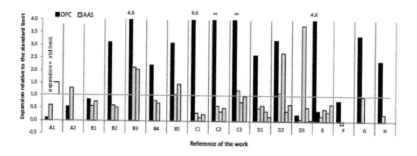

| | Reference | Temperature | Age of test | Explanation of the different AAS mixtures |
|---|---|---|---|---|
| A1, A2 | Bakharev et al. (2001) | 38°C | 1 year | A1: nonreactive aggregate<br>A2: reactive aggregate |
| B1 to B5 | Gifford and Gillott (1996) | 38°C | 1 year | B1: nonreactive aggregate<br>B2: reactive (ASR) aggregate S1<br>B3: reactive (ASR) aggregate S2<br>B4: reactive (ASR) aggregate V<br>B5: reactive (ASR) aggregate B<br>AAS activated with Na$_2$CO$_3$ or Na$_2$SiO$_3$ |
| C1 to C3 | Chen et al. (2002) | 38°C | 180 days | C1, C2 and C3: 2, 3.5, and 5% alkalis, respectively<br>AAS activated with waterglass, NaOH, Na$_2$CO$_3$ or Na$_2$SO$_4$ |
| D1 to D3 | Metso (1982) | 40°C | 70 days | D1, D2 and D3: 3, 8, and 15% of opal, respectively<br>Two types of slag, activated with 1.5 or 2.4% Na |
| E | Al-Otaibi (2007) | 60°C | 1 year | AAS activated with sodium silicate or sodium metasilicate, with Na$_2$O content of 4 or 6%%<br>Two OPC references: doped and non-doped in alkalis |
| F | Fernandez-Jimenez A, Palomo A., Puertas F. (1999) | 80°C | 16 days | |
| G | Puertas et al. (2009) | 80°C | 14 days | |
| H | Wang et al. (2010) | 80°C | 14 days | |

**Figure 9.8** Ratios of the expansion relative to the limit proposed in the standard used for the ASR test, for Portland cement, and slag-based AACB mixtures. *Reprinted from Cyr M, Pouhet R. Resistance to alkali-aggregate reaction (AAR) of alkali-activated binders. In: Pacheco-Torgal F, Labrincha J, Palomo A, Leonelli C, Chindaprasirt P, editors. Handbook of alkali-activated cements, mortars and concretes. Cambridge, UK: WoddHead Publishing; 2014. p. 397–422. Copyright © 2014, with permission from Elsevier.*

### 9.4.3 Corrosion of Steel Reinforcement in AACBs

The corrosion of steel reinforcement is one of the causes that influences the structural capability of concrete elements. As concrete attack depends on its high volume and therefore is not of great concern, an attack on the steel reinforced bars is a serious threat eased by the fact that steel bars are very near to the concrete surface and are very corrosion sensitive. In OPC binders, steel bars are protected by a passivity layer, due to the high alkalinity of calcium hydroxide. The steel bars' corrosion may happen if pH decreases, thus destroying the passivity layer, due to carbonation phenomenon or chloride ingress. The steel corrosion occurs due to an electrochemical action, when metals of different nature are in electrical contact in the presence of water and oxygen. The process consists of the anodic dissolution of iron when the positively charged iron ions pass into the solution and the excess of negatively charged electrons goes to steel through the cathode, where they are absorbed by the electrolyte constituents to form hydroxyl ions. These in turn combine with the iron ions to form ferric hydroxide, which then converts to rust. The volume increase associated with the formation of the corrosion products will lead to cracking and spalling of the concrete cover. For AACB, the literature is scarce concerning its capability to prevent reinforced steel corrosion. Aperador et al. [64] mention that AACB slag concrete (AAS) is associated with poor carbonation resistance, a major cause for corrosion of steel reinforcement. The calculated carbonation rate coefficients were 139 and 25 mm $(year)^{-1/2}$ for AAS and OPC concretes. Fig. 9.9 shows the low corrosion resistance of AACB slag concretes. Other authors also confirmed the low carbonation resistance of AACB mixtures [65,66].

Lloyd et al. [67] showed that AACB is prone to alkali leaching, which could lead to a rapidly and disastrous reduction in the pH, causing steel corrosion. They stated that it is not certain how long a steel-reinforced AACB concrete structure would be able to resist corrosion. They also mention that the presence of calcium is crucial for having durable steel-reinforced AACB concrete because calcium-rich mixtures have much lower diffusion coefficients and a more tortuous pore system that hinders the movements of ions through the paste. Law et al. [68] recently recognized that for chloride-induced attack the long-term protection provided by AACB concrete may be lower than for OPC and blended cement concretes. It is true that as Criado [69] recommends, the use of stainless steel reinforcement could overcome the corrosion problems of AACB concrete;

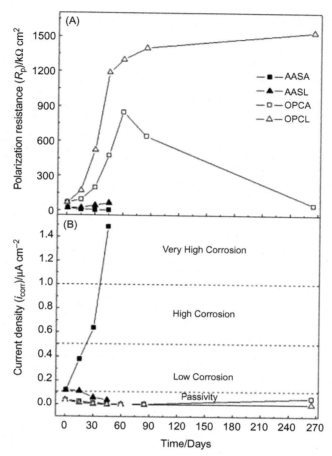

**Figure 9.9** Polarization resistance ($R_p$) and current density ($i_{corr}$) versus time for steel rebars embedded in AAS and OPC concretes with and without exposure to carbonation: AASA and OPCA were exposed to accelerated testing in a cabinet with 3% CO$_2$, 65% RH, and 25°C while AASL and OPCL remain in a laboratory environment with 0.03% CO$_2$, 65% RH, and 25°C. *Reprinted from Aperador W, de Gutiérrez R, Bastidas D. Steel corrosion behaviour in carbonated alkali-activated slag concrete. Corros Sci 2009;51:2027–33. Copyright © 2009, with permission from Elsevier.*

however, since stainless steel is much more expensive than current steel this option will damage the AACB concrete cost competitiveness against OPC concrete. The use of corrosion inhibitors or even the use of concrete coatings may be a more cost-effective option than stainless steel. Still, further studies are needed to confirm this.

## 9.5 CONCLUSIONS AND FUTURE TRENDS

Research carried out so far in the development of AACB showed that much has already been investigated and also that an environmentally friendly alternative to Portland cement is becoming more viable. However, AACBs still show some shortcomings that need to be addressed so that they can effectively compete against Portland cement. This chapter reviewed some AACB shortcomings, including its costs and carbon dioxide emissions, and also some durability issues like efflorescences, ASR, and corrosion of steel reinforcement. Currently the cost of AACB concretes is located midway between OPC concretes and high-performance concretes. These materials only start to become economically competitive compared to OPC concretes with a strength class. In the short term, the above-cited disadvantage means that the study of AACB applications should focus only on high-cost construction materials. The use of activators based on less-expensive waste feedstocks may reduce AACB costs. This constitutes a research area that deserves priority attention. AACBs have been advertised as low-carbon footprint materials; still, no study has ever confirmed the very low emissions (0.184 t of $CO_2$ per ton of binder) found by Davidovits. Some studies even found that OPC and AACB have similar carbon footprints. However, and since Davidovits has mentioned that the carbon footprint calculations of sodium silicate used in those studies are wrong, further studies are needed to confirm the real carbon footprint of AACB. The durability of AACB is the most important issue in determining the success of these new materials; still, some durability issues show some worrying results. So far, very few authors have investigated the efflorescence of AACB, which is originated by the fact that Na, K is bounded only weakly in the nanostructure of these materials and is therefore almost completely leachable. Efflorescence can be greatly reduced by the use of hydrothermal curing and the addition of calcium aluminate. However, the use of hydrothermal curing has serious limitations for onsite concrete-placement operations. On the other hand, the use of calcium-based mixtures reduces the acid resistance and raises the chances for the occurrence of the deleterious ASRs. Besides, the use of calcium reduces the global warming emissions advantage over Portland cement. Although ASR has been intensively studied for OPC concrete, the chance that it also may take place in AACB concrete is still scarcely studied. Since calcium plays a significant role in ASR expansion this could mean that studies on how to prevent ASR in calcium-based AACB are needed. The corrosion of steel reinforcement is one of the causes that influences the structural capability of concrete elements. In OPC binders, steel bars are protected by a passivity layer, due to the high alkalinity of calcium hydroxide. Some studies show that since AACB is prone to alkali leaching, that could lead to a rapidly and disastrous reduction in the pH causing steel corrosion. They also show that the presence of calcium is crucial for having durable steel-reinforced AACB.

## REFERENCES

[1] Juenger M, Winnefeld F, Provis J, Ideker J. Advances in alternative cementitious binders. Cem Concr Res 2011;41:1232–43.

[2] Gartner E. Industrially interesting approaches to low-CO$_2$ cements. Cem Concr Res 2004;34:1489–98.

[3] Damtoft J, Lukasik J, Herfort D, Sorrentino D, Gartner E. Sustainable development and climate change initiatives. Cem Concr Res 2008;38:115–27.

[4] Ali M, Saidur R, Hossain M. A review on emission analysis in cement industries. Renewable Sustainable Energy Rev 2011;15:2252–61.

[5] Taylor M, Gielen D. Energy efficiency and CO$_2$ emissions from the global cement industry. International Energy Agency; 2006.

[6] Pacheco-Torgal F. Introduction. In: Pacheco-Torgal F, Labrincha J, Palomo A, Leonelli C, Chindaprasirt P, editors. Handbook of alkali-activated cements, mortars and concretes. Cambridge, UK: WoddHead Publishing; 2014. p. 1–16.

[7] Provis JL. Geopolymers and other alkali activated materials: why, how, and what? Mater Struct 2014;47:11–25.

[8] Van Deventer JSJ, Provis J, Duxson P. Technical and commercial progress in the adoption of geopolymer cement. Miner Eng 2012;29:89–104.

[9] Zheng D, Van Deventer J.S.L, Duxson P. (2007) The dry mix cement composition, methods and systems involving same. International Patent WO 2007/109862 A1.

[10] Heidrich C, Sanjayan J, Berndt M, Foster S, Sagoe-Crentsil K. (2015) Pathways and barriers for acceptance and usage of geopolymer concrete in mainstream construction. World of Coal Ash (WOCA) conference in Nashville, TN. May 5–7.

[11] Fernandez-Jimenez A, Puertas F. The alkali-silica reaction in alkali-activated slag mortars with reactive aggregate. Cem Concr Res 2002;32:1019–24.

[12] Fernandez-Jimenez A, Palomo J, (2002b) Alkali activated fly ash concrete: alternative material for the precast industry. In proceedings of 2002 geopolymer conference. Melbourne, Australia.

[13] Pacheco-Torgal F. Development of alkali-activated binders using tungsten mine waste mud from Panasqueira mine. PhD Thesis. : University of Beira Interior; 2007.

[14] ERMCO (2014) Statistics of the year 2013. Boulevard du Souverain 68, B-1170 Brussels, Belgium.

[15] Pacheco Torgal F, Gomes JP, Jalali S. Investigations on mix design of tungsten mine waste geopolymeric binders. Constr Build Mater 2008;22:1939–49.

[16] Pacheco Torgal F, Gomes JP, Jalali S. Properties of tungsten mine waste geopolymeric binder. Constr Build Mater 2008;22:1201–11.

[17] Pacheco Torgal F, Gomes J, Jalali S. Adhesion characterization of tungsten mine waste geopolymeric binder. Influence of OPC concrete substrate surface treatment. Constr Build Mater 2008;22:154–61.

[18] Pacheco Torgal F, Gomes JP, Jalali S. Tungsten mine waste geopolymeric binders. Preliminary hydration products. Constr Build Mater 2009;23:200–9.

[19] Habert G, d'Espinose de Lacaillerie JB, Roussel N. An environmental evaluation of geopolymer based concrete production: reviewing current research trends. Journal of Cleaner Production 2011;19(2011):1229–38.

[20] McLellan B, Williams R, Lay J, Van Riessen A, Corder G. Costs and carbon emissions for geopolymer pastes in comparison to ordinary Portland cement. J Clean Prod 2011;19:1080–90.

[21] Puertas F, Torres-Carrasco M, Alonso A. Reuse of urban and industrial waste glass as novel activator for alkali-activated slag. In: Pacheco-Torgal F, Labrincha J, Palomo A, Leonelli C, Chindaprasirt P, editors. Handbook of alkali-activated cements, mortars and concretes. Cambridge, UK: WoddHead Publishing; 2014. p. 75–109.

[22] Abdollahnejad Z, Pacheco-Torgal F, Félix T, Tahri W, Aguiar A. Mix design, properties and cost analysis of fly ash-based geopolymer foam. Constr Build Mater 2015;80:18–30.

[23] Cristelo N, Miranda T, Oliveira D, Rosa I, Soares E, Coelho P, et al. Assessing the production of jet mix columns using alkali activated waste based on mechanical and financial performance and CO2 (eq) emissions. J Clean Prod 2015;102:447–60.

[24] Ribeiro A. Techniques for soil stabilization-jet grouting. Assessment of a real case study-S.Apolonia water front and Jardim do Tabaco. Master dissertation. Technical University of Lisbon; 2010. (only in Portuguese).

[25] Provis JL, Palomo A, Shi C. Advances in understanding alkali-activated materials. Cem Concr Res 2015;78A:110–25.

[26] Davidovits J, Comrie D, Paterson J, Ritcey D. Geopolymeric concretes for environmental protection. ACI Concr Int 1990;12:30–40.

[27] Duxson P, Provis J, Luckey G, Van Deventer JSJ. The role of inorganic polymer technology in the development of "Green Concrete". Cem Concr Res 2007;37(2007):1590–7.

[28] Duxson P, Van Deventer JSJ. Provis J, Van Deventer J, editors. Geopolymers, structure, processing, properties and applications. Cambridge, UK: Woodhead Publishing Limited Abington Hall; 2009.

[29] ACM (2012) LCA of geopolymer concrete (E-Crete). Final report, Aurora Construction Materials (ACM). START2SEE Life Cycle Assessments.

[30] Azevedo F, Pacheco-Torgal F, Jesus C, Barroso de Aguiar JL. Properties and durability of HPC with tyre rubber waste. Constr Build Mater 2012;34:186–91.

[31] Weil M, Dombrowski K, Buchawald A. Life-cycle analysis of geopolymers. In: Provis J, Van Deventer J, editors. Geopolymers, structure, processing, properties and applications. Cambridge, UK: Woodhead Publishing Limited Abington Hall; 2009. p. 194–210.

[32] Turner L, Collins F. Carbon dioxide equivalent (CO2-e) emissions: a comparison between geopolymer and OPC cement concrete. Constr Build Mater 2013;43:125–30.

[33] Ouellet-Plamondon C, Habert G. Life cycle analysis (LCA) of alkali-activated cements and concretes. In: Pacheco-Torgal F, Labrincha J, Palomo A, Leonelli C, Chindaprasirt P, editors. Handbook of alkali-activated cements, mortars and concretes. Cambridge, UK: WoddHead Publishing; 2014. p. 663–86.

[34] Kolousek D, Brus J, Urbanova M, Andertova J, Hulinsky V, Vorel J. Preparation, structure and hydrothermal stability of alternative (sodium silicate free) geopolymers. J Mater Sci 2007;42:9267–75.

[35] Peng M-X, Wan Z-H, Shen SH, Xiao QG. Synthesis, characterization and mechanisms of one-part geopolymeric cement by calcining low-quality kaolin with alkali. Mater Struct 2014;48:699–708.

[36] Ke X, Bernal S, Ye N, Provis J, Yang J. One-part geopolymers based on thermally treated red mud/NaOh blends. J Am Ceram Soc 2015;98:5–11.

[37] Abdollahnejad Z, Hlavacek P, Miraldo S, Pacheco-Torgal F, Aguiar A. Compressive strength, microstructure and hydration products of hybrid alkaline cements. Mater Res 2014;17(4).

[38] Abdollahnejad Z, Pacheco-Torgal F, Aguiar A, Jesus C. Durability of fly ash based one-part geopolymer mortars. Key Eng Mater 2015;634(2015):113–20.

[39] Poowancum A, Horpibulsuk S. Development of low cost geopolymer from calcined sedimentary clay. In: Scrivener, K.Favier A, editors. Calcined clays for sustainable concrete, vol. 10. : RILEM Bookseries; 2015. p. 359–64.

[40] Davidovits, J. Environmental implications of geopolymers. Materials Today, http://www.materialstoday.com/polymers-soft-materials/features/environmental-implications-of-geopolymers/; 2015 [accessed on 15.07.15].

[41] Provis JL, Muntingh Y, Lloyd RR, Xu H, Keyte LM, Lorenzen L, et al. Will geopolymers stand the test of time? Ceram Eng Sci Proc 2008;28:235–48.

[42] Reed M, Lokuge W, Karunasena W. Fibre-reinforced geopolymer concrete with ambient curing for in situ applications. J Mater Sci 2014;49:4297–304.

[43] Scrivener, K. (2015) Practical option for more sustainable cement based building materials. Proceedings of the 1st symposium Knowledge Exchange for Young Scientists (KEYS) 9-23, Dar es Salaam Tanzania.

[44] Rowles MR, Hanna JV, Pike KJ, Smith ME, O'Connor BH. $^{29}$Si, $^{27}$Al, $^{1}$H and $^{23}$Na MAS NMR study of the bonding character in aluminosilicate inorganic polymers. Appl Magn Reson 2007;32:663–89.

[45] Skvara F, Kopecky L, Smilauer V, Alberovska L, Bittner Z. Material and structural characterization of alkali activated low-calcium brown coal fly ash. J Hazard Mater 2008;168:711–20.

[46] Skvara F, Kopecky L, Smilauer V, Alberovska L, Vinsova L. Aluminosilicate polymers—influence of elevated temperatures, efflorescence. Ceramics—Silikaty 2009;53:276–82.

[47] Škvára F, Šmilauer V, Hlaváček P, Kopecký L, Cílová Z. A weak alkali bond in (N, K)–A–S–H gels: evidence from leaching and modelling. Ceramics—Silikáty 2012;56(4):374–82.

[48] Temuujin J, Van Riessen A, Williams R. Influence of calcium compounds on the mechanical properties of fly ash geopolymer pastes. J Hazard Mater 2009;167:82–8.

[49] Pacheco Torgal F, Jalali S. Influence of sodium carbonate addition on the thermal reactivity of tungsten mine waste mud based binders. Constr Build Mater 2010;24:56–60.

[50] Kani E, Allahverdi A, Provis J. Efflorescence control in geopolymer binders based on natural pozzolan. Cem Concr Compos 2011;34:25–33.

[51] Zhang, Z. and Wang, H. and Provis, J.L. and Reid, A. (2013) Efflorescence: a critical challenge for geopolymer applications? In: Concrete Institute of Australia's Biennial National Conference (Concrete 2013): Understanding Concrete, 16–18 Oct 2013, Gold Coast, Australia.

[52] Zhang Z, Provis J, Reid A, Wang H. Fly ash-based geopolymers: The relationship between composition, pore structure and efflorescence. Cem Concr Res 2014;64:30–41

[53] Allahverdi A, Kani N, Hossain K, Lachemi M. Methods to control efflorescence in alkali-activated cement-based materials. In: Pacheco-Torgal F, Labrincha J, Palomo A, Leonelli C, Chindaprasirt P, editors. Handbook of alkali-activated cements, mortars and concretes. Cambridge, UK: WoddHead Publishing; 2014. p. 463–83.

[54] Stanton TE. Influence of cement and aggregate on concrete expansion. Eng News Record 1940;1:50–61.

[55] Sims I, Brown B. (1998) Concrete aggregates. In: Hewlett PC, editor. Lea's Chemistry of cement and concrete. 4th ed. London. p. 903–989.

[56] Wood JG, Johnson RA. The appraisal and maintenance of structures with alkali-silica reaction. Struct Eng 1993;71(2).

[57] Davies D, Oberholster RE. Alkali-silica reaction products and their development. Cem Concr Res 1988;18:621–35.

[58] Davidovits J. Geopolymers: inorganic polymeric new materials. J Therm Anal 1991;37(1991):1633–56.

[59]  Puertas F. Cementos de escórias activadas alcalinamente. Situacion actual y perpectivas de futuro. Mater Constr 1995;45:53–64.

[60]  Bakharev T, Sanjayan JG, Cheng YB. Resistance of alkali-activated slag to alkali-aggregate reaction. Cem Concr Res 2001;31:331–4.

[61]  García-Lodeiro I, Palomo A, Fernández-Jiménez A. Alkali–aggregate reaction in activated fly ash systems. Cem Concr Res 2007;37(2007):175–83.

[62]  Puertas F, Palacios M, Gil-Maroto A, Vázquez T. Alkali-aggregate behaviour of alkali-activated slag mortars: effect of aggregate type. Cem Concr Compos 2009;31:277–84.

[63]  Cyr M, Pouhet R. Resistance to alkali-aggregate reaction (AAR) of alkali-activated binders. In: Pacheco-Torgal F, Labrincha J, Palomo A, Leonelli C, Chindaprasirt P, editors. Handbook of alkali-activated cements, mortars and concretes. Cambridge, UK: WoddHead Publishing; 2014. p. 397–422.

[64]  Aperador W, de Gutiérrez R, Bastidas D. Steel corrosion behaviour in carbonated alkali-activated slag concrete. Corros Sci 2009;51:2027–33.

[65]  Bernal S, de Gutiérrez R, Pedraza A, Provis J, Rose V. Effect of silicate modulus and metakaolin incorporation on the carbonation of alkali silicate-activated slags. Cem Concr Res 2010;40:898–907.

[66]  Bernal S, de Gutiérrez R, Pedraza A, Provis J, Rodriguez E, Delvasto S. Effect of binder content on the performance of alkali-activated slag concretes. Cem Concr Res 2011;41:1–8.

[67]  Lloyd R, Provis J, Van Deventer SJS. Pore solution composition and alkali diffusion in inorganic polymer cement. Cem Concr Res 2010;40:1386–92.

[68]  Law D, Adam A, Molyneaux T, Patnaikuni I, Wardhono A. Long term durability properties of class F fly ash geopolymer concrete. Mater Struct 2015;48:721–31.

[69]  Criado M. The corrosion behaviour of reinforced steel embedded in alkali-activated mortar. In: Pacheco-Torgal F, Labrincha J, Palomo A, Leonelli C, Chindaprasirt P, editors. Handbook of alkali-activated cements, mortars and concretes. Cambridge, UK: WoddHead Publishing; 2014. p. 333–72.

# CHAPTER 10

# Progress in the Adoption of Geopolymer Cement[*]

## J.S.J. Van Deventer[1,2]
[1]University of Melbourne, Melbourne, VIC, Australia
[2]Zeobond Pty. Ltd., Melbourne, VIC, Australia

## 10.1 INTRODUCTION

Under the 2009 Copenhagen Accord, more than 120 countries have agreed to keep global average temperature increase below 2°C. This maximum acceptable temperature increase is based on recommendations from numerous scientific studies, warning that increases in excess of 2°C can trigger dangerous anthropogenic interference with the climate system, including climate-tipping points, with unmanageable consequences to water supply, agricultural productivity, sea-level rise, human habitability, and global security [1]. This is an urgent goal, because many scientists estimate that the concentrations of $CO_2$ and other climate-forcing substances in the atmosphere already exceed the safe level [2]. In contrast, there remain many skeptics, including in the minerals industry, which do not accept that climate change is a result of human activity [3]. This chapter does not aim to contribute to such debate, but instead assumes that there is economic and social benefit in reducing $CO_2$ emissions and valorizing waste materials at the same time, and develops discussions from this viewpoint.

Concrete made from Portland cement, including its blends with mineral admixtures, is second only to water as the commodity most used by mankind today. Global cement production in 2008 was around 2.6 billion tons [4], contributing conservatively 5–8% of global anthropogenic $CO_2$ emissions [5,6], and the rapidly increasing demand for advanced civil infrastructure in China, India, the Middle East, and the developing world is expected to expand the cement and concrete industries significantly [7]. $CO_2$ emissions are due mainly to the decomposition of limestone and combustion of fossil fuels during cement production; grinding and

---

[*]This chapter is an updated version of the article: Van Deventer JSJ, Provis JL, Duxson P. Technical and commercial progress in the adoption of geopolymer cement. *Miner Eng* 2012;29:89–104.

*Handbook of Low Carbon Concrete.*
DOI: http://dx.doi.org/10.1016/B978-0-12-804524-4.00010-5

transport are also significant contributors to the environmental footprint of the cement industry.

The cement industry has started to acknowledge the role of alternative binders in a carbon-constrained industry, given that there are significant reductions in $CO_2$ emissions and also advantages in performance only offered by these alternative binding systems [8,9]. Historically, the driver for competition in the construction-materials industry has been cost reduction, in which case alternative binders, starting from a low-volume basis, could never compete against large-scale Portland cement production. However, $CO_2$ abatement and technical features are now playing a major role in the growth of alternative binder systems. Juenger et al. [10] recently presented a review of potential alternatives to Portland cement technology, including calcium sulfoaluminate cements, magnesium cements, the magnesium phosphate system, and alkali-activated materials or geopolymers. In this chapter, the term "geopolymer," which was coined by Joseph Davidovits [11] for a certain class of alkali-activated aluminosilicates, will be used here for all alkali-activated materials.

Calcium sulfoaluminate cements are made from clinkers that include ye'elimite as a primary phase, which requires a lower amount of limestone and lower fuel consumption. Their commercial use in expansive cements and ultrahigh early-strength cements has been pioneered in China. Magnesium-based cements and magnesium phosphate cements have been used in niche applications and can also give superior fire resistance, with much lower $CO_2$ emissions than Portland cement. The use of magnesium silicate hydrates together with magnesium carbonate as a "carbon-negative concrete" is also attracting commercial attention at present [12]. However, many of these alternative binders require a new supply chain for raw materials, the development of new chemical admixtures, regulatory approval, the development of new durability testing protocols, and an in-service track record before they would be adopted widely by industry.

Geopolymer cement faces the same obstacles, but has a longer in-service track record [13–16], supported by an expanding body of fundamental research relating gel chemistry and nanostructure to durability. In geopolymer chemistry, the reactive aluminosilicate phases present in precursors including fly ash from coal combustion, metallurgical (including blast furnace) slag, calcined clays, volcanic ash, and/or reactive natural materials are reacted with alkaline reagents including alkali metal silicates, hydroxides, carbonates, and/or aluminates [17,18] to form aluminosilicate gel phases with varying (but generally low) degrees of crystalline zeolite formation. Geopolymer concrete has been widely reported to display high resistance to fire and acids, and does not produce the high evolution of reaction heat associated with Portland concrete, reducing cost and potential cracking issues when the material is placed in large volumes. The benefits of geopolymerization, when compared with Portland technology, are largely based around the ability to valorize high-volume industrial waste streams into high-performance concretes, with a highly significant reduction in $CO_2$ emissions [19]. Fly ash and slag appear at present to be the most promising precursors for large-scale industrial production of geopolymer cement due to the more favorable rheological properties and lower water demand achievable when compared to mixes based on calcined clays [20].

The history, chemical principles, reaction phenomena, and engineering properties of geopolymer concrete have been reviewed extensively [11,17,18,21–27]. With the core focus of these recent reviews being laboratory research on geopolymer binders rather than the production of concrete, it is not the aim of this review to duplicate such analysis or discussion. Instead, this chapter will fill a gap in the current literature by explaining the relevance of particle technology in geopolymer concrete design, linking synchrotron-based and other nanostructural characterization of geopolymers to durability and engineering properties, identifying the technical, commercial and regulatory barriers to industrial adoption, and reviewing progress made in Australia along the path of commercialization. This will be addressed from a joint research–commercialization viewpoint, with a particular focus on the areas of research that are specifically beginning (or continuing) to impact developments in the commercial arena. The underlying assumption in research papers and grant applications is that good research will necessarily lead to adoption in industry, which is far from the situation in reality. This chapter will use an experiential and tutorial style to show that the interplay between research problem identification, technical development work, and commercial strategy is as important as high-quality fundamental research, and that both are required in the implementation of a new class of binders on a large scale.

## 10.2  THE ROLE OF CHEMICAL RESEARCH IN THE COMMERCIALIZATION OF GEOPOLYMERS

The commercial implementation of geopolymer technology in Australia is currently being driven by multiple teams operating in different parts of the country. Curtin University in Perth has made advances in this area over the past decade and conducted a number of trial pours in recent years. The Melbourne-based company, Zeobond, developed its own pilot-scale production facilities in 2007 and has now licensed its E-Crete technology for application in major civil infrastructure projects including freeway expansion works, bridge construction, railway structures, and structural elements in buildings. Significant advances have also been made recently in South Africa, where Murray & Roberts Construction has built several structures, including a high-rise building in Cape Town utilizing the principles of geopolymerization. E-Crete utilizes for its binder mostly a blend of fly ash and ground blast furnace slag, with combinations of proprietary alkaline-activating components that are tailored for specific raw materials and products.

There are various technical and commercial factors driving the commercial adoption of geopolymer technology; demand pull, led by a carbon-conscious end-user market, continues to be the key driver for the short-term adoption of geopolymer concrete in Australia. Counterbalancing this demand driver is the inherent resistance of the civil construction industry to the large-scale adoption of new products, where time and demonstration on an industrial scale are prerequisites for practical credibility, as well as the cost implications of the absence of economies of scale. Nevertheless, there is now an increasing openness in the market internationally to consider alternative binders, especially when the driver is to utilize large volumes of waste such as fly ash in India, thereby reducing production cost, or where there is a need for construction materials with advanced properties like superior fire resistance.

A detailed chemical understanding of the properties of geopolymer binders, in particular in areas such as control of setting time, workability, and durability, also plays an enabling role in the commercialization process. There is a growing volume of scientific literature exploring the properties of geopolymeric materials at the laboratory scale. Unfortunately, much of this information has limited direct value in commercial adoption; geopolymer concrete that performs adequately according to all standards can be synthesized easily in a laboratory, while it is substantially more difficult to reproduce such performance in a commercially and practically feasible form in real-world applications. Indeed, closing this apparent disparity and gap between the laboratory and the real-world has been the focus of much research conducted by the author and associates. Being able to achieve this challenging goal has thus unlocked the commercial value of geopolymer technology.

## 10.3 DEVELOPMENTS IN GEOPOLYMER GEL-PHASE CHEMISTRY

This section summarizes selected research advances in geopolymer gel chemistry and also identifies topics that require further work, as depicted in Fig. 10.1.

### 10.3.1 Precursor Design

Based on data obtained from the literature, Duxson and Provis [28] proposed an ideal composition range for glassy aluminosilicate precursors containing network–modifying cations (particularly calcium, magnesium,

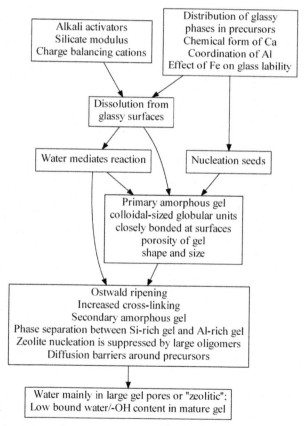

**Figure 10.1** Conceptual diagram showing geopolymer gel-phase processes. *Reproduced from Van Deventer JSJ, Provis JL, Duxson P. Technical and commercial progress in the adoption of geopolymer cement. Miner Eng 2012;29:89–104.*

sodium, and potassium) in order to give sufficiently high solubility to supply the necessary aluminum into the growing geopolymer gel. This control of aluminum availability has been highlighted as being necessary to enable the growth of geopolymer gel particulates, leading to cross-linking, hardening and strength development [29,30]. The need for addition of a separate alkali source may be either greatly reduced or even eliminated if the correct glass, or combination of glasses, can be selectively synthesized [31]. Duxson and Provis [28] postulated that this may be achieved by the addition of components into pulverized coal prior to combustion, or by the selective manufacture of a highly reactive raw material [32] that can be blended with less-reactive raw materials to provide a geopolymer cement to a given specification. Provis et al. [17] reviewed recent work on the use of alternative precursors in geopolymers.

Keyte [33] determined that the aluminosilicate glass particles present within Class F fly ashes predominantly consisted of two intimately intermixed amorphous phases, one silica-rich and one very alumina-rich, with composition close to $Al_6Si_2O_{13}$ and resembling mullite. By applying devitrification to geopolymers, Keyte [33] showed that the poor performance of certain fly ashes in geopolymer synthesis is related to the low aluminum content of the amorphous phases present in the precursor, resulting in a high silicon-to-aluminum ratio in the formed geopolymer. The coordination environment of aluminum in geopolymer precursors is a complex issue, and plays a key role in controlling the availability of Al in geopolymerization; this is an area of active research and debate [34].

Clearly, there is a need for a workable one-part ("just add water") mix design if geopolymeric cements and concretes are to achieve widespread market penetration, as this would greatly simplify the process of handling and distributing the components of a geopolymer binder compared to the use of two-part systems involving large volumes of alkaline liquids. Various methodologies for designing the necessary precursors have been proposed [28,32,35,36], but its successful commercial implementation still remains to be demonstrated.

## 10.3.2 Binder-Phase Chemistry

By studying the remnant fly ash and slag particles left embedded in a hardened geopolymer binder using scanning electron microscopy (SEM), the authors of Refs. [37,38] observed that calcium is active during alkali activation, and that, if discrete high-calcium binder regions are formed (rather than the calcium being incorporated into the main aluminosilicate gel binder), it is on a nanometer rather than a micron-length scale. Yip et al. [39,40] suggested that both geopolymeric sodium aluminosilicate hydrate (N-A-S-H) gel and calcium silicate hydrate (C-S-H) gel coexist at low alkalinities, while geopolymeric gel appeared to be the dominant product at high alkalinities, while Buchwald et al. [41] also showed gel coexistence at relatively high alkalinity. Recent work from García-Lodeiro et al. [42] shows different trends with respect to alkalinity, with C-A-S-H gel proposed to be more stable than N-A-S-H at high pH and further work is certainly required in this area to resolve the remaining disagreements. Provis et al. [43] did identify discrete Ca-rich regions within a hydroxide-activated Class F fly ash binder using 80-nm-resolution X-ray fluorescence microscopy; these regions were not observed in silicate-activated samples, and were proposed to be related to the precipitation of $Ca(OH)_2$ in poorly crystalline form during the very early stages of reaction of the fly ash particles in highly alkaline environments, in contrast to its wider distribution throughout the gel formed at lower alkalinity.

Using the results of synchrotron X-ray diffraction, Oh et al. [44] proposed that the geopolymeric gel contains zeolitic precursors related to disordered forms of the ABC-6 family of zeolites, depending on the pH environment of the paste. Despite a high calcium content, only very weak C-S-H(I) peaks and no $Ca(OH)_2$ phase were found in their Class C fly ash paste when activated with sodium hydroxide solution. This implies that the calcium in Class C fly ash did not dissolve in the activator solution as readily as the calcium in the slag; slag-based systems yielded much higher levels of crystalline C-S-H(I). The high calcium content in the Class C fly ash seemed to reduce the strength of the matrix, while the calcium in the slag appeared to increase strength. Oh et al. [44] postulated that the calcium in the slag is available to form C-S-H(I) while it is not available in the Class C fly ash "due to the different chemical forms of calcium in these raw materials." In summary, the precise role of calcium in geopolymer formation remains poorly defined, although it is clearly pivotal in determining the engineering properties and durability of the product.

Iron is also likely to be important in geopolymers. Lloyd et al. [37] observed that a high iron content appears to render precursor particles relatively unreactive, and noted that phase segregation between iron-rich and iron-poor glasses within particles can mean that compositional information obtained even at an individual particle level is not necessarily able to describe reactivity in geopolymer formation. During alkali activation, iron does not appear to move much from its original position within fly ash particles [38,43]. Bernal et al. [45] recently used synchrotron nanoprobe X-ray fluorescence maps to show discrete iron-rich, titanium-rich and manganese/silicon-rich particles present in blast furnace slag grains, and that these particles remain intact when the slag is alkali activated. These particles appear to be entrained during slag production, and remain stable under the reducing conditions prevailing during alkali activation. There is no evidence of chemical interaction between these particles and the C-A-S-H gel. However, the effect of iron on the lability of precursor glassy phases and the role that iron may play during geopolymer reactions remain largely undescribed; Mössbauer spectroscopy does appear to be giving some initial answers in this area, at least for the case of very Fe-rich precursors [46].

According to Provis and Bernal [26], magnesium incorporation into C-A-S-H-type structures is very limited because the ionic radius of $Mg^{2+}$ is not a good match for the $Ca^{2+}$ sites in the tobermorite-type structure. Nevertheless, it appears that the presence of higher levels of MgO in alkali activated blast furnace slag can enhance the strength of the binder [47,48], which appears to be linked to the formation of hydrotalcite-type phases. Bernal et al. [49] showed that these hydrotalcite phases increase resistance to carbonation. This beneficial effect of hydrotalcite, which itself is not believed to be a particularly strength-giving phase, that may be related to a reduced level of aluminum incorporation into the C-A-S-H gel [26]. Increased $Al_2O_3$ content in the slag reduces the extent of reaction at early times of curing and consequently decreases the compressive strength of alkaliactivated slag binders [26], although the specific effect of aluminum incorporation on the mechanical properties of C-A-S-H-type gels is not yet clear.

The mechanistic distinction between silicate-activated and hydroxideactivated geopolymer formation can be attributed to some extent to the differences in the sites at which gel precipitation takes place, with hydroxideactivated gels forming predominantly on fly ash particle surfaces rather than by polymerization in the bulk region [38]. In silicate-activated fly ash, the gel consisted of roughly similar colloidal-sized, globular units closely bonded together at their surfaces. The gel appeared to be homogenous throughout the sample, whether at the surface of the ash particles or relatively far away in the interstitial space. It is, however, possible to influence this through nanoparticle seeding of hydroxide-activated binders to manipulate the nucleation location and enhance gel growth [50]. This is an area that deserves further research with potentially important commercial outcomes.

The reason why this geopolymer gel is formed instead of large zeolitic crystallites is related to the degree of polymerization of silicate species, as well as the influence of temperature and water content [18,30,51,52]. White et al. [53] used neutron pair distribution function (PDF) analysis to show that, with increasing reaction time, the geopolymer gel derived from metakaolin transitions to a more ordered state via an increase in cross-linking. Following the initial nuclear magnetic resonance and thermal analysis data of Duxson et al. [54,55], a neutron PDF study of heated geopolymer gels [56] confirmed that the water in metakaolin geopolymers was present mainly as free water in large pores, with only a small percentage (approximately <5%) of water either physically bound in small pores or chemically bound as hydroxyl groups attached to the aluminosilicate framework structure.

Coarse-grained Monte Carlo simulations, based on interaction energies derived from density functional modeling, showed that, as the activator silica content of a metakaolin geopolymer system is increased, more species (both aluminate and silicate) participate in geopolymerization, leading to a denser nanostructure and less monomeric species existing in the pore solution [57]. The precipitates formed become larger with increasing silica content, which indicates that different structural transformation mechanisms occur depending on the type of activator used. The results of these simulations have confirmed a previously proposed hypothesis [52] regarding nucleation of precipitates in metakaolin-based systems, where in the presence of silicate-activating solutions the early release of aluminum from metakaolin was suggested to lead to localized nucleation close to the surface of the partially dissolved metakaolin particles; this behavior was reflected in the simulation results. On the other hand, in hydroxide-activated systems there is no localized nucleation taking place, and therefore the precipitates form throughout the system. This differs from the nucleation behavior observed in fly ash systems, as outlined above; the aluminum release rates and dissolution mechanisms differ significantly between the two systems, and this leads to differences in the influence of dissolved silica in the nucleation processes. The simulation results for silicate-activated systems also provide direct evidence of Ostwald ripening of geopolymer gel particles, which has never before been explicitly shown to occur in geopolymer systems [57].

These differences in particle-dissolution mechanisms, and their influence on geopolymer gel development, are essential in determining the processes that control the rate and location of geopolymer gel formation, and thus the microstructure and performance of the final geopolymer binder. By using attenuated total reflectance Fourier transform infrared (ATR-FTIR) spectroscopy, Rees et al. [58,59] observed that in a sodium hydroxide–activated Class F fly ash system selective leaching of Al from the fly ash produces in the first instance a loose, Al-rich, "primary" gel, as proposed earlier by Fernández-Jiménez and Palomo [60]. The Al-deficient surface layer on the fly ash particles then dissolves, followed by later stoichiometric release of Al and Si species. During an induction period, the gel slowly comes to pseudoequilibrium with the surrounding solution via depolymerization/repolymerization reactions [57]. Gel nuclei (particles that are sufficiently stable to resist depolymerization) begin to form, and the growth of a new gel phase begins. This new gel is the phase predominantly responsible for strength development and durability in geopolymers.

When nanoparticle seeds are added to the geopolymer system, no induction period occurs, as the nanoparticles immediately catalyze the formation of nuclei [50]. In this case, when the first Al-rich species are released into the solution from the fly ash particles, their immediate addition to the nuclei forms an Al-rich gel. Structural reorganization of this gel will later lead to the formation of zeolites, which can differ in crystal structure from the zeolites that develop in unseeded systems [50]. Dissolution of the remnant siliceous layer on the fly ash particles releases Si-rich species, which also add rapidly to the growing nuclei, forming a high-silica gel region. Congruent ash dissolution then releases Al and Si species, which add to the growing nuclei in a similar way to the unseeded system, creating a bulk geopolymer gel with similar composition and structure to that observed in the absence of seeds. The authors of [61–63] continued with this work by applying synchrotron radiation-based FTIR microscopy to the spatially resolved analysis of both one-part and two-part geopolymer systems, showing that the release rates of both Si and Al are critical in determining strength development and microstructural evolution in growing geopolymer gels. The enhancement of the nucleation process through seeding led to the formation of an additional silica-rich phase in the early stages of the reaction, which left more Al available to contribute to bulk gel formation and improved the early strength development of geopolymer binders. The later release of more silica from the Si-rich phase then also enhanced final strength.

## 10.3.3 Modeling of Phase Assemblage

Provis et al. [17] explain that substantial progress has been made recently in modeling the phase assemblage in geopolymers. This is beneficial as it is challenging to unravel this complex system by experimentation alone. Also, the availability of appropriate structural models is essential in enabling the correct interpretation of spectroscopic data for C-A-S-H and N-A-S-H gels. Myers et al. [64] implemented a structurally based framework by which cross-linking degrees can be calculated in tobermorite-like C-A-S-H gels, and Richardson [65] also provided a crystal chemical model to describe layer spacings in these gels.

Geochemical-type thermodynamic models have been successful in describing the phase assemblages formed in Portland cement and other Ca-rich systems, but until recently have not fully described the alkali metals and Al, which are known to play important structural roles in the C-(N)-A-S-H gels in geopolymers [17]. This gap has been filled by the thermodynamic model of Myers et al. [66], which describes this phase as an ideal eight-member solid solution, and can accurately describe solubilities in the full quaternary $CaO-Na_2O-Al_2O_3-SiO_2$ aqueous system, as well as the pore-solution chemistry of high-Ca geopolymers. This provides confidence in the long-term phase stability of the C-(N)-A-S-H gel and accompanying secondary phases, as the observed phase assemblages are consistent with predictions made from a thermodynamic basis [17]. Unfortunately, such models are not yet available for N-A-S-H type gels, as the thermodynamics of this type of gel are much less well defined. It is envisaged that further developments in modeling of phase assemblage will aid in the prediction of durability and in-service life for geopolymers.

White et al. [67] used neutron PDF analysis to show differences in ordering between the C-A-S-H gels formed by hydration of tricalcium silicate and in alkali activation of blast furnace slag. Provis et al. [17] ascribe this mainly to the low Ca/Si ratio and the high Al content of the gel produced in slag-based geopolymers, which increases the likelihood of cross-linking between the silicate chains within the tobermorite-like gel [64]. Incorporation of alkali metal cations into the gel structure is important, and this may provide a partial explanation for the differences in PDF analysis results between slag-based geopolymers and tricalcium silicate hydration products. Incorporation of alkali cations into the interlayer space of the C-A-S-H gel is inducing additional structural disorder in the gel [17].

## 10.4 ROLE OF PARTICLE TECHNOLOGY IN THE OPTIMIZATION OF GEOPOLYMER PASTE AND CONCRETE

Fig. 10.2 depicts a conceptual model for the interrelationship between geopolymer concrete mix design, the behavior of wet concrete, and the performance of in-service concrete regarding engineering properties and durability. In this section the principles of particle technology will be used to analyze the behavior of fresh geopolymer cement paste and wet concrete, while in Section 10.5 recent progress on the interrelationship between binder microstructure and durability will be reviewed.

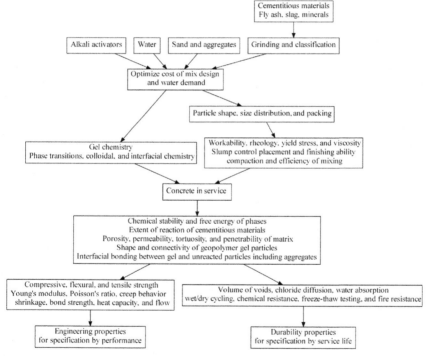

**Figure 10.2** Conceptual diagram showing interrelationships between mix design, gel chemistry, matrix characterization, engineering properties, and durability of geopolymer concrete. *Reproduced from Van Deventer JSJ, Provis JL, Duxson P. Technical and commercial progress in the adoption of geopolymer cement. Miner Eng 2012;29:89–104.*

## 10.4.1 Particle-Shape Effects in Fresh Pastes

The grinding of clinker and blast furnace slag into cementitious powders results in nonspherical particle morphologies. It has been hypothesized that the spherical particle shape of fly ash reduces viscosity and yield stress of fresh paste when fly ash is added to Portland cement [68]. Provis et al. [20] used a packing model to demonstrate the "ball-bearing" effect of spherical ash particles in a paste, by the reduction of particle interlocking when the paste is sheared. Kashani et al. [69] showed that addition of fly ash to a concrete mix can reduce the yield stress due to its broad particle-size distribution. Palomo et al. [70] observed that ash chemistry and variability largely affected the rheology of fly ashes activated by sodium hydroxide solutions. Various studies have focused on the effect of ash fineness on early-age properties of Portland and geopolymer pastes. However, many of these investigations used milling to obtain different particle-size distributions [71–73], which adds the complication of a change in particle shape as the glassy spherical ash particles are shattered into fragments during the milling process.

Kumar et al. [72] showed that slightly higher geopolymer strengths were achievable by use of the finer particle-size fractions obtained by classification or attrition milling, when compared to the use of raw fly ash, but that vibratory milling was able to give an improvement in strength of as much as 50%. These effects could not be attributed solely to particle-size reduction, given that the vibratory-milled fly ash was not the finest of the ash samples studied, and a mechanochemical activation process was postulated to enhance reactivity. In contrast, Keyte [33] did not show a significant effect of either classification or ring milling on the compressive strength of geopolymers.

Clearly, the effect of particle geometry on the behavior of fresh geopolymer pastes is complex, so there is a need to deconvolute the effects of particle shape, particle size, mechanochemical phenomena, and fly ash chemistry. Some detailed studies in this area have used synthetic aluminosilicate glasses [33,74]. However, to fully simulate the geopolymerization of waste materials, it is also necessary to incorporate nonframework cations into these synthetic glasses to better reflect the main reactive phases participating in alkaline activation [28]. Some of the knowledge of mixed alkali/alkaline earth aluminosilicate glasses developed recently through the study of synthetic slags [75,76] should also be relevant in future developments in this area.

## 10.4.2 Water–Binder Ratio and Rheology of Geopolymer Pastes

The pore volume of calcium silicate-based (Portland cement), aluminate-based (high-alumina cement) or sulfate-based (supersulfated or calcium sulfoaluminate cement) binders decreases significantly as they hydrate over the first few days to weeks after mixing, with this reduction in pore volume attributable to the effective consumption of water as it partici-pates in hydration reactions. A water to binder (w/b) ratio of approxi-mately 0.2, where the exact value depends on the cement composition [77], is converted during the hydration of Portland cement to "non-evaporable" water in hydration products, in addition to the consider-able amount of water bound in "gel pores" (less than $\sim$2.7 nm diameter), which is also not readily removed from the gel. Thus, hydrated Portland cement with a nominal water/binder ratio of 0.5 has a final pore volume that is markedly less than the volume of the water that was initially added into the mix [78].

In contrast, geopolymeric binders are primarily aluminosilicate-based and do not form hydrate products, as discussed in Section 10.3.2. Hence, geopolymers do not normally have the same pore-volume reduction ben-efit through the conversion of water into a solid via its incorporation into reaction products. This distinction has significant implications for the development of geopolymer concrete. While in geopolymer binders the pore-size distribution is much finer, and is greatly refined throughout the hardening process [79–81], the absolute pore volume is not affected by the formation of hydration products and there is no significant amount of water bound into the solidified geopolymer gel [82]. Consequently, the min-imization of w/b ratio in geopolymer concrete is a particularly significant operational factor in ensuring quality low-permeability concrete.

## 10.4.3 Particle Packing and Mix Design in Geopolymer Concretes

Besides particle shape, particle-size distribution also has a marked effect on the mixing, workability, and rheology of wet concrete. This is an active area of research in which many fundamental questions remain unanswered [83], and the development of advanced analytical and simulation tech-niques is continuing to provide important advances. Because this is based around the analysis of the locations and interactions of particles on every length scale from a few nanometers (the basic building units in C-S-H or geopolymer gel) to more than a centimeter (coarse aggregate), it is not a problem that has an easy or straightforward answer. It is very difficult to develop a unified discussion of particle–fluid interactions across such a range of length scales, as the effects, relative importance and interactions of parameters such as surface tension, chemical admixtures, and solution ionic strength will differ across length scales [20]. Therefore, most studies consider in detail either the packing of larger particles (particularly the "fine" and "coarse" aggregates; usually sand and crushed rock) [84,85] or the packing of binder components [86], with the combination of these two types of components presenting a much more challenging set of experimental and theoretical problems [87,88].

The term "mix design" in concrete is used mainly to describe the process of proportioning of binder constituents, chemical and mineral admixtures, rock, and sand, with the aim of optimizing some specified combination of technical (mechanical and durability) properties, placement and finishing, and cost [89]. In geopolymer concrete mix design, it is important to tailor the content and composition of the alkali activator and aluminosilicate precursor materials, as well as the aggregate components. This does not necessarily mean that the change from Portland cement to a geopolymer binder requires wholesale changes to overall mix designs, as the optimization of the aggregate blends and aggregate–binder ratio is driven by the order of magnitude cost differentials between cementitious materials and aggregate. However, with a fundamental understanding of aluminosilicate chemistry, significant improvements in geopolymer concrete quality and economy can be achieved, particularly with regard to control of curing conditions, activator dosage, and efflorescence. Successful geopolymer concrete mix designs have been published by various researchers with some of these mix designs incorporating chemical admixtures (particularly superplasticizers). However, analysis of the effectiveness of many of the admixtures that are commonly used in Portland cement mixes under the highly alkaline conditions of geopolymerization has shown mixed outcomes [90,91]. Kashani et al. [92] outlined the complex surface chemistry of the alkali-activated slag system and its relationship with paste rheology, which explains why specially designed polymer architectures are required to plasticize geopolymer concrete, as suggested by Kashani et al. [93].

## 10.5 LINKING GEOPOLYMER BINDER STRUCTURE AND DURABILITY

### 10.5.1 Factors Affecting the Service Life of Reinforced Concrete

The fundamental basis of the durability of reinforced concrete is the ability to develop and maintain a dense, impermeable binder gel that creates and stabilizes a highly alkaline environment, with appropriate chemical (in particular electrochemical) conditions to enable the stabilization of embedded steel in a passive state. Steel corrosion is, on a worldwide basis, the predominant cause of premature failure of reinforced concrete elements [94]. Weather-induced mechanisms of attack on the binder and concrete (including freeze–thaw damage and/or salt scaling), and alkali–aggregate reactions involving some siliceous or carbonate aggregates, can also be problematic under given circumstances and in some climatic conditions, but the focus of the review presented here will be on mechanisms of degradation that are explicitly related to ionic transport. The loss of alkalinity (via leaching, carbonation, or other mechanisms) and the ingress of chloride are the primary causes of steel corrosion within concrete, and the resistance of the material to degradation by these mechanisms will obviously depend intrinsically on its mass transport properties. In particular, the width, connectivity, and degree of water saturation of the pores and cracks that are present throughout the binder will play a highly significant role in determining the resistance to ingress by aggressive agents. Cracks and pores with relevance to binder structure and durability performance are present on length scales from nanometers to millimeters in most concrete structures, making this another true multiscale problem, similar to the problems of mix design and particle packing as discussed above. The application of both traditional and advanced techniques in the analysis of the microstructure of the binder therefore again becomes imperative to understanding and designing for durability, particularly when introducing a new binder system such as the alkali-activated family of binders.

Van Deventer et al. [15], Provis and Van Deventer [27], and Provis et al. [17] outlined the physicochemical factors which can determine the service life of a concrete and emphasized the complexity of the system. The discussion below will present a brief overview of selected physicochemical factors, with a specific focus on the design and characterization of durable geopolymer binders and concretes.

## 10.5.2 Microcracking Phenomena

In broad terms, cracking of concretes can be caused by chemomechanical (shrinkage or expansion of gel phases, autogenous heating during curing) or physicomechanical (applied load, freeze–thaw) processes. Concrete is well known to be strong in compression but weak in flexion and tension; the use of steel reinforcing, often in combination with techniques such as pretensioning, and careful structural design, will often be intended specifically to compensate for this weakness by ensuring that the concrete itself bears minimal tensile load. However, an advantage of geopolymer binders is that they tend to demonstrate a flexural (and presumably also tensile, although this is rarely tested directly) strength significantly higher than is specified by the standard relationships that apply for Portland concretes of similar compressive strength [95]. Mortar compressive strengths in excess of 100 MPa, and concrete strengths higher than 70 MPa, have been relatively widely reported for laboratory samples. Achieving such strengths consistently in large-scale production is more complex, but the large-scale production of "high-performance" concretes by alkaline activation of suitably chosen precursor blends is certainly possible, and is beginning to be demonstrated on a commercial scale.

It is also important to distinguish load-induced macroscopic cracking of the concrete from microcracking of the binder; both are likely to be deleterious in terms of durability performance, but the length scales on which the cracks form are very different. In an early investigation, Byfors et al. [96] observed significant microcracking in samples of "F-concrete" (superplasticized $NaOH/Na_2CO_3$-activated slag). The authors of Refs. [97,98] observed a tendency towards microcracking in alkali-activated slag concretes, with a corresponding increase in the rate of carbonation and reduction in strength, and found that a high activator content tended to correlate with a high extent of microcracking. Collins and Sanjayan [99] conducted a detailed investigation in this area and concluded that drying effects during curing were a primary cause of microcracking of alkali-activated slags.

More recently, Bernal et al. [100] proposed, via capillary-suction measurements, that the extent of microcracking in silicate-activated slag-based concretes depends significantly on the paste content of the concrete mix design; excessive binder content was suggested to lead to heat generation during curing of concrete specimens, leading to thermally induced microcracking of the concrete at early ages. However, in this case, microcracking was not observed to have any significant influence on the rate of carbonation of the concretes, indicating that the higher binder content was able to provide sufficient densification of the matrix to compensate for additional carbonation taking place via transport along cracks. The fundamental cause of microcracking is the partially restrained chemical shrinkage of binder phases after hardening, which has been proposed from a thermodynamic basis to be intrinsic to the chemistry of alkali-activated binders and related (pozzolanic) systems [101,102]. Almost all cementlike binders show either shrinkage or expansion during or after hardening as a result of the process of crystallographic (including gel) phase evolution, which results in strength generation and development [77]. Maintaining dimensional stability is thus a key challenge across the field of cement and concrete technology.

The observed trends in microcracking intensity in alkali-activated concretes as a function of paste content and curing regime thus highlight the value of understanding interactions between the binder and aggregate, and effects related to heat generation and heat and moisture transport during curing, in mitigating the effects of microcracking on concrete performance and durability. The key interactions take place in the region known as the interfacial transition zone, and the microstructural characteristics of this region of the concrete are critical in terms of both strength and durability performance.

## 10.5.3 Interfacial Transition Zone Effects

In the area where the binder comes into contact with aggregate particles, there are often microstructural differences when compared with bulk binder regions, meaning that these specific regions can exert a disproportionately high (and usually negative) influence on the mass transport, tensile, and flexural properties of the geopolymer concrete. This is an area in which geopolymer binders are believed to provide significant advantages over Portland cement. The chemistry of the high-calcium Portland systems tends to lead to the formation of a porous zone containing large, mechanically weak crystals surrounding aggregate particles [103,104], and this is a key pathway for both mechanical failure and mass transport in concrete. The interfacial transition zone in geopolymer concretes has been identified as being dense and much less microstructurally distinct from the bulk of the binder region [105–110], which provides much higher tensile and flexural performance, as well as removing the possibility for these regions to lead to the formation of a percolated porous pathway for mass

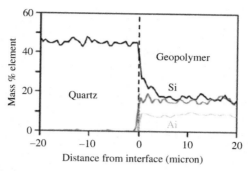

**Figure 10.3** Elemental concentrations (from SEM-EDS). *Data from Lee WKW, Van Deventer JSJ. Chemical interactions between siliceous aggregates and low-Ca alkaliactivated cements. Cem Concr Res 2007;37(6):844–55.*

transport through the binder, thus enhancing the durability of the geopolymer material.

Fig. 10.3 shows an example, presented in detail by Lee and Van Deventer [109] and reproduced in simplified form here, of the elemental compositions observed across the interface between a (K, Na) silicateactivated fly ash geopolymer binder and a quartz particle within a section of siltstone. The elemental compositions were determined by scanning electron microscopy-energy dispersive spectroscopy (SEM-EDS), and show a uniform binder region spanning the distance from the bulk binder to the aggregate particle surface. Similar results were also obtained for a range of phases present within a basalt specimen; an iron-rich augite region in the basalt did generate a distinguishable region less than $2\,\mu m$ in thickness and rich in Fe and Si at the interface, but there was not a distinct and chemically/crystallographically different region at the interface in any case when comparing to the bulk binder structure. The SEM-EDS data in Fig. 10.3 obviously do not provide detailed microstructural information; however, imaging of the same specimens did not show any differences in microstructure or any apparent additional porosity in these regions [109], consistent with the general consensus in the literature regarding geopolymer binders.

## 10.5.4 Microporosity in the Bulk of the Geopolymer Binder

There are a variety of techniques available for the analysis of micronscale pores within a geopolymer (or any cement-type binder), which can be classified roughly into the categories of one-dimensional, two-dimensional, and three-dimensional analysis. Among these categories, there is no universally applicable technique that can provide a full multi-scale characterization of a complex material such as a geopolymer binder; a fuller toolkit of techniques is required to obtain detail across the length scales of interest.

One-dimensional analysis of porosity in the range of nanometers to microns is usually conducted by gas sorption (most commonly the Barrett–Joyner–Halenda, BJH, technique), direct penetration measurements (air, water, and chloride penetration being the most commonly applied), or mercury intrusion porosimetry (MIP). The limitations and inaccuracies of MIP in application to cements are well known [111], but it is widely used due to its apparent simplicity and the ready availability of instrumentation. MIP largely fails in the case of samples with complex pore geometries (the "ink-bottle" effect), as the full volume of a complex-shaped pore is registered as having an effective pore size equivalent to that of the narrowest part of its entry. This results in an underestimation of mean pore size. The high pressures used to intrude mercury into the smallest pores in a sample are also problematic, as the applied pressure will often exceed the strength (compressive or tensile) of the material to which it is being applied, which can result in significant crushing effects. Multicycle MIP has been proposed as a method by which the ink-bottle effect can be minimized or avoided [112], and appears to provide some significant advances in this area with good agreement with $N_2$ sorption data. Gas sorption is believed to provide good characterization of very small pores, but cannot accurately measure the larger pores which are essential to mass transport in samples with pores spanning a very wide range of length scales.

Direct water and air permeability measurements of geopolymer and other alkali–activated binders have shown a range of performance, depending mainly on the mix designs tested. Binders with well-cured alkali-activated binders with low water/binder ratios perform acceptably in these tests [113–115], but generally do not provide results that could be considered particularly outstanding, most likely due to the low levels of space-filling bound water associated with these gels. On the other hand, chloride permeability testing of geopolymer mortars and concretes has shown a wide range of performance, with the outcomes depending to a significant extent on the details of the testing methodology selected. The ASTM C1202 test methodology involves the application of a current to the sample and uses measurement of the charge passed as a proxy for the movement of chloride ions into the sample under the electrical field gradient. The outcomes of this test are strongly dependent on pore-solution chemistry [116,117], and so it sometimes registers geopolymer binders as showing very good resistance to mass transport [100,114,118], and sometimes as performing rather poorly. Direct measurement of alkali cation movements from the pore solutions of geopolymer binders [119] provides insight into the cause of this behavior. The mobility of alkalis will certainly display a strong influence on the total charge passed by the samples during the test. It has been proposed that alternative methodologies which provide a more direct measurement of the progress of chloride migration into the binder; for example ponding tests, or the NordTest Build 492 accelerated test, will provide a more valid comparison that is relatively independent of the pore-solution chemistry of the binder, and work in this area is ongoing [27].

Two-dimensional analysis of pore structures is predominantly conducted using microscopic methods, in particular SEM. The primary challenge associated with the observation of porosity by SEM (designed to detect solid matter) is that the aim of the measurement is to detect the regions that contain no solid matter. This process of "trying to see the parts that are not there" provides challenges in pore identification. It can be overcome to some extent for calcium-rich systems, such as Portland or alkali-activated slags, by the use of image-analysis algorithms [120,121]. However, for aluminosilicate geopolymer systems with lower electron density contrast between the solid and pore regions, the application of Wood's metal intrusion prior to the collection and analysis of SEM images has proven to be of value [81]. In this technique, the pores of the sample are filled with a high elemental-number, low melting-point alloy, which is intruded into the pore space of the sample under moderate pressure (lower than the pressures used in MIP, to prevent damage to delicate parts of the microstructure) and at temperatures below 100°C but still above the melting point of the alloy. The sample is then cooled, and the alloy solidifies within the pore network, to provide a high degree of elemental contrast and make the pores visible against the low-elemental number binder regions [122,123]. In the recent application of this technique to geopolymer binders [81], it was calculated that pores as small as 11 nm—well below the spatial resolution achievable by standard SEM techniques—are able to be filled with the molten alloy, while using an intrusion pressure lower than the compressive strength of the binder material to minimize microstructural damage.

The primary three-dimensional characterization technique which is able to be applied to the analysis of cements (geopolymer and traditional) is X-ray microtomography. There have been a number of studies of Portland-based materials by this technique over the past decade or more [124]. The systematic analysis of geopolymer binders by X-ray microtomography and hard X-ray nanotomography has recently been presented for the first time [82,125].

In the analysis of the geopolymer sample set discussed here, the samples were analyzed using beamline 2-BM at the Advanced Photon Source synchrotron at Argonne National Laboratory, using 22.5 keV X-ray radiation and achieving a 750-nm voxel resolution. Porosity was calculated during segmentation of the sample into "pore" and "solid" regions using slightly modified versions of freely available data processing scripts [126] based on the full greyscale tomographic reconstructions, and tortuosity was computed by random walker simulations utilizing the segmented pore network, again based on the algorithms of Nakashima and Kamiya [126]. An example showing the greyscale and segmented images for a given region of a fly ash–Na silicate geopolymer binder is given in Fig. 10.4.

Fig. 10.4 also demonstrates some of the challenges associated with microtomographic characterization of fly ash geopolymer samples [15].

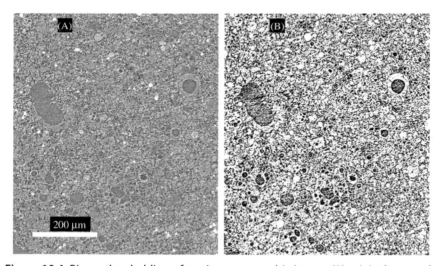

**Figure 10.4** Binary thresholding of a microtomographic image: (A) original greyscale image and (B) binary segmented image. Scale bar represents 200 m. *Reproduced from Van Deventer JSJ, Provis JL, Duxson P. Technical and commercial progress in the adoption of geopolymer cement. Miner Eng 2012;29:89–104.*

The large dark regions represent the interior of hollow or partially hollow (cenosphere or plerosphere) fly ash particles; these regions are not connected to the pore volume within the gel, and so will not contribute to the transport properties of the material. However, in measurements or calculations of total porosity, these regions must be excluded via the application of algorithms which considers "connected porosity" only. As an additional complicating factor, the tomographic reconstruction process (if sample alignment was not on a single vertical axis during rotation for scanning in the instrument) can introduce streaking artifacts that effectively break through the shell of a fly ash particle and connect its interior to the binder pore network. Thus, the collection of high-resolution, high-quality tomography data is of particular importance in enabling the correct analysis of pore network transport parameters for samples containing fly ash. This consideration is less problematic for samples based on slag (or Portland cement) because the precursor particles do not contain significant inaccessible pore volumes.

The data collected by Provis et al. [82] as shown in Fig. 10.5, demonstrate that the porosity of the more calcium-rich geopolymer binders (the systems containing 50% or more slag) decreases as a function of curing

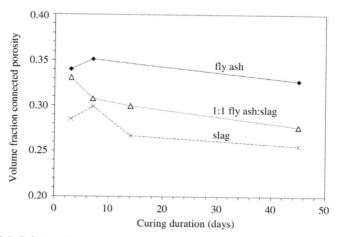

**Figure 10.5** Relationship between segmented porosity and curing duration for a range of sodium silicate-activated fly ash/slag systems. *Reproduced from Van Deventer JSJ, Provis JL, Duxson P. Technical and commercial progress in the adoption of geopolymer cement. Miner Eng 2012;29:89–104, with data selected from Provis JL, Myers RJ, White CE, Rose V, Van Deventer JSJ. X-ray microtomography shows pore structure and tortuosity in alkali-activated binders. Cem Concr Res 2012;42(6):855–64.*

duration, indicating that there is some chemical binding of water taking place in this system, consistent with the known formation and coexistence of N-A-S-H and C-A-S-H gels in mixed fly ash/slag geopolymer binders.

The results of drying tests conducted on a series of mortar samples synthesized from the same binders as were studied by tomography [127] show further that the C-(A)-S-H phase is binding some water, but to a lesser extent than in the case of Portland hydrate products. As the gel evolves, and its porosity decreases over time, the tortuosity of the pore network also increases. The sample set presented in Fig. 10.5 shows close agreement with an inverse relationship between connected porosity and diffusion tortuosity [82]. The diffusion tortuosity describes the effect on mass transport of the constriction due to the presence of the pore network. An increase in this parameter by almost a factor of two between early age (<7 days) and 45 days of age indicates that the rate of diffusion through a well-cured binder would be halved when compared to a poorly cured binder. This further highlights the importance of adequate curing in achieving high performance and durability in geopolymer concretes. A criticism of geopolymer technology is that when mixes are poorly designed, thermal curing is often required for adequate strength development. This is not the case for a well-designed mix with sufficiently well-controlled activation conditions, as has been demonstrated for systems based on fly ash, on slag, and on mixtures of these two precursors. However, it is clear from the tomography results that, regardless of the rate of early-strength development, an extended period of curing will provide marked advantages in service life and overall durability performance once the binder is placed in service and exposed to aggressive environments.

The substitution of fly ash for slag is seen in Fig. 10.5 to lead to higher porosity. This is consistent with the results of nitrogen sorption analysis of a range of similar binders [81], and the porosities obtained from tomography are also within 2 vol.% of the values obtained in that study for samples of similar mix design and curing duration. The slag-rich systems form predominantly a C-(A)-S-H gel, which appears to bind a larger amount of water into its nanostructure, while the fly ash–rich systems form N-A-S-(H) gels with a low bound-water content [10].

The discussion presented here therefore highlights the importance of understanding and controlling the porosity of geopolymer binders, particularly in the case of fly ash–rich systems. If the higher porosity of the geopolymer-type (sodium aluminosilicate) gel when compared with calcium silicate gels such as those produced by Portland hydration were to lead to high diffusivity (and thus rapid mass transport) of aggressive ions through the binder to reach the embedded steel reinforcing, this would mean that the durability of these materials would in all likelihood be unacceptably poor. Evidence from the in-service performance of geopolymer binders [13,16,128], as well as from laboratory chloride penetration testing as discussed above, shows that the observed performance is significantly better than would be expected from raw permeability or carbonation rate data [13,100,110,129,130].

Bernal et al. [131] showed that the depth of carbonation in a set of slag-geopolymers exposed to ambient conditions for 7 years was much lower than would be predicted through accelerated carbonation testing of specimens formulated to the same mix designs, demonstrating that the exposure conditions used in accelerated testing do not replicate the phenomena that take place under natural service conditions. In a recent study of in-service geopolymer concrete in the Netherlands it was shown that after 2 years the carbonation depths were comparable with those of blended Portland cement concrete containing slag or fly ash [132].

Carbonation of alkali-activated binders is an open and active area of research, and much remains to be explained in this area, particularly with regard to the relationship between carbonation and steel degradation, which may or may not be similar to the corresponding relationship in Portland cement–based concretes. This suggests that there are additional effects that compound with, or mitigate, the direct influence of porosity on permeability (particularly ionic permeability, which also relates to gel chemistry-specific effects and interactions) and durability.

## 10.6 TECHNICAL CHALLENGES

In developing the technology of novel cements and concretes, it is often important to apply expertise from other disciplines such as mineral processing, and in particular fluid–particle interactions. For example, the phase chemistry of cement and other mineral components (particularly fly ash and slag) is critical in controlling reactivity. As was discussed previously, an understanding of the different glassy phases present in geopolymer precursor materials is required to determine the optimal processing method and composition of binder components in order to manipulate concrete properties. Classified fly ash and ground slag available through the cement supply chain are not always able to be used straightforwardly with existing commercially available alkaline activators to prepare optimal geopolymer binders, so special preparation of alkaline activators may also be warranted. In addition, fly ash and slag (being waste materials) are subject to far more variability, especially when they are obtained from a range of sources, than Portland cement as a quality controlled product. Therefore, online monitoring and optimization of activator types and binder components is necessary in the production of consistent, high-quality geopolymer cements.

As was mentioned briefly above, it is important to control the rheology of fresh concrete to enable it to be placed and finished, and the ability to do this without adversely affecting the final properties of the hardened concrete depends primarily on the manipulation of colloidal interactions by the use of chemical admixtures such as superplasticizers [90,91]. The existing range of commercially available superplasticizers has been developed specifically to suit the complex series of chemical reactions that take place in the Portland system, and are usually not effective in the geopolymer system. Moreover, most of the various admixtures used to control slump, air dispersion, water retention, and other properties of the Portland system are less effective in the geopolymer system. Consequently, there is a need to develop a whole set of new admixtures for the geopolymer system, which presents a significant challenge for an emerging industry with a lack of scale, but which is still required to compete with the well-established Portland cement industry.

Geopolymerization also provides the potential for the utilization of non–blast furnace slags; alternative materials such as ferronickel, steel, and phosphorus slags have also been alkali activated to form usable binders and concretes, some on a laboratory scale and some in larger-scale applications in China and the former Soviet Union [13]. The leaching of toxic metal components from some of these slags during activation may prove to be a cause for concern, but—as is the case for fly ashes—the selection of appropriate waste materials for use as geopolymer precursors is both important and possible.

At present, geopolymer concrete is most commonly produced in a "two-part" mix format, by blending coal fly ash, slag, and alkali activators together in a concrete-batching plant. This requires a high level of skill from the operators of the batching plant, which is rarely available in the premix concrete industry. Ash and slag are waste materials of variable composition, so that the concrete mix designs and the added alkalis need to be varied to compensate for these variations in the ash and slag. Such a technology, where the quality control is mainly at the batching plant level, is not scalable on an industry-wide level and has limited appeal in the market, despite the growing market pull for "green" construction materials.

In contrast with this situation, Zeobond has developed a process whereby the various solid materials and proprietary activators are processed together to produce a dry cement binder that behaves in a similar way to Portland cement [32]. The quality control is hence centralized in the cement binder plant, so that the dry powder can be distributed to various concrete-batching plants, as is the case with Portland cement. To a large extent this addresses the supply chain challenges and difficulties in price competition which are inherent when sourcing materials from Portland-related suppliers. However, the process of establishing such dry binder-processing facilities is capital intensive and requires significant market drivers for geopolymer concretes (synthesized via the existing process) to justify investment.

## 10.7 REDUCTION IN CARBON EMISSIONS

$CO_2$ emissions from cement production are incurred through the consumption of fossil fuels, the use of electricity, and the chemical decomposition of limestone during clinkerization, which can take place at around 1400°C. The decarbonation of limestone to give the calcium required to form silicates and aluminates in clinker releases roughly 0.53 t $CO_2$ per ton of clinker [8]. In 2005, cement production (total cementitious sales including ordinary Portland cement (OPC) and OPC blends) had an average emission intensity of 0.89 with a range of 0.65–0.92 t $CO_2$ per ton of cement binder [133]. Therefore, the decarbonation of limestone contributes about 60% of the carbon emissions of Portland cement, with the remaining 40% attributed to energy consumption, most of which is related to clinker kiln operations; the WWF-Lafarge Conservation Partnership [6] estimated that the production of clinker is responsible for over 90% of total cement production emissions.

In view of the fact that the requirement for decarbonation of limestone presents a lower limit on $CO_2$ emissions in clinker production, and that there exist technical issues associated with the addition of supplementary cementitious materials (SCMs, including fly ash and ground granulated blast furnace slag), which restrict the viability of direct Portland cement supplementation by SCM above certain limits, the possibility to reduce $CO_2$ emissions using Portland chemistry is limited. The WWF-Lafarge Conservation Partnership [6] expects that the emissions intensity of cement, including SCM, could be reduced to 0.70 t $CO_2$ per ton of cement by 2030, which still amounts to around 2 billion tons of $CO_2$ per annum worldwide, even if cement production does not increase from its current level.

Fig. 10.6 shows the $CO_2$ emissions of various binder designs as a function of Portland cement content. There have been a limited number of life-cycle analyses (LCA) of geopolymer technology. One reasonably extensive research program carried out in Germany [134] has provided information regarding the selection of precursors and mix designs for a range of geopolymer-based materials. However, geographic specificity plays a significant role in a full LCA, so there is the need for further studies considering different locations in addition to a wider range of mix designs spanning the broader spectrum of geopolymers. The main carbon-intensive and also the most expensive ingredient in geopolymer cement is the alkali activator, which should be minimized in mix design. McLellan et al. [135] provided further detail, while Habert et al. [136] concluded that geopolymer cement does not offer any reduction in carbon emissions; such a conclusion needs to be drawn with caution.

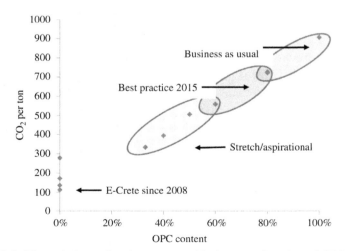

**Figure 10.6** $CO_2$ emissions of various cement binders as a function of OPC content. *Reproduced from Van Deventer JSJ, Provis JL, Duxson P. Technical and commercial progress in the adoption of geopolymer cement. Miner Eng 2012;29:89–104.*

Sodium carbonate is the usual Na source for the production of sodium silicate. The different processes for conversion of $Na_2CO_3$ (or NaOH) and $SiO_2$ to sodium silicate, via either furnace or hydrothermal routes, differ by a factor of 2–3 in $CO_2$ emissions, and up to a factor of 800 in other emissions categories [137]. It is therefore essential to state which of these processes is used as the basis of any LCA. Moreover, the best available data for emissions due to sodium silicate production were published in the mid-1990s [137], so improvements in emissions since that time have not been considered. Sodium carbonate itself can be produced via two main routes, which vary greatly in terms of $CO_2$ emissions. The Solvay process, which converts $CaCO_3$ and $NaCl$ to $Na_2CO_3$ and $CaCl_2$, has emissions between 2 and 4t $CO_2$ per ton of $Na_2CO_3$, depending on the energy source used. Conversely, the mining and thermal treatment of trona for conversion to $Na_2CO_3$ has emissions of around 0.14t $CO_2$ per ton of $Na_2CO_3$ produced plus a similar level of emissions attributed to the electricity used. This indicates an overall factor of 5–10 difference in emissions between the two sources of $Na_2CO_3$ [138].

A commercial LCA was conducted by the NetBalance Foundation, Australia, on Zeobond's E-Crete geopolymer cement, as reported in the "Factor Five" report published by the Club of Rome [139]. This LCA compared the geopolymer binder to the standard Portland blended cement available in Australia in 2007 on the basis of both binder-to-binder comparison and concrete-to-concrete comparison. The binder-to-binder comparison showed an 80% reduction in $CO_2$ emissions, whereas the comparison on a concrete-to-concrete basis showed slightly greater than 60% savings, as the energy cost of aggregate production and transport was identical for the two materials. However, this study was again specific to a single location and a specific product, and it will be necessary to conduct further analyses of new products as they reach development and marketing stages internationally. Fig. 10.6 shows a comparison of the $CO_2$ emissions of four different E-Crete products against the "Business as Usual," "Best Practice 2011," and a "Stretch/Aspirational" target for OPC blends. It is noted that in some parts of the world (particularly Europe), some of the blends shown here in the "Stretch/Aspirational" category are in relatively common use for specific applications, particularly CEM III-type Portland cement/slag blends, but this is neither achievable on a routine scale worldwide at present, nor across the full range of applications in which Portland cement is used in large volumes.

## 10.8 STANDARDS FRAMEWORK

Fig. 10.7 shows that the strategic development of standards is pivotal to the commercialization of geopolymer cement. The regulatory framework governing concrete to be utilized in various applications relies on a typical cascade of standards, with application standards referring to concrete standards, and concrete standards referring to standards covering cement

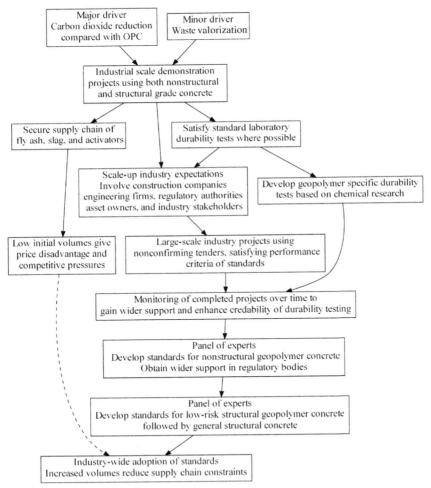

**Figure 10.7** Flow diagram for commercialization of geopolymer cement and concrete. *Reproduced from Van Deventer JSJ, Provis JL, Duxson P. Technical and commercial progress in the adoption of geopolymer cement. Miner Eng 2012;29:89–104.*

and other raw materials. Hence, when considering the regulatory framework for a concrete binder system such as geopolymer cement or concrete, most of the attention to regulatory aspects should be focused on the cement standards, although some aspects of concrete standards also need to be considered. In general, all of the world's concrete and cement standards are based on two "super" standards, i.e., European Union EN 197 and United States ASTM C150/C595/C1157. For instance, Chinese cement and concrete standards are based largely on European Union Standards, while Australian standards are based mainly on the American standards.

These standards have been developed over many years, and in collaboration with input from Portland manufacturing companies, with the chemistry and behavior of Portland-based concretes intrinsically in mind. However, prescriptive standards containing constraints such as "minimum cement content" are increasingly being viewed as excessively prohibitive, even for Portland-based systems. Products such as geopolymer concrete may not simply be an evolution of existing Portland technology, but instead may require an entirely different chemical paradigm to understand their behavior, and may perform entirely acceptably, but without conforming exactly to the established regulatory standards, particularly with regard to rheology and chemical composition [140]. This is a significant obstacle to the acceptance of geopolymer technology [141]. However, by working with all stakeholders (Fig. 10.7), these barriers can be overcome, provided that the intent of regulatory standards is met.

There is general agreement that a performance-based standards framework must be developed if different binder types and proprietary cementitious mixes are to be used. The question remains which performance testing and durability testing methods should be used in order to specify performance criteria. It is no simple task to develop testing methods for durability that are independent of initial binder-phase assemblage. In a critical review of performance-based approaches, Alexander and Thomas [142] explained that it is possible to relate service-life prediction models to durability testing, even when it is known that the diffusion parameters in concrete are complicated by several factors, including interaction between the diffusing species and the matrix, and the reduction of diffusion coefficients with age. It is noteworthy that South Africa has developed a suite of durability index tests, i.e., oxygen permeability, sorptivity, and chloride conductivity, and these are linked to service-life models for the relevant deterioration mechanisms in reinforced concrete structures [142].

VicRoads, the state roads authority in Victoria, Australia, has recognized geopolymer concrete as being equivalent to Portland concrete for nonstructural applications in a 2010 update to their design specification Section 703 [143]; this specification requires acceptable binder content, strength development, water content and quality, aggregate properties, consistency, mixing and placement methodologies, curing, finishing, and material supplier competency guarantees. Zeobond is working with VicRoads to also recognize geopolymer concrete for structural applications in Section 620. Zeobond's E-Crete is already used in VicRoads structural projects, and by local councils and housing developers in subdivisional works and slabs, applications that represent approximately 70% of all concrete usage. These large-scale applications are pivotal to the process of gradually convincing standards authorities to accept geopolymer concrete.

## 10.9 TESTING FOR DURABILITY

The question of whether geopolymer concretes are durable remains the major obstacle to recognition in standards for structural concrete, and hence to their commercial adoption. A material such as geopolymer, which has been subjected to detailed investigation only recently, cannot possibly have the availability of decades of in-service testing and durability data to prove its long-term stability. Most standard methods of testing cement and concrete durability involve exposing small samples to very extreme conditions—in particular, highly concentrated acid or salt solutions, with or without the application of electrical field gradients—for short periods of time. The data obtained from these tests are then used to predict how the material will perform under normal environmental conditions over a period of decades or more. In some of these predictive models, engineering concepts including mass transport through porous media, reaction kinetics, and particle packing are used, although usually in a simplified and semiempirical form to enable utilization of the derived equations by nonspecialists in the field. However, the key shortcoming of this approach to "proving" durability is that it can only provide indications of the expected performance, rather than definitive proof. Therefore, there has been a very slow process of adoption of new materials, as it is considered necessary to wait up to 20–30 years for "real-world" verification. Adoption of fly ash and slag in Portland concretes is the prime example of this, where the use of these SCMs was resisted for decades. It is the author's experience that asset owners and their insurance companies are willing to use geopolymer concrete in low-risk applications based on accelerated durability testing. Higher-risk applications such as high-rise buildings, which constitute a smaller fraction of the total concrete market, will follow only when the market is comfortable with the real-world track record of the material in low-risk applications. Therefore, a staged approach towards the development of standards and commercial adoption needs to be followed, as outlined in Fig. 10.7.

Zeobond also works closely with various roads authorities in monitoring the performance of in situ geopolymer concrete structures and also with researchers in developing appropriate test methods for durability (Fig. 10.7). The state-of-the-art report by RILEM Technical Committee 224-AAM on Alkali-Activated Materials [27] and the current work by RILEM Technical Committee 247-DTA on the Durability Testing of Alkali-Activated Materials provide essential advice to standards authorities about the structure of performance-based standards for geopolymers and the associated testing methods for durability.

As outlined by Provis et al. [144], there have been a number of studies of chloride diffusion in slag geopolymer concretes, with the performance of these materials in accelerated chloride penetration tests generally observed to be at least comparable to that of Portland concrete. Zeobond's E-Crete has been shown repeatedly to have significantly lower chloride diffusion and acceptable freeze–thaw performance compared with Portland concrete. As outlined in Section 10.5.4, the excellent performance of geopolymer concrete is related to the highly refined pore network forming a dense low-calcium C-A-S-H phase [144]. It is envisaged that advanced techniques such as synchrotron-based nanotomography will ultimately be used to compute transport properties of the geopolymer gel generated from various mix designs (such as water/binder ratio), and hence changes in durability performance.

## 10.10  SUPPLY CHAIN RISKS

The presence of a competitive and efficient supply chain is essential for the successful scale-up of geopolymer technology. For instance, granulated blast furnace slag (GBFS) is produced during the production of pig-iron. Although there is a risk that production methods for pig-iron will change and reduce the availability of GBFS, this is not currently viewed as a substantial risk. As demand for GBFS increases, blast furnace operators will increasingly invest in rapid-cooling slag-handling equipment in order to increase the production of GBFS. The further development and industrial adoption of the new Commonwealth Scientific and Industrial Research Organization air-cooled slag-granulation method [145] is important in regions with a shortage of clean water, which is the standard chilling medium used at present. The main risk facing GBFS supply for geopolymer use is preferential utilization of GBFS in Portland blends rather than in geopolymers.

There is a risk that a substantial reduction in worldwide carbon emissions will result in reduced coal-fired energy production, and hence reduced availability of coal ash. While there is a global determination to reduce energy dependency on thermal coal, it is more likely that coal combustion will continue to increase in the coming decades, but with some level of carbon capture through sequestration or mineralization. In any event, only a very small fraction of coal ash is currently used in concrete. A more substantial risk is the well-intended control of coal ash as a hazardous material by, for example the US Environment Protection Agency, which leads to uncertainty in investment in the supply chain of all SCMs. Laboratory studies have shown that natural pozzolans could be used as geopolymer precursors, but no large-scale application has been attempted and a supply chain is not in place, except in countries like Indonesia. In the event of a shortage of fly ash or slag, natural pozzolans could possibly be used as an alternative precursor.

In economically developed and structurally segmented markets for cementitious materials, distribution networks for slag and fly ash are established, making it uncompetitive or practically impossible to invest in new channels to market. In some cases, even voluminous byproduct streams such as slag and fly ash are largely utilized or earmarked via long-term option contracts. Due to the rapid increase in demand and production in markets such as China and India, new opportunities for both Portland and geopolymer supply channels are foreseeable. However, in established markets such as the United Kingdom, where fly ash production does not exceed its consumption in cement, the existing market for Portland cement utilizes all of the available fly ash. Despite the increasing number of research papers on geopolymers, there has been little development of the supply chain channels necessary for scale production, which will continue to limit the wider adoption of geopolymer technology until this bottleneck is resolved.

## 10.11  PERSPECTIVES ON COMMERCIALIZATION

Apart from the challenge to relate accelerated durability testing data to in-service life predictions, there is also the question whether existing engineering design methods for Portland-based structural concrete are applicable to geopolymer concrete. From laboratory research and monitored industrial applications it indeed appears to be the case. However, more needs to be done to calibrate design codes for the specific characteristics of geopolymer concrete.

Each concrete market is different, as price structures vary greatly, logistics and supply chains are different, and the precursors available for geopolymer concrete will be different; for example, fly ash varies greatly in reactivity from one location to another. Therefore, it is necessary to build confidence in geopolymer concrete from scratch in each new market. Small low-risk projects, where the cost of replacement is low if performance is not met, must be completed first to build confidence before more complex projects are taken on. The key challenge is often the availability of suitable precursors at the right price for small demonstration projects in a new location, which is a more challenging situation technically and commercially than when there is a suitable supply chain at scale. This is the equivalent of building a car from components compared with delivery from an assembly line. Nevertheless, commercial adoption of geopolymer concrete has been achieved in selected applications [27,141]. Recently, there has been renewed interest in commercial application of geopolymers in new markets, with the drivers being valorization of fly ash and waste metallurgical slag, reduction of $CO_2$ emissions, improved technical properties, and cost reduction.

Fig. 10.8(A) shows a small section of E-Crete paving completed as part of the Victorian Government upgrade of the Westgate Freeway in Port Melbourne. While small in volume, this project required that the concrete meet all of the technical specifications of the local road authority, VicRoads, and the ultimate asset owner, the Port Melbourne City Council, and was approved for use by a construction consortium which included numerous national and multinational construction companies and engineering firms. This small project demonstrates the entire process of commercialization of geopolymer concrete (Fig. 10.7).

Fig. 10.8B shows the installation of 55-MPa E-Crete precast panels for VicRoads. This concrete was required to meet the structural concrete code (VicRoads Section 610), which is far more stringent than the requirements faced by the nonstructural grade concrete in Fig. 10.8A.

**Figure 10.8** Examples of E-Crete: (A) Footpath along Westgate Freeway extension, Port Melbourne; (B) Precast panels across Salmon Street bridge, Port Melbourne; (C) Embankment at Swan Street bridge, Melbourne; (D) Retaining walls for Regional Rail Link, Melbourne; (E) Retaining walls along M80 freeway, Victoria; (F) Precast bridge deck, Queensland; (G) and (H) Precast structural panels and in situ concrete for Melton library, Victoria; (I) Vertically cast, roller compacted pipes for partly treated sewerage, Victoria; (J) Portside building in Cape Town utilizing some Portland cement and the principles of geopolymerization.

Here the level of concrete specification and scrutiny was consistent with bridge design, with the aim of using the highest-grade concrete specified by VicRoads as a "stretch target" for trial purposes.

Structural grade concrete (40 MPa) was used for the southern embankment at the Swan Street bridge in the city of Melbourne, as shown in Fig. 10.8C. This structure was instrumented by VicRoads and has been monitored for durability and steel corrosion, with good effect. Fig. 10.8D shows 40-MPa structural retaining walls for the Regional Rail Link in the city of Melbourne. This project involved several authorities and took a long time to get approved, even after a recommendation by VicRoads. Despite the obstacles to get such projects approved, it is essential to pursue such signature projects as they build confidence in the technology in the wider market. Fig. 10.8E shows a section of the 40-MPa retaining walls as part of the VicRoads project to upgrade the M80 freeway to the

**Figure 10.8** (Continued)

northwest of Melbourne. Although a large geopolymer concrete bridge has not been built, several smaller precast bridge decks have been made, such as the one depicted in Fig. 10.8F.

Zeobond's E–Crete was used for the in situ cast concrete and architectural panels in the Melton Library and Learning Hub project, as depicted in Figs. 10.8G and H. This project won several architectural and sustainability awards and was the first completed building in Australia to achieve a Five-Star Green Star–Public Building Design Pilot rating from the Green Building Council of Australia. Fig. 10.8I shows vertically cast, roller-compacted pipes for partly treated sewerage in Victoria. It is significant that Murray and Roberts in South Africa constructed the Portside high-rise building (Fig. 10.8J) in Cape Town using minimum Portland cement by applying the principles of geopolymerization. Although this project should contribute substantially to the building of confidence in geopolymers worldwide, the market remains hesitant to transfer positive experience from one location to the next.

As opposed to focusing on the technical detail of these projects, attention is drawn to the fact that these case studies demonstrate something that is rarely seen in geopolymer technology, i.e., the vast regulatory, asset management, liability, and industry stakeholder engagement process that has been undertaken by Zeobond to commercialize its E-Crete (Fig. 10.7). The shift from the laboratory to the real world is not a process of scale-up of technology alone. While such technical challenges are sufficient to be insurmountable to many, it is the ability to manage the scale-up of industry participation and acceptance of geopolymer concrete that provides the core challenge to the future of a geopolymer cement/concrete industry.

## 10.12 FINAL REMARKS

Significant progress has been made in developing an understanding of the colloid and interface science, gel chemistry, phase formation, reaction kinetics, and transport phenomena underlying geopolymerization of aluminosilicates. Despite much research, the role of calcium and the location of alkali metals in geopolymeric tobermorite-type C-A-S-H phases remain poorly understood. Analysis of the nanostructure of geopolymer gels has enabled the tailored selection of geopolymer precursors and the design of alkali–activator composition; such studies have established the relationship between geopolymer gel microstructure and durability. Geopolymer concrete has performed well in accelerated durability tests, which is supported by its tortuous pore structure and the specific details of its gel chemistry, although detailed analytical work in this area is ongoing. In the absence of a long in-service track record, microstructural research is essential to validate durability testing methodology and improve geopolymer cement technology. Although the principles of particle technology have been applied to the mix design and rheology of fresh geopolymer concrete, this area is in its infancy and could benefit from the knowledge base of particle technology and surface chemistry. For example, suitable superplasticizers for the geopolymer system have been developed only recently, so more progress can be expected in this area.

Demand pull by a carbon-conscious market continues to be the key driver for the adoption of geopolymer cement in Australia. While the scale-up from the laboratory to the real world is technically challenging, the core challenge is the scale-up of industry participation and acceptance of geopolymer cement. High-profile demonstration projects in Australia and South Africa have highlighted the vast regulatory, asset management, liability, and industry stakeholder engagement process required to commercialize geopolymer cement.

The fact that all premixed and precast concrete standards are based on an assumption of the use of Portland cement remains a major obstacle to the commercial adoption of geopolymer cement. Even when asset owners and specifiers such as government, architects, and design engineers accept the results of durability testing of geopolymer concrete, the main barrier to the entry of geopolymers into an established market is access to a suitable supply of source materials including fly ash, GBFS and alkaline activators. Significant development of supply chains to accommodate geopolymer cement, predominantly in developing markets, is essential in overcoming this barrier. This crucial aspect of commercialization is seldom appreciated by the research community, governments, or the ultimate users of concrete. It is important for commercial geopolymer concrete producers to work closely with research partners to develop testing methods for accelerated durability, especially as longer-term in-service testing data become available. Substantial progress has been made in Australia, where the local road authority has recognized geopolymer concrete for non-structural applications.

## ACKNOWLEDGMENTS

This work has been funded by Zeobond Pty Ltd and the Australian Research Council (ARC) through an ARC Linkage Project grant. The author holds a financial interest in Zeobond Pty Ltd, a producer of geopolymer cements and concretes.

# REFERENCES

[1] Rockström J, Steffen W, Noone K, Persson Å, Chapin III FS, Lambin EF, et al. A safe operating space for humanity. Nature 2009;461:472–5.

[2] Richardson K, Steffen W, Schellnhuber HJ, Alcamo J, Barker T, Kammen DM, et al. Synthesis report from climate change: global risks, challenges & decisions. Copenhagen, Denmark: University of Copenhagen; 2009 (http://climatecongress.ku.dk/pdf/synthesisreport/).

[3] Plimer I. Heaven and earth Global warming: the missing science. Ballan, Australia: Connor Court Publishing; 2009, 503 pp.

[4] Freedonia Group. World cement to 2012, http://www.freedoniagroup.com/World-Cement.html; 2009.

[5] Scrivener KL, Kirkpatrick RJ. Innovation in use and research on cementitious material. Cem Concr Res 2008;38(2):128–36.

[6] WWF-Lafarge Conservation Partnership. A blueprint for a climate friendly cement industry: how to turn around the trend of cement related emissions, http://assets.panda.org/downloads/cement_blueprint_climate_fullenglrep_lr.pdf; 2008.

[7] Taylor M, Tam C, Gielen D. Energy efficiency and $CO_2$ emissions from the global cement industry. International Energy Agency; 2006.

[8] Damtoft JS, Lukasik J, Herfort D, Sorrentino D, Gartner E. Sustainable development and climate change initiatives. Cem Concr Res 2008;38(2):115–27.

[9] Juenger MCG, Siddique R. Recent advances in understanding the role of supplementary cementitious materials in concrete. Cem Concr Res 2015;78:71–80.

[10] Juenger MCG, Winnefeld F, Provis JL, Ideker J. Advances in alternative cementitious binders. Cem Concr Res 2011;41:1232–43.

[11] Davidovits J. Geopolymer chemistry & applications, 2nd ed. Saint-Quentin, France: Institut Géopolymère; 2009. 590 pp.

[12] Gartner EM, Macphee DE. A physico-chemical basis for novel cementitious binders. Cem Concr Res 2011;41(7):736–49.

[13] Shi C, Krivenko PV, Roy DM. Alkali-activated cements and concretes. Abingdon, UK: Taylor & Francis; 2006. 376 pp.

[14] Van Deventer JSJ, Provis JL, Duxson P, Brice DG. Chemical research and climate change as drivers in the commercial adoption of alkali activated materials. Waste Biomass Valorization 2010;1(1):145–55.

[15] Van Deventer JSJ, Provis JL, Duxson P. Technical and commercial progress in the adoption of geopolymer cement. Miner Eng 2012;29:89–104.

[16] Xu X, Provis JL, Van Deventer JSJ, Krivenko PV. Characterization of aged slag concretes. ACI Mater J 2008;105(2):131–9.

[17] Provis JL, Palomo A, Shi C. Advances in understanding alkali-activated materials. Cem Concr Res 2015;78:110–25.

[18] Provis JL, Van Deventer JSJ, editors. Geopolymers: structures, processing, properties and industrial applications. Cambridge, UK: Woodhead Publishing, CRC Press; 2009. 454 pp.

[19] Duxson P, Provis JL, Lukey GC, Van Deventer JSJ. The role of inorganic polymer technology in the development of 'Green concrete'. Cem Concr Res 2007;37(12):1590–7.

[20] Provis JL, Duxson P, Van Deventer JSJ. The role of particle technology in developing sustainable construction materials. Adv Powder Technol 2010;21(1):2–7.

[21] Shi C, Fernández-Jiménez A, Palomo A. New cements for the 21st century: the pursuit of an alternative to Portland cement. Cem Concr Res 2011;41(7):750–63.

[22] Duxson P, Fernández-Jiménez A, Provis JL, Lukey GC, Palomo A, Van Deventer JSJ. Geopolymer technology: the current state of the art. J Mater Sci 2007;42(9):2917–33.

[23] Komnitsas K, Zaharaki D. Geopolymerisation: a review and prospects for the minerals industry. Miner Eng 2007;20(14):1261–77.

[24]  Pacheco-Torgal F, Castro-Gomes J, Jalali S. Alkali-activated binders: a review. Part 1. Historical background, terminology, reaction mechanisms and hydration products. Constr Build Mater 2008;22(7):1305–14.

[25]  Pacheco-Torgal F, Castro-Gomes J, Jalali S. Alkali-activated binders: a review. Part 2. About materials and binders manufacture. Constr Build Mater 2008;22(7):1315–22.

[26]  Provis JL, Bernal SA. Geopolymers and related alkali-activated materials. Ann Rev Mater Res 2014;44:299–327.

[27]  Provis JL, Van Deventer JSJ, editors. Alkali-activated materials: state-of-the-art report. Dordrecht: RILEM TC 224-AAM, Springer/RILEM; 2014.

[28]  Duxson P, Provis JL. Designing precursors for geopolymer cements. J Am Ceram Soc 2008;91(12):3864–9.

[29]  Fernández-Jiménez A, Palomo A, Sobrados I, Sanz J. The role played by the reactive alumina content in the alkaline activation of fly ashes. Microporous Mesoporous Mater 2006;91(1–3):111–9.

[30]  Provis JL, Van Deventer JSJ. Geopolymerisation kinetics. 2. Reaction kinetic modeling. Chem Eng Sci 2007;62(9):2318–29.

[31]  Buchwald A, Wiercx J. ASCEM cement technology—alkali-activated cement based on synthetic slag made from fly ash. In: Shi C, Shen X, editors. Advances in chemically activated materials CAM'2010. Jinan, China: RILEM Publications; 2010. p. 15–21.

[32]  Van Deventer JSJ, Feng D, Duxson P. Dry mix cement composition, methods and systems involving same. US Patent 7,691,198 B2. 2010b.

[33]  Keyte LM. What's wrong with Tarong? The importance of coal fly ash glass chemistry in inorganic polymer synthesis, Ph.D. Thesis. Australia: University of Melbourne; 2008.

[34]  White CE, Perander LM, Provis JL, Van Deventer JSJ. The use of XANES to clarify issues related to bonding environments in metakaolin: a discussion of the paper S. Sperinck et al., 'Dehydroxylation of kaolinite to metakaolin—a molecular dynamics study,' J. Mater. Chem., 2011, 21, 2118–2125. J Mater Chem 2011;21(19):7007–10.

[35]  Koloušek D, Brus J, Urbanova M, Andertova J, Hulinsky V, Vorel J. Preparation, structure and hydrothermal stability of alternative (sodium silicate-free) geopolymers. J Mater Sci 2007;42(22):9267–75.

[36]  O'Connor SJ, MacKenzie KJD. Synthesis, characterisation and thermal behaviour of lithium aluminosilicate inorganic polymers. J Mater Sci 2010;45(14):3707–13.

[37]  Lloyd RR, Provis JL, Van Deventer JSJ. Microscopy and microanalysis of inorganic polymer cements. 1: Remnant fly ash particles. J Mater Sci 2009;44(2):608–19.

[38]  Lloyd RR, Provis JL, Van Deventer JSJ. Microscopy and microanalysis of inorganic polymer cements. 2: The gel binder. J Mater Sci 2009;44(2):620–31.

[39]  Yip CK, Lukey GC, Van Deventer JSJ. The coexistence of geopolymeric gel and calcium silicate hydrate at the early stage of alkaline activation. Cem Concr Res 2005;35(9):1688–97.

[40]  Yip CK, Lukey GC, Provis JL, Van Deventer JSJ. Effect of calcium silicate sources on geopolymerisation. Cem Concr Res 2008;38(4):554–64.

[41]  Buchwald A, Hilbig H, Kapps C. Alkali-activated metakaolin-slag blends—performance and structure in dependence on their composition. J Mater Sci 2007;42(9):3024–32.

[42]  García-Lodeiro I, Palomo A, Fernández-Jiménez A, Macphee DE. Compatibility studies between N-A-S-H and C-A-S-H gels. Study in the ternary diagram $Na_2O$-$CaO$-$Al_2O_3$-$SiO_2$-$H_2O$. Cem Concr Res 2011;41(9):923–31.

[43]  Provis JL, Rose V, Bernal SA, Van Deventer JSJ. High resolution nanoprobe X-ray fluorescence characterization of heterogeneous calcium and heavy metal distributions in alkali activated fly ash. Langmuir 2009;25(19):11897–904.

[44]  Oh JE, Monteiro PJM, Jun SS, Choi S, Clark SM. The evolution of strength and crystalline phases for alkali-activated ground blast furnace slag and fly ash-based geopolymers. Cem Concr Res 2010;40(2):189–96.

[45] Bernal SA, Rose V, Provis JL. The fate of iron in blast furnace slag particles during alkali-activation. Mater Chem Phys 2014;146:1–5.

[46] Gomes KC, Lima GST, Torres SM, De Barros S, Vasconcelos IF, Barbosa NP. Iron distribution in geopolymer with ferromagnetic rich precursor. Mater Sci Forum 2010;643:131–8.

[47] Douglas E, Brandstetr J. A preliminary study on the alkali activation of ground granulated blast furnace slag. Cem Concr Res 1990;20:746–56.

[48] Ben Haha M, Lothenbach B, Le Saout G, Winnefeld F. Influence of slag chemistry on the hydration of alkali-activated blast-furnace slag. Part I. Effect of MgO. Cem Concr Res 2011;41:955–63.

[49] Bernal SA, San Nicolas R, Myers RJ, Mejía de Gutiérrez R, Puertas F, Van Deventer JSJ, et al. MgO content of slag controls phase evolution and structural changes induced by accelerated carbonation in alkali-activated binders. Cem Concr Res 2014;57:33–43.

[50] Rees CA, Provis JL, Lukey GC, Van Deventer JSJ. The mechanism of geopolymer gel formation investigated through seeded nucleation. Colloids Surf A 2008;318(1-3):97–105.

[51] Fernández-Jiménez A, Monzó M, Vicent M, Barba A, Palomo A. Alkaline activation of metakaolin–fly ash mixtures: obtain of Zeoceramics and Zeocements. Microporous Mesoporous Mater 2008;108(1–3):41–9.

[52] Provis JL, Lukey GC, Van Deventer JSJ. Do geopolymers actually contain nanocrystalline zeolites? A reexamination of existing results. Chem Mater 2005;17(12):3075–85.

[53] White CE, Provis JL, Llobet A, Proffen T, Van Deventer JSJ. Evolution of local structure in geopolymer gels: an in-situ neutron pair distribution function analysis. J Am Ceram Soc 2011;94(10):3532–9.

[54] Duxson P, Lukey GC, Separovic F, Van Deventer JSJ. The effect of alkali cations on aluminum incorporation in geopolymeric gels. Ind Eng Chem Res 2005;44(4):832–9.

[55] Duxson P, Lukey GC, Van Deventer JSJ. Physical evolution of Na–geopolymer derived from metakaolin up to 1000°C. J Mater Sci 2007;42(9):3044–54.

[56] White CE, Provis JL, Proffen T, Van Deventer JSJ. The effects of temperature on the local structure of metakaolin-based geopolymer binder: a neutron pair distribution function investigation. J Am Ceram Soc 2010;93(10):3486–92.

[57] White CE, Provis JL, Proffen T, Van Deventer JSJ. Molecular mechanisms responsible for the structural changes occurring during geopolymerization: multiscale simulation. AIChE J 2012;58(7):2241–53.

[58] Rees CA, Provis JL, Lukey GC, Van Deventer JSJ. Attenuated total reflectance Fourier transform infrared analysis of fly ash geopolymer gel aging. Langmuir 2007;23(15):8170–9.

[59] Rees CA, Provis JL, Lukey GC, Van Deventer JSJ. In situ ATR-FTIR study of the early stages of fly ash geopolymer gel formation. Langmuir 2007;23(17):9076–82.

[60] Fernández-Jiménez A, Palomo A. Mid-infrared spectroscopic studies of alkali-activated fly ash structure. Microporous Mesoporous Mater 2005;86(1–3):207–14.

[61] Hajimohammadi A, Provis JL, Van Deventer JSJ. The effect of alumina release rate on the mechanism of geopolymer gel formation. Chem Mater 2010;22(18):5199–208.

[62] Hajimohammadi A, Provis JL, Van Deventer JSJ. The effect of silica availability on the mechanism of geopolymerisation. Cem Concr Res 2011;41(3):210–6.

[63] Hajimohammadi A, Provis JL, Van Deventer JSJ. Time-resolved and spatially-resolved infrared spectroscopic observation of seeded nucleation controlling geopolymer gel formation. J Colloid Interface Sci 2011;357(2):384–92.

[64] Myers RJ, Bernal SA, San Nicolas R, Provis JL. Generalized structural description of calcium-sodium aluminosilicate hydrate gels: the crosslinked substituted tobermorite model. Langmuir 2013;29:5294–306.

[65] Richardson IG. Model structures for C-(A)-S-H(I). Acta Crystallogr 2014;B70:903–23.

[66] Myers RJ, Bernal SA, Provis JL. A thermodynamic model for C-(N)-A-S-H gel: CNASH_ss. Derivation and validation. Cem Concr Res 2014;66:27–47.

[67] White CE, Daemen LL, Hartl M, Page K. Intrinsic differences in atomic ordering of calcium (alumino)silicate hydrates in conventional and alkali-activated cements. Cem Concr Res 2015;67:66–73.

[68] Laskar AI, Talukdar S. Rheological behavior of high performance concrete with mineral admixtures and their blending. Constr Build Mater 2008;22(12):2345–54.

[69] Kashani A, San Nicolas R, Qiao GG, Van Deventer JSJ, Provis JL. Modelling the yield stress of ternary cement-slag-fly ash pastes based on particle size distribution. Powder Technol 2014;266:203–9.

[70] Palomo A, Banfill PFG, Fernández-Jiménez A, Swift DS. Properties of alkali-activated fly ashes determined from rheological measurements. Adv Cem Res 2005;17(4):143–51.

[71] Bouzoubaâ N, Zhang MH, Malhotra VM, Golden DM. Blended fly ash cements—a review. ACI Mater J 1999;96(6):641–50.

[72] Kumar S, Kumar R, Alex TC, Bandopadhyay A, Mehrotra SP. Influence of reactivity of fly ash on geopolymerisation. Adv Appl Ceram 2007;106(3):120–7.

[73] Bouzoubaâ N, Zhang MH, Bilodeau A, Malhotra VM. The effect of grinding on the physical properties of fly ashes and a Portland cement clinker. Cem Concr Res 1997;27(12):1861–74.

[74] Hos JP, McCormick PG, Byrne LT. Investigation of a synthetic aluminosilicate inorganic polymer. J Mater Sci 2002;37(11):2311–6.

[75] Rajaokarivony-Andriambololona Z, Thomassin JH, Baillif P, Touray JC. Experimental hydration of two synthetic glassy blast furnace slags in water and alkaline solutions (NaOH and KOH 0.1 N) at 40°C: structure, composition and origin of the hydrated layer. J Mater Sci 1990;25(5):2399–410.

[76] Shimoda K, Tobu Y, Kanehashi K, Nemoto T, Saito K. Total understanding of the local structures of an amorphous slag: perspective from multi-nuclear ($^{29}$Si, $^{27}$Al, $^{17}$O, $^{25}$Mg, and $^{43}$Ca) solid-state NMR. J Non-Cryst Solids 2008;354(10–11):1036–43.

[77] Brouwers H. The work of Powers and Brownyard revisited: Part 1. Cem Concr Res 2004;34(9):1697–716.

[78] Lothenbach B, Le Saout G, Gallucci E, Scrivener K. Influence of limestone on the hydration of Portland cements. Cem Concr Res 2008;38(6):848–60.

[79] Sindhunata Provis JL, Lukey GC, Xu H, Van Deventer JSJ. Structural evolution of fly ash-based geopolymers in alkaline environments. Ind Eng Chem Res 2008;47(9):2991–9.

[80] Duxson P, Provis JL, Lukey GC, Mallicoat SW, Kriven WM, Van Deventer JSJ. Understanding the relationship between geopolymer composition, microstructure and mechanical properties. Colloids Surf A 2005;269(1–3):47–58.

[81] Lloyd RR, Provis JL, Smeaton KJ, Van Deventer JSJ. Spatial distribution of pores in fly ash-based inorganic polymer gels visualised by Wood's metal intrusion. Microporous Mesoporous Mater 2009;126(1–2):32–9.

[82] Provis JL, Myers RJ, White CE, Rose V, Van Deventer JSJ. X-ray microtomography shows pore structure and tortuosity in alkali-activated binders. Cem Concr Res 2012;42(6):855–64.

[83] Hunger M, Brouwers HJH. Flow analysis of water-powder mixtures: Application to specific surface area and shape factor. Cem Concr Compos 2009;31(1):39–59.

[84] Amirjanov A, Sobolev K. Optimization of a computer simulation model for packing of concrete aggregates. Part Sci Technol 2008;26(4):380–95.

[85] Goltermann P, Johansen V, Palbol L. Packing of aggregates: an alternative tool to determine the optimal aggregate mix. ACI Mater J 1997;94(5):435–43.

[86] Wong HHC, Kwan AKH. Packing density of cementitious materials: measurement and modelling. Mag Concr Res 2008;60(3):165–75.

[87] Jones MR, Zheng L, Newlands MD. Comparison of particle packing models for proportioning concrete constituents for minimum voids ratio. Mater Struct 2002;35(5):301–9.

[88] Jones MR, Zheng L, Newlands MD. Estimation of the filler content required to minimise voids ratio in concrete. Mag Concr Res 2003;55(2):193–202.

[89] Ji T, Lin TW, Lin XJ. A concrete mix proportion design algorithm based on artificial neural networks. Cem Concr Res 2006;36(7):1399–408.

[90] Puertas F, Palomo A, Fernández-Jiménez A, Izquierdo JD, Granizo ML. Effect of superplasticisers on the behaviour and properties of alkaline cements. Adv Cem Res 2003;15(1):23–8.

[91] Criado M, Palomo A, Fernández-Jiménez A, Banfill PFG. Alkali activated fly ash: effect of admixtures on paste rheology. Rheol Acta 2009;48(4):447–55.

[92] Kashani A, Provis JL, Qiao GG, Van Deventer JSJ. The interrelationship between surface chemistry and rheology in alkali activated slag paste. Constr Build Mater 2014;65:583–91.

[93] Kashani A, Provis JL, Xu J, Kilcullen AR, Qiao GG, Van Deventer JSJ. Effect of molecular architecture of polycarboxylate ethers on plasticizing performance in alkali activated slag paste. J Mater Sci 2014;49(7):2761–72.

[94] Hobbs DW. Concrete deterioration: causes, diagnosis, and minimising risk. Int Mater Rev 2001;46(3):117–44.

[95] Sofi M, Van Deventer JSJ, Mendis PA, Lukey GC. Engineering properties of inorganic polymer concretes (IPCs). Cem Concr Res 2007;37(2):251–7.

[96] Byfors K, Klingstedt G, Lehtonen HP, Pyy H, Romben L. Durability of concrete made with alkali-activated slag. In: Malhotra VM, editor. 3rd international conference on fly ash, silica fume, slag and natural pozzolans in concrete. ACI SP 114; 1989. p. 1429–44.

[97] Häkkinen T. The permeability of high strength blast furnace slag concrete. Nordic Concr Res 1992;11(1):55–66.

[98] Häkkinen T. The microstructure of high strength blast furnace slag concrete. Nordic Concr Res 1992;11(1):67–82.

[99] Collins F, Sanjayan JG. Microcracking and strength development of alkali activated slag concrete. Cem Concr Compos 2001;23(4–5):345–52.

[100] Bernal SA, Mejía de Gutierrez R, Pedraza AL, Provis JL, Rodríguez ED, Delvasto S. Effect of binder content on the performance of alkali-activated slag concretes. Cem Concr Res 2011;41(1):1–8.

[101] Chen W, Brouwers H. The hydration of slag, part 1: reaction models for alkali-activated slag. J Mater Sci 2007;42(2):428–43.

[102] Justnes H, Ardoullie B, Hendrix E, Sellevold EJ, Van Gemert D. The chemical shrinkage of pozzolanic reaction products. In: Malhotra VM, editor. 6th CANMET conference on fly ash, silica fume, slag, and natural pozzolans in concrete. Malhotra, Bangkok, Thailand; 1198. p. 191–205.

[103] Mehta PK, Monteiro PJM. Concrete: microstructure Properties and materials, 3rd ed. New York: McGraw-Hill; 2006, p. 659.

[104] Ollivier JP, Maso JC, Bourdette B. Interfacial transition zone in concrete. Adv Cem Based Mater 1995;2(1):30–8.

[105] San Nicolas R, Bernal SA, Mejía de Gutiérrez R, Van Deventer JSJ, Provis JL. Distinctive microstructural features of aged sodium silicate-activated slag concretes. Cem Concr Res 2014;65(201):41–51.

[106] Shi C, Xie P. Interface between cement paste and quartz sand in alkali-activated slag mortars. Cem Concr Res 1998;28(6):887–96.

[107]  Zhang JX, Sun HH, Wan JH, Yi ZL. Study on microstructure and mechanical property of interfacial transition zone between limestone aggregate and sialite paste. Constr Build Mater 2009;23(11):3393–7.

[108]  Lee WKW, Van Deventer JSJ. The interface between natural siliceous aggregates and geopolymers. Cem Concr Res 2004;34(2):195–206.

[109]  Lee WKW, Van Deventer JSJ. Chemical interactions between siliceous aggregates and low-Ca alkali-activated cements. Cem Concr Res 2007;37(6):844–55.

[110]  Provis JL, Muntingh Y, Lloyd RR, Xu H, Keyte LM, Lorenzen L, et al. Will geopolymers stand the test of time? Ceram Eng Sci Proc 2007;28(9):235–48.

[111]  Diamond S. Mercury porosimetry. An inappropriate method for the measurement of pore size distributions in cement-based materials. Cem Concr Res 2000;30(10):1517–25.

[112]  Kaufmann J, Loser R, Leemann A. Analysis of cement-bonded materials by multi-cycle mercury intrusion and nitrogen sorption. J Colloid Interface Sci 2009;336(2):730–7.

[113]  Sagoe-Crentsil K, Brown T, Yan S. Medium to long term engineering properties and performance of high-strength geopolymers for structural applications. Adv Sci Technol 2010;69:135–42.

[114]  Shi C. Strength, pore structure and permeability of alkali-activated slag mortars. Cem Concr Res 1996;26(12):1789–99.

[115]  Olivia M, Nikraz H, Sarker P. Improvements in the strength and water penetrability of low calcium fly ash based geopolymer concrete. In: Uomoto T, Nga TV, editors. 3rd ACF international conference—ACF/VCA 2008. Ho Chi Minh City, Vietnam; 2008. p. 384–91.

[116]  Shi C. Effect of mixing proportions of concrete on its electrical conductivity and the rapid chloride permeability test (ASTM C1202 or ASSHTO T277) results. Cem Concr Res 2004;34(3):537–45.

[117]  Liu Z, Beaudoin JJ. The permeability of cement systems to chloride ingress and related test methods. Cem Concr Aggregates 2000;22(1):16–23.

[118]  Roy DM, Jiang W, Silsbee MR. Chloride diffusion in ordinary, blended, and alkali-activated cement pastes and its relation to other properties. Cem Concr Res 2000;30(12):1879–84.

[119]  Lloyd RR, Provis JL, Van Deventer JSJ. Pore solution composition and alkali diffusion in inorganic polymer cement. Cem Concr Res 2010;40(9):1386–92.

[120]  Brough AR, Atkinson A. Automated identification of the aggregate-paste interfacial transition zone in mortars of silica sand with Portland or alkali-activated slag cement paste. Cem Concr Res 2000;30(6):849–54.

[121]  Wong HS, Buenfeld NR, Head MK. Estimating transport properties of mortars using image analysis on backscattered electron images. Cem Concr Res 2006;36(8):1556–66.

[122]  Willis KL, Abell AB, Lange DA. Image-based characterization of cement pore structure using Wood's metal intrusion. Cem Concr Res 1998;28(12):1695–705.

[123]  Nemati KM. Preserving microstructure of concrete under load using the Wood's metal technique. Int J Rock Mech Min Sci 2000;37(1–2):133–42.

[124]  Gallucci E, Scrivener K, Groso A, Stampanoni M, Margaritondo G. 3D experimental investigation of the microstructure of cement pastes using synchrotron X-ray microtomography (CT). Cem Concr Res 2007;37(3):360–8.

[125]  Provis JL, Rose V, Winarski RP, Van Deventer JSJ. Hard X-ray nanotomography of amorphous aluminosilicate cements. Scr Mater 2011;65(4):316–9.

[126]  Nakashima Y, Kamiya S. Mathematica programs for the analysis of three-dimensional pore connectivity and anisotropic tortuosity of porous rocks using X-ray computed tomography image data. J Nucl Sci Technol 2007;44(9):1233–47.

[127] Ismail I, Provis JL, Van Deventer JSJ, Hamdan S. The effect of water content on compressive strength of geopolymer mortars. In: CD proceedings of 7th AES-ATEMA 2011 international conference on advances and trends in engineering materials and their applications. Milan, Italy, 4–8 July 2011; 2011.

[128] Deja J. Carbonation aspects of alkali activated slag mortars and concretes. Silic Indus 2002;67(1):37–42.

[129] Adam AA. Strength and durability properties of alkali activated slag and fly ash-based geopolymer concrete. PhD Thesis. Melbourne, Australia: RMIT University; 2009.

[130] Rodríguez E, Bernal S, Mejía de Gutierrez R, Puertas F. Alternative concrete based on alkali-activated slag. Materiales de Construcción 2008;58(291):53–67.

[131] Bernal SA, San Nicolas R, Provis JL, Mejía de Gutiérrez R, Van Deventer JSJ. Natural carbonation of aged alkali-activated slag concretes. Mater Struct 2014;47(4):693–707. 2014.

[132] Vermeulen E, De Vries P. De uitdagingen van geopolymeerbeton (English: The challenges of geopolymer concrete). Betoniek Vakblad (In Dutch) 2015;2:20–7.

[133] International Energy Agency. Tracking industrial energy efficiency and $CO_2$ emissions—executive summary;2007. Retrieved at http://www.iea.org/Textbase/npsum/tracking2007SUM.pdf.

[134] Weil M, Dombrowski K, Buchwald A. Life-cycle analysis of geopolymers. In: Provis JL, van Deventer JSJ, editors. Geopolymers: structures, processing, properties and industrial applications. Cambridge, UK: Woodhead Publishing, CRC Press; 2009. p. 194–212

[135] McLellan BC, Williams RP, Lay J, van Riessen A, Corder GD. Costs and carbon emissions for geopolymer pastes in comparison to ordinary Portland cement. J Clean Prod 2011;19(9–10):1080–90.

[136] Habert G, d'Espinose de Lacaillerie JB, Roussel N. An environmental evaluation of geopolymer based concrete production: reviewing current research trends. J Clean Prod 2011;19(11):1229–38.

[137] Fawer M, Concannon M, Rieber W. Life cycle inventories for the production of sodium silicates. Int J Life Cycle Assess 1999;4(4):207–12.

[138] Provis JL. Green concrete or red herring? Future of alkali-activated materials. Adv Appl Ceram 2014;113(8):472–6.

[139] Von Weizsäcker E, Hargroves K, Smith MH, Desha C, Stasinopoulos P. Factor five: transforming the global economy through 80% improvements in resource productivity. London, UK: Earthscan; 2009.

[140] Hooton RD. Bridging the gap between research and standards. Cem Concr Res 2008;38(2):247–58.

[141] Van Deventer JSJ, Brice DG, Bernal SA, Provis JL. Development, standardization and applications of alkali-activated concretes. In: Leslie Struble, James K. Hicks, editors. ASTM symposium on geopolymer binder systems, special technical paper 1566. West Conshohocken, PA: PAASTM International; 2013. p. 196–212.

[142] Alexander M, Thomas M. Service life prediction and performance testing—current developments and practical applications. Cem Concr Res 2015;78:155–64.

[143] VicRoads. Standard Specification Section 703, General Concrete Paving, <http://www.vicroads.vic.gov.au/Home/Moreinfoandservices/TendersAndSuppliers/StandardDocuments.htm>; 2010.

[144] Provis JL, Ismail I, Myers RJ, Rose V, Van Deventer JSJ. Characterising the structure and permeability of alkali-activated binders. RILEM international conference on advances in construction materials through science and engineering. Hong Kong; 4–8 September 2011; 2011.

[145] Jahanshahi S, Xie D. A new integrated dry slag granulation and heat recovery process. In: McCaffrey R, Edwards P, editors. CD proceedings of 6th global slag conference. Sydney, Australia: PRo Publications Int. Ltd; 2010.

# CHAPTER 11

# An Overview on the Influence of Various Factors on the Properties of Geopolymer Concrete Derived From Industrial Byproducts

W.K. Part, M. Ramli and C.B. Cheah
Universiti Sains Malaysia, Penang, Malaysia

## 11.1 INTRODUCTION

Since the 1980s, the feasibility and the sustainability of utilizing ordinary Portland cement (OPC) as primary construction material has been questioned extensively due to the environmental impact resulting from the production of clinker, in particularly the greenhouse gas emissions and the embodied energy of production [1,2]. In fact, the production of Portland cement clinker from cement production plants worldwide emits up to 1.5 billion tons of $CO_2$ annually. This accounts for around 5% of the total man-made $CO_2$ emissions and if this undesirable trend continues, the figure will rise to 6% by 2015 [3–5]. Apart from OPC, sand and aggregate are also the main constituent source materials in the production of concrete, which originated from quarrying operations, which are both energy intensive and produce a high level of waste materials. A shortage of natural resources for construction materials in many developing countries has also led to long-distance haulage of the materials and thus significantly increased the life-cycle carbon emission and production cost of concrete. All of the above-mentioned issues are against the context of sustainable development in the construction industry. Hence, immediate remedy actions must be taken to ensure sustainability in the construction industry [1].

The aforementioned issues prompted various studies in an attempt to reduce the global carbon footprint, ranging from utilizing supplementary cementitious materials as partial cement replacement materials [6–10] to developing a whole new cementless binder, namely geopolymer [11–15]. Geopolymer is an alternative cementitious material synthesized

*Handbook of Low Carbon Concrete.*
DOI: http://dx.doi.org/10.1016/B978-0-12-804524-4.00011-7

by combining source materials that are rich in silica and alumina such as fly ash (FA), ground granulated blast furnace slags (GGBFS) with strong alkali solutions such as potassium hydroxide (KOH), sodium hydroxide (NaOH). In many circumstances, soluble silicates such as sodium silicate are also used as the primary alkali activator. Geopolymer binder is formed when the dissolved $Al_2O_3$ and $SiO_2$ minerals undergo geopolymerization to form a three-dimensional (3D) amorphous aluminosilicate network with strength similar or higher than that of OPC concrete. Generally, the mechanism of geopolymerization can be divided into three main stages: (1) dissolution of oxide minerals from the source materials (usually silica and alumina) under highly alkaline conditions; (2) transportation/orientation of dissolved oxide minerals, followed by coagulation/gelation; (3) polycondensation to form a 3D network of silicoaluminates structures [16]. Based on the types of resultant chemical bonding, three types of structures can be derived from the 3D aluminosilicate network: poly(sialate) (-Si-O-Al-O-), poly(sialate-siloxo) (Si-O-Al-O-Si-O) and poly(sialate-disiloxo) (Si-O-Al-O-Si-O-Si-O-) [17]. As most of the geopolymer's source materials, e.g., FA and GGBFS, originated from the waste stream of their respective industries, the carbon footprint of these materials is extremely low compared to OPC. This is because processing energy is incurred only during the posttreatment process, namely drying, milling, and separation. This presents fellow researchers and concrete manufacturers a very attractive avenue for reducing the carbon footprint of concrete materials and the construction industry as a whole.

The potential of geopolymer binders to replace the traditional OPC binders was further supported by the fact that there is abundance of industrial byproducts generated in various industries that was found to be suitable to use as geopolymer source materials. These industrial wastes pose a significant challenge in term of finding an ideal solution for disposal purposes. For instance, pulverized fuel ash (PFA) or more commonly known as FA, an industrial byproduct of the coal-burning power plant industry, makes up 75–80% of global annual ash production [18]. With alkali activation, PFA can be used for production of geopolymer concrete with superior mechanical and durability properties as compared to OPC concrete [19–21]. GGBFS, a byproduct of pig-iron manufacture from iron ore, has also found significant use in the production of high-strength geopolymer concrete [22,23]. The use of palm oil fuel ash (POFA), waste materials derived from the burning of empty fruit brunches, oil palm shells and oil palm clinker from the oil palm industry to generate electricity as geopolymer binder, has gathered pace in recent years. POFA is widely used as a geopolymer binder especially in oil palm–rich countries such as Malaysia and Thailand due to its increasing amount, which rendered the disposal method by means of landfilling not economically and technically feasible [4,11,24]. Other industrial byproducts, for example, rice husk ash (RHA) from the rice milling industry, red mud (RM) from the alumina refining industry, and copper and hematite mine tailings (MTs) from the mining industry [14,21,25,26] have also found considerable use in the fabrication of geopolymer concrete.

With the ever-present problem of reducing the carbon footprint in the construction industry, coupled with the problems of disposing of industrial byproducts in various industries, there is a huge potential for geopolymeric binder to completely replace OPC for concrete production. This marks the emergence of a new low-carbon binding material for the production of concrete and ultimately ensuring the sustainability of construction industry in terms of environmental perspective. The current work aims to review the current research and development of geopolymer concrete by identifying the governing factors that affect the various properties of the resultant geopolymer concrete. Besides, the relationship of these strength–determining factors with the eventual carbon footprint of the end products is also assessed. This chapter is also constructed with the aim to identify the various challenges presents in the geopolymer field that must be overcome to ensure full-scale industrial implementation of geopolymer technology. Some proposed solutions and recommendations to address the various challenges are also included in the current chapter.

## 11.2 EFFECT OF CHEMICAL ACTIVATORS AND CURING REGIME ON THE MECHANICAL, DURABILITY, SHRINKAGE, MICROSTRUCTURE, AND PHYSICAL PROPERTIES OF GEOPOLYMER

The efficiency of the alkali activation process of geopolymers is very much dependent on the addition of chemical activators (sodium/potassium hydroxide, soluble silicates, etc.) and also the curing regime (heat treatment) employed on the hardened geopolymer concrete. Strength development of geopolymers fabricated without the addition of chemical activators or subsequent heat treatment is very slow, particularly during the early stages. This renders the industrial application of geopolymers, especially in precast industries, impractical. The following sections discuss in detail the importance or the effect of chemical activators' addition and curing regime employed towards the mechanical, durability, shrinkage, microstructure, and physical properties of geopolymers.

## 11.2.1 Mechanical Properties

Chemical-activator or alkali–activator solution plays a vital role in the initiation of the geopolymerization process. Generally, a strong alkaline medium is necessary to increase the surface hydrolysis of the aluminosilicate particles present in the raw material while the concentration of the chemical activator has a pronounced effect on the mechanical properties of the geopolymers [27,28]. On the other hand, the dissolution of Si and Al species during the synthesis of geopolymer is very much dependent on the concentration of NaOH, where the amount of Si and Al leaching is mostly governed by the NaOH concentration and also the leaching time [29]. Somna et al. [30] studied the compressive strength of ground fly ash (GFA) cured at ambient temperature by varying the NaOH concentration from 4.5 to 16.5 M. Results showed that by increasing NaOH concentrations from 4.5 to 9.5 M, a significant increase in the compressive strength of paste samples can be observed, while the variation of NaOH concentrations from 9.5 to 14 M also increases the compressive strength of paste samples, but in a much lesser extent. The increase in compressive strength with the increasing NaOH concentrations is mainly due to the higher degree of silica and alumina leaching. The compressive strength of GFA-hardened pastes starts to decline at the NaOH concentrations of 16.5 M. This decrease in compressive strength is mainly attributed to the excess hydroxide ions that caused the precipitation of aluminosilicate gel at very early stages, thus resulting in the formation of lower-strength geopolymers.

Gorhan and Kurklu [19] investigated the influence of the NaOH solution on the 7 days compressive strength of ASTM Class F FA geopolymer mortars subjected to different NaOH concentrations. Three different concentrations of NaOH (3, 6, and 9 M) were used throughout the laboratory work while other parameters such as sand/ash ratio and sodium silicate/NaOH (SS/SH) ratio was maintained constant. Based on the compressive strength results acquired, the optimum NaOH concentration that produced the highest 7 days compressive strength of 22.0 MPa is 6 M. In the aforementioned concentration, an ideal alkaline environment was provided for proper dissolution of FA particles and at the same time the polycondensation process was not hindered. When the NaOH concentration is too low at 3 M, it is not sufficient to stimulate a strong chemical reaction, while the excessively high concentration of NaOH (9 M) resulted in premature coagulation of silica that in both cases culminated in lower-strength mortars.

Eco-friendly geopolymer bricks were fabricated by Ahmari and Zhang [25] using solely copper MTs and NaOH as the chemical activators. NaOH concentrations were varied from 10 to 15 M to study the effect of NaOH concentration on the unconfined compressive strength (UCS) of cured geopolymer bricks. The authors reported that the UCS of 15 M NaOH specimens is higher than the 10 M NaOH counterparts for all the mixtures due to the higher NaOH/MT ratio, which consequently resulted in higher Na/Al and Na/Si ratios. This in turn produced a much thicker geopolymer binder gel that binds the unreacted particles and directly contributes to the UCS of geopolymer bricks.

According to Komljenovic et al. [31], the nature and concentration of alkali activators is the most dominant parameter in the alkali activation process. The authors utilized five different types of alkali activators, i.e., $Ca(OH)_2$, NaOH, NaOH + $Na_2CO_3$, KOH, and $Na_2SiO_3$, with various concentrations to fabricate FA-based geopolymer mortars. The curing condition was made constant in order to effectively study the influence of types of alkali activators on the mechanical properties of the geopolymer mortars. Based on the compressive strength results, the alkali activator that possesses highest activation potential, i.e., highest compressive strength, was $Na_2SiO_3$, followed by $Ca(OH)_2$, NaOH, NaOH + $Na_2CO_3$, and KOH. The lower activation potential of KOH compared to NaOH was due to the difference in ionic diameter difference between sodium and potassium. Regardless of the types of alkali activators used, the compressive strength generally increased with the increase in activator concentration. The authors also concluded the optimum value of $Na_2SiO_3$ modulus was 1.5. Anything higher than the prescribed modulus will have deleterious effects on the compressive strength of geopolymer mortars.

While most of the research papers reported enhancement in compressive strength with the increase in the concentration of chemical activators, particularly NaOH [19,25,30,32], some research shows a total contrast in compressive strength development. For example, a study done by He et al. [14], which focused on RM-/RHA-based geopolymer concluded that a higher concentration of NaOH had resulted in a decrease in the compressive strength of the geopolymer. The possible reasons for the contrasting trends could be attributed to (1) the high viscosity of NaOH solution due to fact that the higher concentration disrupts the leaching of Si and Al ions, (2) excessive $OH^-$ concentration results in premature precipitation of geopolymeric gels, (3) the partially reacted/unreacted RHA particles due to the incomplete dissolution of Si and Al species caused the deterioration in the mechanical properties of the geopolymer produced.

The dissolution, hydrolysis, and condensation reaction of geopolymers are greatly affected by the effective Si/Al ratios. In a low Si/Al ratio geopolymer system, the condensation reaction tends to occur between aluminate and silicate species, thus resulting in mainly poly(sialate) geopolymeric structures. On the other hand, the condensation reaction in a high Si/Al system would result in a predominantly between-the-silicate species itself, forming oligomeric silicates that in turn condense with $Al(OH_4)^{4-}$ and form geopolymeric structures of poly(sialate-siloxo) and poly(sialate-disiloxo) [16,33].

Sukmak et al. [34] studied the effect of sodium silicate/sodium hydroxide ($Na_2SiO_3$/NaOH) and liquid alkaline activator/FA (L/FA) ratios on the compressive strength development of clay-FA geopolymer bricks under prolonged curing ages. The $Na_2SiO_3$/NaOH ratios used were 0.4, 0.7, 1.0, 1.5, and 2.3 while the L/FA ratios used were 0.4, 0.5, 0.6, and 0.7 by dry clay mass. The clay–FA geopolymer bricks were compressed using a hand-operated hydraulic jack at the optimum water content to attain maximum dry unit weight. The bricks were left to set at room temperature for 24 h before being subjected to oven curing at 75°C for 48 h. The compressive strength tests were performed after 7, 14, 28, 60, and 90 days of curing. Results showed that L/FA ratios of less than 0.3 and greater than 0.8 are not suitable for fabrication of clay–FA geopolymer bricks as the strength is null at the aforementioned L/FA ratios. The optimum values of $Na_2SiO_3$/NaOH and L/FA ratios are 0.7 and 0.6, respectively. The optimum $Na_2SiO_3$/NaOH ratio of 0.7 is less than that of FA-based geopolymers as the clay possesses high cation absorption capability and may have absorbed some of the NaOH added. The significant decrease in strength for clay-FA geopolymer bricks with excessive alkali activator (L/FA > 0.6) was claimed to be due to the precipitation of dissolved Si and Al species at the early stages before the initiation of the polycondensation process, resulting in the formation of cracks on the FA particles. The maximum compressive strength attained at the aforementioned optimized specimens was approximately 15 MPa at the age of 90 days. On the other hand, Ridtirud et al. [35] reported the optimum SS/SH ratio for FA-based geopolymer mortar to be 1.5. FA geopolymer mortars with SS/SH ratios of 0.33, 0.67, 1.0, 1.5, and 3.0 achieved compressive strengths of 25.0,

28.0, 42.0, 45.0, and 23.0 MPa, respectively. The increasing trend of compressive strength is mainly attributed to the increasing Na content in the mixture where the $Na^+$ ion plays a critical role in the formation of geopolymer by acting as a charge, balancing ions. However, excessive silicate in the geopolymer system reduces its compressive strength as excess sodium silicate hampered the evaporation of water and also disrupted the formation of 3D networks of aluminosilicate geopolymers.

Various factors such as liquid alkaline/ash ratio, SS/SH ratio, and NaOH concentration on the compressive strength of ground bottom ash (GBA) geopolymer mortars were reported by Sathonsaowaphak et al. [32]. Besides, water and naphthalene-based superplasticizer (NCP) were also incorporated into the mortar mixes in an effort to improve the workability while maintaining the strength of the mortars. Results have shown that the liquid alkaline/ash ratio, SS/SH ratio and NaOH concentration values of 0.4209–0.709, 0.67–1.5 and 10 M, respectively, produced GBA geopolymer mortars with high compressive strength and desired workability. The authors also stressed that the addition of 10 M NaOH solution is essential to the geopolymerization as the $Na^+$ ions act as charge-balancing ions while the NaOH solution increases the dissolution rate of silica and alumina. While for POFA geopolymers, the optimum solid-to-liquid ratios and SS/SH ratios to achieve highest compressive strength were reported to be 1.32 and 2.5, respectively [36]. Higher presence of voids in solid to liquid ratio less than 1.32 will adversely affect the compressive strength. Also, SS/SH ratios more than 2.5 will cause excessive sodium silicate, which hindered the geopolymerization process.

A proper adjustment of $SiO_2/Al_2O_3$ ratio in geopolymers by hybridizing two different sources of aluminosilicate source material and adjustment in hybridization ratios can improve the compressive strength of geopolymers [14,21,37]. Nazari et al. [21] proposed an innovative approach for large-volume recycling of rice husk-bark ash (RHBA), which is a solid waste, generated by biomass power plants using rice husks and eucalyptus bark as source of fuel. The proposed approach requires the blending of RHBA (high silica source) with FA in order to modify the chemical composition of the resultant geopolymer. Sodium silicate and varying concentration of NaOH (4, 8, and 12 M) were used as chemical activators for solidification and stabilization of the RHBA-FA blend. SS/SH ratio and chemical activator to FA-RHBA mixture were fixed at 2.5 and 0.4, respectively. RHBA was added to the mixture at replacement levels of 20%, 30%, and 40% by total binder weight. The specimens were tested after being subjected to 80°C oven curing for 36 h upon precuring duration of 24 h. The authors concluded that at all FA replacement levels, the compressive strength of FA-RHBA geopolymer increased proportionally with the concentration of NaOH. Moreover, RHBA-FA geopolymer with FA replacement level of 30% using RHBA at any concentration of NaOH exhibited the highest compressive strength among the various geopolymer mixes examined. The compressive strength of geopolymer with the aforementioned FA replacement level by RHBA and varying NaOH concentrations was found to range between 20 and 30 MPa.

Besides the SS/SH ratio, it is also very important to study the effect of silica modulus ($M_s$) of the activator and its relationship with the SS/SH ratio in order to maximize the strength and also the economy aspect in the synthesis of alkali-activated binders [38,39]. $M_s$ determines the amount of soluble silicates present and is crucial in controlling the dissolution rate and also the gelation process during geopolymerization, which in turn has a significant effect on the strength development of the hardened geopolymer mixes. However, the appropriate $M_s$ varied for different geopolymer systems comprising of different source materials, i.e., different chemical composition, indicates the need to understand the suitable $M_s$ for each category of geopolymer. For instance, Guo et al. [40] attempted to evaluate the influence of the $M_s$ and the content of alkali activator on the compressive strength of Class C Fly Ash (CFA)-based geopolymer. Mixtures of sodium silicate and sodium hydroxide were used as the alkali activator for the CFA geopolymer (CFAG). The silica–alkali modulus of the alkali activator was varied from 1.0 up to 2.0 while the content of alkali activator was based on the mass proportion of $Na_2O$ to CFA and ranged between 5% and 15%. From the results acquired, both the silica–alkali modulus and content of alkali activator were proven to be equally crucial for the strength development of CFAGs. The optimum modulus and content of alkali activator values were found to be 1.5% and 10%, respectively, which yielded 3, 7, and 28 days compressive strength values of 22.6, 34.5, and 59.3 MPa, respectively when cured under normal room temperature (23°C).

Law et al. [39] suggested that the optimum $M_s$ for class F FA based geopolymer concrete is 1.0, and that further increase in $M_s$ does not bring about any significant increase in compressive strength. The authors suggested that at $M_s > 1.0$, either all the FA particles have been dissolved, or any increase in $M_s$ beyond 1.0 does not result in further dissolution of the protective crust on the FA particles developed from the precipitation of the geopolymerization reaction products. Yusuf et al. [38] found that the strength of alkaline-activated ground steel slag/ultrafine (AAGU)

POFA with various $M_s$ ranging from 0.915 to 1.635 to be insignificant, with $M_s$ = 0.915 yielded compressive strength of 69.13 MPa while $M_s$ = 1.635 yielded compressive strength of 65 MPa.

Generally, compressive strength can be correlated with the modulus of elasticity, where a higher degree of geopolymerization brings about a denser geopolymer matrix, which in turns resulted in higher compressive strength and modulus of elasticity [41]. Chemical activators have a pronounced effect on the compressive strength development of geopolymer concrete. However, other mechanical properties such as modulus of elasticity of the resultant geopolymer concrete do not depend entirely on the chemical activator dosage [42]. It was found that modulus of elasticity was governed by the amount of aggregate present in the geopolymer concrete mixtures. A proper adjustment in the total aggregate content and also the ratio of fine aggregate to total aggregate can result in geopolymer concrete with equal or higher modulus of elasticity compared with OPC concrete [18]. In separate studies, it was reported that the high silicate content might increase the elasticity of geopolymer concrete, subsequently resulting in a lower modulus of elasticity than OPC concrete [43,44]. Topark-Ngarm et al. [41] reported high-calcium FA geopolymer concrete can exhibit similar or higher modulus of elasticity with a lower SS/SH ratio, corresponding to a higher amount of $Na_2O$.

It is well known that conventional geopolymers require heat treatment in order to attain similar or higher compressive strength in comparison with OPC concrete [14,23,24,45,46]. Heat treatment is beneficial towards the dissolution and geopolymerization of aluminosilicate gel, which results in high early-strength gain [47]. It also helps in accelerating the dissolution of silica and alumina species and the subsequent polycondensation process. However, the heat-curing regime applied must be appropriate in such a way that it provides an ideal condition for the proper dissolution and precipitation of dissolved silica and alumina species. Geopolymerization might be hindered upon exceeding certain temperature and heat treatment period, depending on the source material, which in turns adversely affect the mechanical properties of the geopolymers [14,21,25]. Different types of heat treatment and curing environment such as steam and autoclave curing [22], saline-water curing [20] and even microwave-assisted curing regime [48] have been employed in order to maximize the potential and capacity of geopolymerization.

He et al. [14] studied the effect of curing period on the compressive strength of geopolymer paste derived from two industrial wastes, namely

RM and RHA. The geopolymer paste was activated using 4 M NaOH with the constant solution-to-solid weight ratio of 1.2. The RHA/RM ratio was fixed at 0.4. Upon casting, the geopolymer pastes specimens were left to cure at room temperature and atmospheric pressure for the periods of 14, 28, 35, 42, and 49 days before being subjected to compression test. The following conclusions were derived from the experimental study: (1) a near constant compressive strength value of 11.7 MPa was obtained at the testing age of 35 days, which implies that the geopolymer-paste specimens have only achieved complete geopolymerization at that particular time frame; (2) the slower rate of RM and RHA geopolymerization compared with other geopolymer source materials such as FA and metakaolin (MK)-based geopolymer is due to the dominant crystalline solid phases in RM, which acted as unreactive fillers; the larger particle size of RHA, which slows down the rate of dissolution; and also the considerable amount of impurities present in both RM and RHA, which may cause a deleterious effect on the geopolymerization rate.

The effect of curing temperature on the compressive strength of FA-RHBA geopolymer was studied by Nazari et al. [21]. FA was replaced by RHBA at three replacement levels of 20%, 30%, and 40% by total binder weight. The FA-RHBA geopolymer mixtures were activated by sodium silicate and NaOH with sodium silicate/NaOH of 2.5, NaOH concentration of 12 M and alkaline activator/ash ratio of 0.4. A precuring period of 24 h was allowed upon casting to increase the homogeneity of geopolymeric materials before the application of heat curing. After the precuring period, the geopolymer samples were subjected to 50–90°C oven curing for 36 h. From the results of compressive strength the following conclusions were made: (1) the optimum curing temperature for all the mixtures at both 7 and 28 days of curing is 80°C; (2) compressive strength of samples cured at higher temperature i.e., 90°C, start to decrease after certain period of time. This is because prolonged curing at high temperature could destroy the granular structure of geopolymer. High-temperature curing had also resulted in dehydration of the geopolymer matrix and subsequently excessive shrinkage due to contraction of the polymeric gel; (3) compressive strength as high as 58.9 MPa after 28 days of curing was achieved for the geopolymer mixture with FA replacement level of 30% cured at 80°C for 36 h.

An experimental study on the effect of curing temperature on the UCS of copper MTs-based geopolymer bricks showed that the optimum curing temperature that produced the highest UCS geopolymer bricks is

90°C. Any temperature higher than that will cause a drastic drop in UCS of the geopolymer bricks. The authors concluded that too high a temperature will cause a rapid polycondensation process and excessive early formation of geopolymeric gels, which will hinder the dissolution of unreacted silica and alumina species. Also, excessively high curing temperature will cause rapid evaporation of pore solutions and may result in an incomplete geopolymerization. The optimum curing temperature yielded geopolymer bricks with UCS of approximately 15 MPa [25]. On the other hand, Ridtirud et al. [35] concluded that higher curing temperature, i.e., 60°C, will give rise to a rapid strength development during the early ages of curing, i.e., 7–28 days. Upon 28 days of curing, the strength development was deemed insignificant for higher curing temperature. The initial curing at elevated temperature accelerates the geopolymerization reaction and thus the strength of the resultant geopolymer was improved.

Giasuddin et al. [20] discovered the potential of geopolymeric binder in replacing traditional oil-well cements in $CO_2$ geosequestration in saline aquifer. The slag-added FA-based geopolymer was cured in three different curing environments, i.e., water curing, saline–water curing (8% and 15% concentration), and sealed curing. American Petroleum Institute (API)-recommended class G oil-well cement was fabricated for benchmarking purposes. From the 28-days compressive strength results, it can be concluded that (1) saline water cured geopolymer specimens exhibited higher compressive strength as compared to water-cured geopolymer specimens, while the oil-well cement specimens showed a contrasting trend as compared to their geopolymer counterpart specimens; (2) for geopolymer specimens, the compressive strength of specimens cured in sealed condition is consistently higher as compared to water-cured and saline-water-cured specimens; (3) the variation in compressive strength with different curing environments was attributed to the leaching out of alkali-activator solution and other useful reactants from the geopolymer specimen to the surrounding curing medium.

Aydin and Baradan [22] studied the effect of steam and autoclave curing on the compressive strength of alkali-activated slag (AAS) mortars. In this study, GGBFS with low hydration modulus (HM) of 1.33 was used. The slag was activated using the mixture of NaOH and $Na_2SiO_3$ in different proportions to obtain different $M_s$ values and different $Na_2O$ content. Aggregate-to-binder (cement or GGBFS) and water-to-binder ratio were fixed at 2.75 and 0.44 respectively for all the mixtures. One batch of specimens was kept in a humidity cabinet for 5 h before being subjected to steam curing at 100°C for 8 h. Another batch of specimens was kept in humidity cabinet for 24 h and subsequently autoclaved at 210°C and under 2.0 MPa pressure for 8 h. The compressive strength results of AAS mortars were then compared with a standard PC mortar. The following conclusions were derived: (1) Compressive strength values in the range of 15–90 MPa were achieved by steam cured AAS mortars, despite using low-HM GGBFS; (2) compressive strength values of steam-cured AAS mortars were significantly higher than the PC mortars when the $Na_2O$ and $M_s$ values are higher than 4% and 0.4, respectively; (3) high-performance AAS mortars with compressive strength value of 70 MPa can be achieved using a mere 2% $Na_2O$ (by weight of slag) under autoclave curing; (4) compressive strength of autoclaved PC mortar was found to be significantly lower than its steam-cured AAS counterpart due to the formation of crystalline structure α-calcium silicate hydrate (α-$C_2SH$) under high temperature and pressure, which caused the increase in porosity and reduction in compressive strength; (5) autoclave curing was found to be more favorable for activator solutions with low $Na_2O$ concentrations and low $M_s$ ratios while steam curing is more favorable for activator solutions with high $M_s$ ratios.

Chindaprasirt et al. [48] proposed a method of reducing the heat-treatment period of high-calcium FA geopolymer paste. The results showed that by subjecting the paste sample to microwave heating of 5 min plus conventional oven curing for 6 h at 60°C, the compressive strength obtained was higher if compared to paste samples cured at 60°C for 24 h without microwave treatment. Tables 11.1 and 11.2 summarize the effect of the chemical activator and curing regime on the mechanical properties of geopolymers.

## 11.2.2 Dimensional Stability and Durability Properties

The effect of chemical activator (SS/SH ratio, NaOH molarity, etc.) and curing regime on the dimensional stability and durability properties of geopolymers has an indirect relationship with the mechanical properties exhibited by the geopolymers. Generally, geopolymers with excellent mechanical properties will exhibit superior dimensional stability and durability properties if compared with geopolymers with inferior mechanical properties [22,35,39,48].

Ridtirud et al. [35] investigated the effect of NaOH concentration and sodium-silicate-to–NaOH (SS/SH) ratio on the shrinkage of ASTM class C FA geopolymer mortars. In order to determine the influence of NaOH concentration, SS/SH ratio of 0.67 and NaOH concentrations of

**Table 11.1** Effect of chemical activator on the compressive strength of geopolymer concrete

| Types of geopolymer | Chemical activator | | | Compressive strength | Primary findings |
|---|---|---|---|---|---|
| | SS/SH ratio | NaOH concentration | $M_s$ | | |
| **FA-based** | | | | | |
| Gorhan and Kurklu [19] | 0.4–2.3 | 3–9 M | | 12–23 MPa | 6 M NaOH optimal |
| Sukmak et al. [34] | | 10 M | | 4–14 MPa | SS/SH ratio of 0.7 optimal |
| Somna et al. [30] | 0.33–3.0 | 4.5–16.5 M | | 7–25 MPa | Strength increased from 4.5 to 14 M NaOH, but decreased at 16.5 M NaOH |
| Ridtirud et al. [35] | 7.5–12.5 M | | | 25–45 MPa | SS/SH ratio of 1.5 and 7.5 M NaOH optimal |
| Guo et al. [40] | | | 1.0–2.0 | 5.0–63.4 MPa | $M_s$ of 1.5 optimal |
| Law et al. [39] | | 10 M | 0.75–1.25 | 39–57.3 MPa | $M_s$ of 1.0 optimal |
| **RHA/RHBA-based** | | | | | |
| He et al. [14] | 2.5 | 2–6 M | | 8–15 MPa | 2 M NaOH optimal |
| Nazari et al. [21] | | 4–12 M | | 20–30 MPa | 12 M NaOH optimal |
| Songpiriyakij et al. [49] | 0.5–2.5 | 14 and 18 M | | 34–56 MPa | 18 M NaOH optimal |
| Detphan and Chindaprasirt [50] | 1.9–5.5 | | | 15–40 MPa | SS/SH ratio of 4.0 optimal |
| **POFA-based** | | | | | |
| Salih et al. [36] | 0.5–3.0 | 10 M | | 7–32 MPa | SS/SH ratio of 2.5 optimal |
| Yusuf et al. [38] | | 10 M | 0.915–1.635 | 65–69 MPa | $M_s$ of 0.915 optimal but insignificant |
| **MT-based** | | | | | |
| Ahmari and Zhang [25] | | 10–15 M | | 4–34 MPa | 15 M NaOH optimal |

**Table 11.2** Effect of curing regime on the compressive strength of geopolymer concrete

| Types of geopolymer | Curing regime | Compressive strength | Primary findings |
|---|---|---|---|
| **FA based** | | | |
| Giasuddin et al. [20] | 10–12 h room temperature curing upon casting, followed by saline–water, normal-water, and sealed-condition curing | 49–91 MPa | Sealed-condition curing optimal, followed by saline–water and normal-water curing |
| Nazari et al. [21] | 24 h precuring period after casting followed by 36 h 50–90°C oven curing | 49–60 MPa | 80°C oven curing optimal |
| Ridtirud et al. [35] | 25°C, 40°C, and 60°C curing for 24 h after 1 h of precuring period | 22–53 MPa | 60°C oven curing optimal (applicable for 7 and 28 days strength development) |
| Chindaprasirt et al. [48] | 24 h 65°C curing; 5 min microwave curing + 3/6/12 h 65°C curing; ambient temperature curing | 20–42.5 MPa | 5 min microwave curing + 6 h 65°C curing optimal |
| **Slag-based** | | | |
| Aydin and Baradan [22] | – Steam curing (5 h humidity cabinet curing followed by 100°C steam curing for 8 h)<br>– Autoclaved curing (24 h humidity cabinet curing followed by 210°C, 2.0 MPa autoclaved curing for 8 h) | 15–90 MPa | Steam-cured specimens exhibited higher strength than autoclaved specimens |
| **MT-based** | | | |
| Ahmari and Zhang [25] | 7 days 60–120°C oven curing after casting | 4–34 MPa | 90°C oven curing optimal |
| **RHBA based** | | | |
| He et al. [14] | Cured at room temperature after casting for 14, 28, 35, 42, and 49 days | 2–12 MPa | Optimum strength achieved at 35 days of curing |

7.5, 10.0, and 12.5 M were used. While for the influence of SS/SH ratio, geopolymer containing 10 M of NaOH and SS/SH ratios of 0.33, 0.67, 1.0, 1.5, and 3.0 were examined. Sand-to-FA ratio of 2.75:1, liquid-to-ash ratio of 0.6, and curing temperature of 40°C were used for the afore-mentioned series of tests. The authors concluded that the shrinkage of FA geopolymer mortars increased with higher NaOH concentration mainly due to the low strength of the corresponding mortar samples. On the other hand, the increase in the SS/SH ratio produces geopolymer with significantly lower values of shrinkage as the high silica-to-alumina ratio with high SS/SH ratio gives rise to rapid geopolymerization reaction or condensation of geopolymer. Thus, the increase in shrinkage was gener-ally associated with the low strength development of geopolymers. The authors also studied the effect of curing temperature on the shrinkage properties of ASTM class C FA geopolymer mortars. The results show that FA geopolymer mortars subjected to higher curing temperature, i.e., 40°C and 60°C, had undergone a significantly lower degree of shrinkage as compared to geopolymer mortars cured at ambient temperature, i.e., 23°C. The lower shrinkage at higher curing temperature is associated with the higher strength value as a result of accelerated heat curing for 24 h. The authors also highlighted the significantly low shrinkage value after three weeks for samples cured at all three temperatures. This shows that the geopolymer mortars have high potential for commercialization espe-cially for the precast structural element industry.

However, the variations in $M_s$ were found to have no significant effect on the chloride diffusivity of class F FA based geopolymer concrete and are comparable to those of OPC concrete [39]. The authors predicted that the long-term chloride resistance of geopolymer concrete would be lower if compared with OPC concrete owing to the lower strength incre-ment over time of the geopolymer concrete. The attempted investigation of long-term chloride resistance of geopolymer concrete by the authors using rapid chloride permeability test (RCPT) was proven to be futile as the geopolymer specimens exhibited a rapid rise in temperature during testing which is against Ohm's law, implying that RCPT is not a suitable testing method to evaluate the chloride resistance of geopolymer concrete.

Aydin and Baradan [22] studied the effect of steam curing and autoclave curing on the drying shrinkage of AAS mortars. A standard OPC mor-tar was fabricated and used for comparison with all other heat-cured AAS mortars. Besides, a group of OPC and AAS mortars were cured in standard water conditions and used for comparison purposes. Upon heat treatment, the specimens were left to be cooled to room temperature and the length changes of the specimens were measured periodically up to 6 months. The authors concluded that (1) drying shrinkage values of AAS mortars were higher as compared to PC mortars in all curing conditions; (2) drying shrinkage values of AAS and PC mortars were reduced upon heat curing; (3) reduction of drying shrinkage values of AAS mortars upon being subjected to heat curing is more significant as compared to heat-cured PC mortars; (4) generally, autoclave curing was found to be more effective in reducing the drying shrinkage of AAS and PC mortars than steam curing.

The sulfate and acid resistance of high-calcium FA geopolymer paste was greatly enhanced by using the microwave-assisted heat-curing method to cure the paste sample. Microwave radiation is thought to enhance the dissolution rate of Si and Al species from the FA particles by the rapid and uniform heating of the aqueous alkaline solution by microwave energy, leading to multiple gel formation, and subsequently a denser, stronger, and durable matrix was formed if compared to the conventional oven heat curing [48].

## 11.2.3 Microstructure of Geopolymer Matrix

Mineral phases of GFA using different NaOH concentrations (4.5–16.5 M) as chemical activator were investigated by Somna et al. [30]. The authors observed that at low NaOH concentrations, i.e., 4.5 and 7.0 M, the X-ray diffraction (XRD) pattern is somewhat similar to GFA paste without any addition of NaOH, suggesting that the degree of geopolymerization is low. The presence of crystalline silicate and aluminosilicate compounds was detected at higher NaOH concentrations (9.5–14.0 M). At NaOH concentrations of 16.0 M, crystalline products of aluminosilicate compound at around 34° and 38° ($2\theta°$) disappeared. The observation is probably due to the excess hydroxide ion concentration, which caused the aluminosilicate gel to precipitate at very early stages, hindering the polycondensation process.

Hanjitsuwan et al. [12] reported similar mineralogical phases in high-calcium lignite FA where the amorphous phase was indicated by the broad hump between 20° and 38° while the crystalline phases were indicated by sharp peaks mainly consisted of quartz ($SiO_2$), hematite ($Fe_2O_3$), anhydrite ($CaSO_4$), magnesioferrite ($MgFe_2O_4$), and calcium oxide ($CaO$). Upon alkali activation by NaOH (8, 10, 12, 15, and 18 M), the broad hump was shifted to around 25°–38°, which is indicative of the existence of alkaline aluminosilicate gel and C-S-H gels due to the high calcium content of the FA. The new phases occurring in the alkali-activated FA include portlandite $(Ca(OH)_2)$, sodium sulfate $(Na_2SO_4)$ and hydrosodalite $(Na_4\text{-}Al_3Si_3O_{12}(OH))$. The peak intensity of C-S-H and hydrosodalite increased as the NaOH concentration increases indicating that both the aforementioned phases are the governing factors in the strength development of high-calcium lignite FA geopolymers. Following the optimization of the modulus $(SiO_2/Na_2O)$ and content $(Na_2O/CFA)$ of alkali activator which were determined to be 1.5% and 10%, respectively, the mineralogical phases of the optimized CFAG specimen were evaluated using XRD and compared against the pure CFA sample. The major components of pure CFA consisted of mullite, quartz, anhydrite, and f-CaO. Upon geopolymerization, a broad and amorphous hump between 20° and 40° $(2\theta)$ appeared in the XRD pattern of CFAG, which included both geopolymeric gels and calcium silicate hydrate (C-S-H). This finding suggested that the geopolymeric and hydrate reaction occurred concurrently in a single CFAG system [40]. Similar findings where the coexistence of geopolymer gel and C-S-H gel were also observed in high-calcium FA-based geopolymer system can be found in Refs. [12,13,51,52].

Komljenovic et al. [31] analyzed the X-ray diffractograms of different sources of FA geopolymer pastes activated using sodium silicate with different $M_s$ and concentrations. All the geopolymer pastes were cured for 1 day at 20°C followed by 6 days at 55°C. The $M_s$ used were 0.5 and 1.5 while the concentration of sodium silicate used was 8% and 10%. The $M_s$ of sodium silicate was adjusted accordingly by adding NaOH. Zeolite morphological phases of faujasite was found in the geopolymer specimens with low $M_s$ of 0.5 but disappeared in the 1.5 $M_s$ specimens, regardless of sodium silicate concentrations, as can be seen in Fig. 11.1. With the increase in the concentration of sodium silicate, the crystallization of zeolites were slowed down as the alkali-activation process was accelerated due to the higher amount of dissolved silicon in the solution, resulting in the formation of amorphous phases as the only reaction products. The authors also concluded that the reduction or absence of crystalline product had contributed to the higher compressive strength of the geopolymers. Similar findings were observed on the alkali-activated POFA paste where the authors concluded that the intensity of main crystalline peaks originated from the grounded POFA was found to be reduced upon alkali-activation process which corresponded to the higher-strength POFA geopolymer paste [36].

Somna et al. [30] performed Fourier transform infrared spectroscopy (FTIR) analysis on GFA with various NaOH concentrations (4.5–16.5 M)

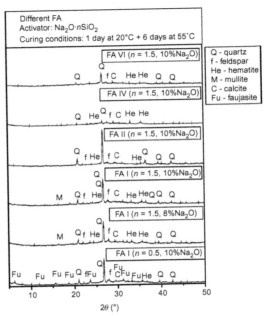

**Figure 11.1** X-ray diffractogram of various FA activated by sodium silicate with various modulus (*n*) and concentrations [31].

cured at ambient temperature. The IR spectrum of GFA consists of an intense band at $450\,cm^{-1}$ (Si-O-Si bending vibration) and $1180\,cm^{-1}$ (Si-O-Si and Si-O-Al asymmetric stretching vibration). A downward shift of intensity band at $1180\,cm^{-1}$ to around $972–990\,cm^{-1}$ was observed for all the samples upon the addition of NaOH. This downward-shifting pattern was due to the rise in the tetrahedrally positioned Al atom present in the geopolymer system and is also indicative that geopolymerization of NaOH-activated GFA has taken place. There is also the existence of a newly formed broad band around $1650\,cm^{-1}$ and $3480\,cm^{-1}$ in the NaOH-activated GFA samples, which is associated with the stretching vibration of –OH and bending vibration of O-H-O due to the geopolymerization of NaOH-activated GFA into geopolymer pastes. Similar adsorption bands were reported by other authors using class C FA (CFA) [40].

The effect of curing temperature and the initial heat-curing durations on the morphology of clay-FA geopolymer bricks were evaluated by Sukmak et al. [34]. The formation of microcracks was clearly seen in specimens cured at high temperature, i.e., 85°C, even at a relatively short

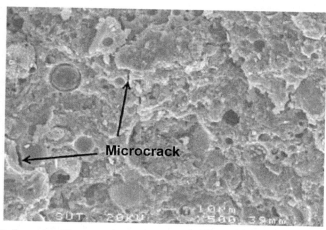

**Figure 11.2** Scanning electron microscopy (SEM) micrograph of 28 days clay-FA geopolymer brick cured at 85°C for 24 h [34].

curing duration of 24 h (Fig. 11.2). Similar microcrack development was observed for the specimens cured at lower temperature (75°C) but much longer curing durations (72 h) and is illustrated in Fig. 11.3. Both microcrack formations were very much related to the substantial loss of moisture and pore fluid from the geopolymer matrix, which induced excessive shrinkage during drying and subsequent loss of structural integrity of the clay-FA geopolymer matrix [53,54].

SEM analysis was performed by Aydin and Baradan [22] to study the microstructure characteristic of steam- and autoclave-cured AAS mortars. NaOH and $Na_2SiO_3$ were used as alkali–activation agents. $Na_2O$ content was fixed at 6% while silicate modulus ($M_s$) values varied between 0 and 1.2. The microstructure of steam-cured AAS mortars was transformed from a porous structure into a well-packed and homogeneous structure with the increasing $M_s$ value. However, the crack intensity of the matrix phase is higher for higher $M_s$ value due to the tension generated during shrinkage [55]. As opposed to steam curing, autoclave-cured specimens exhibited a dense and well-packed structure especially for $M_s$ values of 0 and 0.4. No significant microcracks were observed at the aggregate matrix interface for all the autoclave-cured specimens. The main reaction products for both steam- and autoclave-cured AAS mortars are Na-substituted C-S-H, though a lower Ca/Si ratio for the C-S-H was observed for the autoclave-cured AAS mortars.

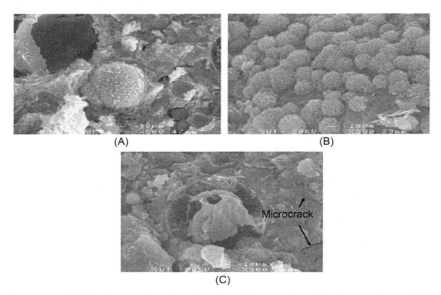

**Figure 11.3** SEM micrographs of 28 days clay-FA geopolymer bricks cured at 75°C for (A) 24 h, (B) 48 h, and (C) 72 h [34].

Guo et al. [40] analyzed the micrographs (Fig. 11.4) of pure class C FA (CFA) and also the optimized CFAG. The optimized CFAG was based on the $M_s$ and content ($Na_2O/CFA$) of alkali activators of 1.5% and 10%, respectively. The morphology of CFA (Fig. 11.4A) showed a series of spherical vitreous particles with varying sizes, similar to that of class F FA. Upon geopolymerization, the partial dissolution of CFA particles can be clearly seen in Fig. 11.4B and the cavities of the broken CFA particles seem to be filled with a large amount of microparticles of the reaction products (Fig. 11.4C). Also, the energy-dispersive X-ray analysis confirmed the main geopolymeric gel existed in CFAG is (Na)-poly(sialate-disiloxo-), i.e., $Na_n-(Si-O-Al-O-Si-O-Si-O-)_n-$. The geopolymeric gels were also found to be coexistent with C–S–H gels and some unreacted CFA spheres (Fig. 11.4D).

Komljenovic et al. [31] studied the effect of different $M_s$ on the elementary composition of sodium silicate–activated FA geopolymers. When the modulus of sodium silicate increased, the Si/Al atomic ratio of the reaction products also increased, while the Na/Si and Na/Al decreased as can be seen in Table 11.3. Based on the compressive strength results obtained in the study, the authors concluded that higher compressive

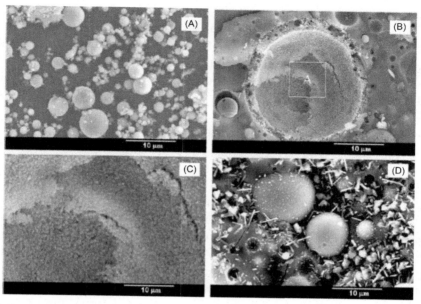

**Figure 11.4** SEM images of (A) pure CFA powder, (B) the reactive CFA sphere, (D) the reactive area A of the CFA sphere in SEM image (B), and (D) CFAG cured at 75°C for 8h followed by curing at 23°C for 28 days [40].

**Table 11.3** Average content of different elements (atomic %) and their ratios in the reaction products of FA + $Na_2O \cdot nSiO_2$ (10% $Na_2O$) [31]

| Elements and ratios | $SiO_2/Na_2O$ | |
|---|---|---|
| | 0.5 | 1.5 |
| Si | 10.25 | 17.40 |
| Al | 3.95 | 5.91 |
| Na | 11.66 | 4.41 |
| Ca | 2.32 | 2.62 |
| Fe | 0.87 | 0.98 |
| Si/Al | 2.64 | 3.16 |
| Na/Si | 1.98 | 0.26 |
| Na/Al | 4.47 | 0.86 |

strength is directly related to higher modulus value, which in turns is also directly related to the higher Si/Al atomic ratio of the reaction products. Yusuf et al. [38] reported that $M_s$ has a significant effect on the bond characteristic, structural units of reaction products, amorphosity, and also the morphology of the products. The authors concluded that (1) the high-silica modulus (HSM) system contained more polymerized aluminosilicate structure due to attachment of more Al than the low-silica modulus (LSM) system, thus the HSM system has the tendency to form products resembling calcium-(alumino) silicate hydrate (C-(A)-S-H) while the LSM system tends to form calcium silicate hydrate (C-S-H); (2) the HSM system yielded products with higher amorphosity than the LSM products; and (3) the HSM paste sample exhibited morphology that was denser, more compact, and segmental than LSM paste sample due to the lower volume of residual water present in the latter system.

The effect of curing temperature and duration on the apparent porosity of ASTM class F FA geopolymer mortars activated by water glass and NaOH were evaluated by Gorhan and Kurklu [19]. Geopolymer mortars were subjected to 65°C and 85°C oven curing for 2, 5, and 24 h after being molded. Results show that both the curing temperature and time have significant effect on the apparent porosity of geopolymer mortars. Similar findings where water absorption has a direct influence on the apparent porosity of geopolymer have been reported [56,57]. From the results acquired, the following conclusions were made: (1) the increase in curing time consistently reduces the apparent porosity of geopolymer mortars cured at constant temperature of 85°C; (2) the samples cured at 65°C have a more porous structure as compared to their counterparts, which were cured at 85°C, with the former exhibiting apparent porosity range of 26.1–29.2% while the latter have the apparent porosity range of 25.3–29.8%; (3) sufficient while not excessive curing temperature and time play an important role in the pore structure development in geopolymer mortars.

The Mercury Intrusion Porosimetry (MIP) analysis carried out to study the pore-size distribution of steam- and autoclave-cured AAS mortars showed that steam curing caused a coarser pore-size distribution (>25 nm). Meanwhile, autoclave-cured specimens exhibited a finer pore-size distribution. This is due to the formation of C-S-H with lower Ca/Si ratio, which has denser matrix structure. Moreover, the higher degree of slag grains hydration and the stronger aggregate–matrix interface of autoclave-cured specimens had also contributed to the finer pore-size distribution [22].

## 11.2.4 Rheological and Physical Properties of Geopolymer

Rheological properties (initial and final setting time and flow test) and physical properties of geopolymers were found to be influenced by the nature and complex of geopolymerization process, which in turn was governed mostly by the chemical activator and also the curing regime employed [12,58–61]. The setting time of high-calcium FA geopolymer pastes was found to increase proportionally with increasing the NaOH concentration up to 18 M of NaOH concentration. The leaching of $Ca^{2+}$ to the pore solution at low NaOH concentration was not disrupted significantly which enables sufficient dissolved $Ca^{2+}$ in the system for the formation of calcium silicate hydrate (C-S-H) and calcium aluminate hydrate (C-A-H) gels. While at higher NaOH concentration, the hardening and setting of the paste is governed by the geopolymerization process, which usually occurred in a slower rate as compared with CSH- and CAH-dependent cementitious systems, thus resulting in a higher setting time [12].

Sathonsaowaphak et al. [32] performed flow tests on fresh GBA geopolymer mortars with various liquid alkaline/ash ratio, sodium silicate/NaOH ratio, and NaOH concentration. The workable range of liquid alkaline/ash ratios was found to be in between 0.429 and 0.709. Higher liquid alkaline/ash ratios give rise to a more workable geopolymer mix due to lower particle interference and also larger interparticle distance. On the other hand, the workable range of sodium silicate/NaOH ratios and NaOH concentration lies between 0.67 and 1.5 and 7.5 and 12.5 M, respectively. The increase in sodium silicate/NaOH ratio and NaOH concentration resulted in less workable mortar mixes owing to the higher viscosity of sodium silicate and NaOH. It is also recommended that the amount of sodium silicate in the mortar mixes should be kept as low as possible for economic reasons, while at the same time does not compromise the workability and also the strength of the geopolymer mortars.

A higher amount of efflorescence was observed in POFA geopolymer paste samples that contained lower solid-to-liquid ratios and lower sodium-silicate-to-sodium-hydroxide ratios [36]. Higher Na ions in lower sodium-silicate-to-sodium-hydroxide ratios and lower solid-to-liquid ratios samples have higher tendency of alkaline leaching-out phenomena on the specimen's surface due to the weakly bound Na ions in the nano-structure of geopolymer gel.

## 11.3 EFFECT OF PARTICLE-SIZE DISTRIBUTION OF BINDER PHASE AND ADDITIVES ON THE PROPERTIES OF GEOPOLYMER

Besides the adjustment in the chemical activator and curing regime, the nature and also the fineness of the geopolymer source materials play a crucial role in the strength development, durability properties, and microstructure of the resultant geopolymer matrices. Several authors have come to an agreement that the variation of particle-size distribution of the binder phase poses a significant effect on the compressive strength, physical properties, durability properties and microstructure of the resultant geopolymer paste [14,62]. Generally, the binder phase with finer particle-size distribution will have a higher reactivity and subsequently produce geopolymer paste which has denser microstructure, higher compressive strength, and refined physical properties [21,50]. For instance, Chindaprasirt et al. [51] reported an improvement in drying shrinkage for geopolymer mortars fabricated using fine high-calcium FA as geopolymer source material. The authors also suggested that high-calcium geopolymer mortars exhibited 1000% improvement in term of drying shrinkage in comparison with OPC mortars, proving the excellent dimensional stability of high-calcium FA geopolymers.

In order to achieve an efficient geopolymer synthesis, one must find or achieve an ideal balancing between the essential elements during the geopolymerization, i.e., $SiO_2$, $Al_2O_3$, $Na_2O$, and most recently CaO. One of the ways of finding the ideal balancing of the aforementioned elements is by the addition of commercially available additives such as calcium hydroxide ($Ca(OH)_2$), aluminum hydroxide ($Al(OH)_3$), silica fume (SF), nano-$SiO_2$, nano-$Al_2O_3$, etc. [11,13,63,64]. Some researchers utilized a "waste-to-waste" stabilization technique by incorporating waste materials such as flue gas desulfurization gypsum (FGDG) and Al-rich waste sludge to achieve a more efficient geopolymerization [52,65]. The addition of different types of fibers, polymer resin, superplasticizers (SPs), and nano-materials was also found to greatly enhance the properties of geopolymers, particularly the mechanical properties such as flexural strength, splitting tensile strength and also the modulus of elasticity [13,66–69].

The influence of particle-size distribution of binder phase and additives on the mechanical, durability, physical, and microstructure properties of geopolymers derived from industrial byproducts will be deliberated in detail in the following sections.

## 11.3.1 Mechanical Properties

Nazari et al. [21] investigated the effect of particle-size distribution on the compressive strength of FA-RHBA based geopolymers. The FA and RHBA were sieved and ground into two different particle sizes. The average particle sizes obtained for FA were 75 and 3 μm while for RHBA, average particle sizes of 90 and 7 μm were obtained. The finer FA and RHBA were denoted fF and fR while the coarser FA and RHBA were denoted cF and cR. A total of four series of geopolymer samples based on the different particle sizes of FA and RHBA as illustrated in Table 11.4 were fabricated and subjected to compressive strength tests at the curing ages of 7 and 28 days. Generally, specimens made with fine FA and fine RHBA particles (fF–fR series) exhibited the highest compressive strength regardless of curing ages. This is due to the fact that finer particles are more capable of filling pores, and hence, resulted in a denser and more compact geopolymer paste structure that can sustain higher applied load prior to ultimate failure. This finding was further strengthened when the fF–cR series yielded higher compressive strength as compared with the corresponding cF–fR series. Although fine RHBA was used in cF–fR series, the total percentage of finer particles in fF–cR series was much higher due to the higher percentage of FA in the FA–RHBA geopolymer mixtures and this has proven to be the key factor in determining the strength of FA–RHBA geopolymers.

In a separate study involving RHA, the compressive strength of blended geopolymer pastes fabricated from RM and RHA was tested at the age of 60 days. Two different gradations of RHA samples were used, one in "as–received" condition and another ground RHA with 100% particles passing through a #100–mesh (150 μm opening) sieve. A constant RHA/RM ratio of 0.4 was used throughout the study. The geopolymer specimens containing ground RHA exhibited a compressive strength value of 16.08 MPa, an increase of 37.43% if compared with an equivalent unground RHA geopolymer. The enhancement in compressive strength was attributed to the higher degree of geopolymerization achieved by the fine particle size and the high specific surface area of ground RHA, resulting in a stronger geopolymer [14]. Similar findings using ground RHA in RHA/FA geopolymers were reported [50].

Besides FA, the production of coal ash comprised approximately 20% of BA and as of now most of the BA is disposed of as landfill due to limited industrial application value if compared to FA. However, due to its similarity in silica and alumina content with FA, with the exception of excessively high carbon content due to incomplete burning and large particle size, several researchers have started to incorporate BA in either geopolymer production or as cement/aggregate replacement materials [32,62,70,71]. Sata et al. [62] evaluated the effect of different particle sizes of BA on the compressive strength of the BA geopolymer mortars. Three

**Table 11.4** Mix design of FA-RHBA based geopolymer specimens [21]

| Sample designation | Weight percent of fine FA (fF wt%) | Weight percent of coarse FA (cF wt%) | Weight percent of fine RHBA (fR wt%) | Weight percent of coarse RHBA (cR wt%) | SiO$_2$/AlO$_3$ ratio |
|---|---|---|---|---|---|
| fF–fR–1 | 60 | 0 | 40 | 0 | 3.81 |
| fF–fR–2 | 70 | 0 | 30 | 0 | 2.99 |
| fF–fR–3 | 80 | 0 | 20 | 0 | 2.38 |
| fF–cR–1 | 60 | 0 | 0 | 40 | 3.81 |
| fF–cR–2 | 70 | 0 | 0 | 30 | 2.99 |
| fF–cR–3 | 80 | 0 | 0 | 20 | 2.38 |
| cF–fR–1 | 0 | 60 | 40 | 0 | 3.81 |
| cF–fR–2 | 0 | 70 | 30 | 0 | 2.99 |
| cF–fR–3 | 0 | 80 | 20 | 0 | 2.38 |
| cF–cR–1 | 0 | 60 | 0 | 40 | 3.81 |
| cF–cR–2 | 0 | 70 | 0 | 30 | 2.99 |
| cF–cR–3 | 0 | 80 | 0 | 20 | 2.38 |

Alkali activator (sodium silicate + sodium hydroxide) to FA–RHBA mixture ratio is 0.4.

different finenesses of BA were incorporated into the mortar specimens: fine BA (15.7 μm), medium BA (24.5 μm), and coarse BA (32.2 μm). The compressive strength of the geopolymer mortars was tested at the ages of 7, 28, 90, 180, and 360 days. Results are indicative that the finer particle size of BA gives rise to a higher compressive strength of the hardened geopolymer mortars at all the curing ages. The geopolymerization rate was increased with the higher fineness and higher specific surface areas of fine BA. Besides, as the "as-received" BA contains large portions of mesopores on the ash surfaces, and the grinding process indirectly helps to reduce the porosity of the BA particles, thus reducing the water demand of the fresh mortar and contributing to the higher compressive strength of the hardened geopolymer mortar. The highest compressive strength of fine BA geopolymer mortar is 61.5 MPa, attained at the curing age of 180 days.

The effect of the addition of calcium hydroxide ($Ca(OH)_2$), aluminum hydroxide ($Al(OH)_3$) and SF on the compressive strength of treated palm oil fuel ash (TPOFA) based geopolymer mortars were investigated by Mijarsh et al. [11]. Six design factors, namely $Ca(OH)_2$ wt%, $Al(OH)_3$ wt%, SF wt%, NaOH concentration (molarity), $Na_2SiO_3$/NaOH (weight ratio), and alkali–activator solid materials (weight ratio), were examined at five levels using the Taguchi experimental design method to obtain the optimum mix proportion. A total of 25 trial mixes were fabricated and tested in accordance with the L25 array proposed by Taguchi method. The TPOFA was obtained by first separating the incompletely combusted fibers and kernel shells from the raw POFA by using a 300 μm sieve. Then, the POFA was heated at 500°C for 1 h to remove the unburned carbon before being subjected to secondary grinding to obtain the TPOFA. The sand-to-binder-material mass ratio for all the mixes was fixed at 1.5. Immediately after molding, the specimens were wrapped using a cling film and left to cure for 1 h before being subjected to oven curing at 75°C for 48 h. The compressive strength of all the specimens was then tested at 1, 3, and 7 days of curing ages. Compressive strength result of the 25 trial mixes ranged from 15.67 to 44.74 MPa at 1, 3, and 7 days of curing. From the trial mix results, the optimum level of substitutions or ratios of various factors examined, i.e., additive materials (20 wt% $Ca(OH)_2$, 5 wt% SF, and 10 wt% $Al(OH)_3$) and alkaline activators (10 M NaOH, $Na_2SiO_3$/NaOH = 2.5, and alkaline activator/solid = 0.47), was obtained and fabricated to test the resultant compressive strength. The optimum TPOFA geopolymer mortar exhibited a compressive strength of 47.27 ± 5.0 MPa at 7 days of curing, which is higher than all 25 trial mixes for the same curing duration.

The addition of a small amount of nanosized additives such as nano-$SiO_2$ and nano-$Al_2O_3$ was known to effectively enhance the compressive and tensile strength of concrete by the means of additional pozzolanic and filler effects [72]. Phoo-ngernkham et al. [13] incorporated 1–3% of nano-$SiO_2$ and nano-$Al_2O_3$ by binder weight to fabricate FA geopolymer pastes. The 7-, 28-, and 90-day compressive strengths of the geopolymer pastes were compared to an OPC paste. The compressive strength results showed that the addition of nano-$SiO_2$ and nano-$Al_2O_3$ regardless of the additive's dosage give rise to superior compressive strength as compared to the reference OPC paste specimens. The dual functionality of the nano-sized additives in FA-based geopolymer, which provides additional $SiO_2$ and $Al_2O_3$ to the geopolymer system and at the same time acts as a micro-filler, yielded additional calcium silicate hydrate (CSH) or calcium aluminosilicate hydrate (CASH) and sodium aluminosilicate hydrate (NASH) gels in the geopolymer matrix and a dense geopolymer structure.

Similar to OPC concrete, high-range water-reducing admixtures or SPs can be incorporated into a geopolymer system in order to reduce its water content while maintaining the desired workability, thus resulting in a higher-strength geopolymer [73–75]. Nematollahi and Sanjayan [67] investigated the effect of different types of commercial SP to the compressive strength of class F FA geopolymer paste. The SP used in the experimental study consisted of naphthalene (N), melamine (M), and modified polycarboxylate (PC) based SPs, and each of them was added to the fresh geopolymer mixture at a dosage of 1% by mass of FA. Although the compressive strength of all the SP-added geopolymer paste decreased as compared to the control mix (without the addition of SP), the authors concluded that PC-based SP is the most suitable type of commercial SP to be incorporated into class F FA geopolymer paste activated using a multi-compound activator ($Na_2SiO_3$/NaOH = 2.5), showing the least reduction in compressive strength (16–29%). However, Puertas et al. [76] reported the addition of vinyl copolymer and polyacrylate copolymer based SPs into FA-based geopolymer paste and mortar does not bring about any significant changes on the compressive strength nor the workability of the resultant geopolymers.

Nath and Kumar [77] utilized two types of iron-making slags, namely granulated blast furnace slag (GBFS) and granulated Corex slag (GCS) in FA-based geopolymer system in the range of 0–50% by weight of binder. GCS is produced during the Corex process in iron making, and has similar chemistry and phase composition with GBFS. Prior to be used as mix constituents, both GBFS and GCS were milled for 2 h in a ball mill to obtain the desired fineness of $d_{50} = 18.49$ and $18.53 \mu m$, respectively. The compressive strength of hardened geopolymers was tested at 7 and 28 days of curing. The following conclusions can be made based on the compressive strength results obtained: (1) Both geopolymers, GBFS-FA and GCS-FA, exhibited an increase in compressive strength with the increasing slag content; (2) the increase in compressive strength rate is more pronounced at slag addition content above 20% for both geopolymers; (3) GCS addition generally has resulted in higher 7- and 28-days compressive strength in comparison with GBFS addition; (4) the strength increment is mainly due to the formation of C-S-H cementitious gel which occupied the pore space and subsequently improved the density of the resultant geopolymer binder matrix; (5) compressive strength as high as 93.4 and 91.2 MPa was obtained for 50% addition of GCS and GBFS, respectively, after 28 days of curing.

Boonserm et al. [65] attempted to improve the geopolymerization of BA by incorporating ASTM class C FA and FGDG. BA:FA as the blended source materials with ratios of 100:0, 75:25, 50:50, 25:75, and 0:100 were used. The source materials were then replaced by FGDG in 0, 5, 10, and 15%. Sodium silicate/NaOH ratio, liquid/ash ratio, sand/ash ratio were fixed at 1.0, 0.6, and 2.75 respectively for all the geopolymer mixes. The fresh geopolymer mixes were casted in $5 \times 5 \times 5$ cm cubic molds and were subjected to 40°C electric oven curing for 48 h. All the geopolymer mortars were subjected to compressive strength test at the age of 7 days. From the experimental results, it can be concluded that (1) the degree of geopolymerization of FA is higher than that of BA geopolymer owing to the high glassy mineral-phase content of FA and also the additional CSH gel formed as a result of the reaction between $Ca^{2+}$ and silicate from the FA, thus the strength of blended geopolymer mortars increased with the increase in FA content; (2) the incorporation of 5–10% of FGDG showed a significant effect on the blended geopolymer mortars with low-FA replacement level, i.e., 0, 25, and 50%. The aforementioned phenomenon was due to the additional CSH gel formed as a result of increase in the concentration of $Ca^{2+}$ ions and also significantly higher dissolution rate of $Al^{3+}$ in BA due to the presence of $SO_4^{2-}$ ions in the system; (3) the addition of 15% of FGDG caused adverse effects to all the geopolymer mortars. High-FGDG content obstructs the geopolymerization process especially in geopolymer mortars with high-FA content, and as a result a thenardite phase that existed as impurity presence in the geopolymer system and caused all the geopolymer mortars to exhibit very low compressive strength ranging from 0.3 to 1.0 MPa.

Reuse and recycling of industrial wastes are the ideal solution to current waste management problems, and Chindaprasirt et al. [52] found a way to utilize Al-rich waste originated from a waste-water treatment unit of a polymer-processing plant as additive in the fabrication of high-calcium FA-based geopolymer mortar. The raw Al-rich wastes were dried, ground, and calcined at temperature ranging from 400°C to 1000°C before it can be used as additive. Results showed that active $\theta$-Al$_2$O$_3$ can be obtained at high calcination temperature, i.e., 1000°C. Seven-days compressive strength of 34.2 MPa could be obtained by adding 2.5 wt% of Al-rich waste calcined at 1000°C. Further increase in the additive's dosage resulted in reduction in compressive strength due to the excess Al species, which act as nonfunctional filler in the geopolymer system.

The brittle nature of geopolymers can be inhibited or overcome by incorporating different types of fibers such as polyvinyl alcohol (PVA) fibers [78], carbon fibers [79], and plain woven stainless steel mesh [80] to improve the ductility of the geopolymer composite. The aforementioned fibers are produced by an energy-intensive process and are therefore not in line with the purpose of geopolymers as a sustainability and environmentally friendly product. In light of this, Chen et al. [68] investigated the effect of sweet sorghum fiber, which is a natural fiber, on the splitting tensile strength of FA-based geopolymer paste. Prior to being incorporated into the geopolymer mixture, the fibers were first treated with alkaline solution in order to improve the adhesion between the fiber and the matrix. Results showed that the inclusion of 2% sweet sorghum fiber increased the splitting tensile strength of the geopolymer paste sample by a massive 36% as compared to an unreinforced paste sample. The authors also observed a change of failure mode from brittle to ductile failure with the incorporation of sweet sorghum fibers and the effects are similar to those geopolymers incorporated by synthetic fibers. In a separate study, the splitting tensile strength of FA-based self-compacting geopolymer concrete (SCGC) increased by 12.8% by the addition of 10 wt% SF, in comparison with the control mix in which no SF is added into the FA-based geopolymer concrete mixtures [81].

The addition of 2% nano-SiO$_2$ and nano-Al$_2$O$_3$ by weight was found to significantly enhance the flexural strength in a particularly high-calcium FA-based geopolymer paste system [13]. The increase in reaction products such as CSH, CASH, and NASH due to the addition of nanoparticles led to remarkable enhancement in the interfacial transition zone (ITZ) within the geopolymer matrix. The authors found that the flexural strength of the resultant geopolymer paste samples increased linearly with the square root of ultimate compressive strength and the flexural strength values obtained were generally higher than that of OPC concrete as given by ACI 318. Similar findings were also reported where the addition of 10 wt% SF in FA-based SCGC was found to increase the flexural strength by 11.09% if compared to the non–SF added FA–based SCGC [81].

Flexural strength of geopolymers can be greatly enhanced by incorporating different types of short synthetic fibers such as PVA fibers, polypropylene, and carbon fibers through a bridging effect during the micro- and macrocracking of the geopolymer matrix under flexure stresses [69]. However, the use of natural fibers to reinforce geopolymers is gaining wide interest due to their environmentally friendly and cost-efficient characteristics [82]. In one of those studies, cotton fabric was used to reinforce ASTM class F FA-based geopolymer composites [66,83,84]. Results showed that the addition of cotton fabric greatly enhanced the flexural strength of the geopolymer composites. The flexural strength was enhanced by almost threefold with the optimum cotton fabric addition, which is 8.3 wt%, in comparison with the unreinforced geopolymer composites. On the other hand, the addition of 2 wt% of sweet sorghum fiber enhanced the flexural strength of ASTM class F FA-based geopolymer paste samples by almost 40%. Higher fiber content will result in a significant decrease in flexural strength due to entrapment of air bubbles in the geopolymer composite resulting from the poor workability and fiber agglomeration [68].

The addition of water-soluble organic polymers was found to improve the mechanical properties of geopolymers through the modification of microstructure and pore-size distribution of the resultant geopolymer matrix. For instance, Zhang et al. [69] improved the mechanical performance, i.e., flexural strength of MK/GBFS based geopolymer composites by the incorporation of 1–15 wt% of polymer resin. The authors concluded that the addition of a mere 1% resin by weight percentage improved greatly, i.e., 41%, the flexural strength of the geopolymer composites. Higher dosage of polymer–resin incorporation resulted in decrease in flexural strength due to the coating effect that resin has on MK and GBFS fine particles, thus resulting in significant reduction in the binder's reactivity. The authors have also evaluated the effect of elevated temperature on the resin-reinforced MK/GBFS based geopolymer composites in a separate study [85]. The geopolymer composites exhibited an increase in flexural strength when exposed to temperatures between 150°C and 300°C due to the enhancement in polycondensation reaction. However, when exposed to elevated temperature ranging from 450°C to 850°C, the flexural strength significantly reduced due to the dehydration of the geopolymer matrix, thermal decomposition of the added resin, and some phase transformation.

The addition of 2% nano-$SiO_2$ and nano-$Al_2O_3$ by weight of binder was found to increase the modulus of elasticity of high-calcium FA geopolymer paste sample cured at ambient temperature by a massive 30%. $E$-value as high as 17.65 GPa is obtained after 90 days of curing, comparable with that of OPC concrete. The observed enhancement in $E$-values of the geopolymer paste samples was due to the denser and stronger matrix formed with the addition of the aforementioned nanoparticles. The $E$-values obtained in the study though are slightly lower if compared with OPC paste and blended cement paste samples and it was concluded that the absence of heat treatment in the current study is the main factor contributing to the observed lower $E$-value [13].

## 11.3.2 Rheological and Physical Properties of Geopolymer

Generally finer geopolymer source materials bring about a higher degree of reactivity and vigorous geopolymerization due to the higher specific surface area attained in finer source materials, which in turn will have a shorter setting time and higher early strength development. However, significant enhancement in mechanical, durability, and microstructure properties can only be achieved if the balancing between the corresponding higher water requirement and source materials fineness is achieved. In one particular study, original FA (CFA), medium fineness FA (MFA), and fine FA (FFA) with corresponding Blaine fineness of 2700, 3900, and 4500 cm$^2$/g were utilized in fabricating high-strength geopolymer using high-calcium FA as the source material [51]. FFA yielded shorter setting time compared to CFA and MFA due to the higher specific surface area and the presence of larger content of amorphous phase, which increased the reactivity of FFA. The relatively long setting time behavior of the three fineness of high-calcium FA suggested that the aforementioned geopolymer products can be suitably used for industrial application for the ease of handling, transporting, placing, and compaction prior to curing. The authors also concluded that the particle shape of FA regardless of its fineness plays an important role in improving the workability of the fresh geopolymer paste samples. It has been agreed upon that FFA with a spherical shape and a smooth surface would yield the best ball-bearing effect and thus increase the flow of the subsequent fresh mixes without the need of additional water or water-reducing admixtures.

ASTM class F FA has been widely used as geopolymer source material [19,34,67]. However, the aforementioned ash material requires heat treatment to achieve high early strength and faster setting time of the resultant geopolymer pastes, which is a drawback as heat treatment induced additional energy requirements for the fabrication of geopolymer with competitive strength if compared to OPC concrete. Several researchers incorporated additional additives such as $Ca(OH)_2$ [11], nano-$SiO_2$, and nano-$Al_2O_3$ [13] or by using or hybridizing high-calcium source materials such as GGBFS and ASTM class C FA in order to achieve fast setting time and strength development while not compromising the strength potential of the geopolymer system [22,35,52,77]. For instance, in high-calcium FA-based geopolymer paste, the addition of nano-$SiO_2$ up to 3% by binder weight has resulted in a decrease of initial and final setting time as compared to OPC paste while the addition of nano-$Al_2O_3$ has resulted in a slight reduction in initial and final setting time in reference with the OPC paste. The reduced setting time by the addition of nano-$SiO_2$ was due to the much faster rate of activation with the readily available free calcium ions in the high-calcium FA and formed additional CSH gels [13]. Similar findings were also reported by Chindaprasirt et al. [61].

Nematollahi and Sanjayan [67] studied the effect of different commercial SPs' addition to the relative slump of class F FA geopolymer pastes. Naphthalene (N), melamine (M), and modified polycarboxylate (PC) based SPs were used at 1% by mass of FA throughout the experimental study. Two types of chemical activators were used in the study, i.e., 8 M concentration of NaOH and multicompound activator consisted of $Na_2SiO_3$ and 8 M NaOH with $Na_2SiO_3/NaOH$ = 2.5. The relative slump of the fresh geopolymer pastes were measured using the mini slump test method in accordance with ASTM C1437. For geopolymer paste activated using only 8 M NaOH, N-based SP was found to be the most effective high-range water-reducing admixture as the addition of 1% of N-based SP increased the relative slump of fresh geopolymer pastes by 136% in reference with the control mix (without SP). For multicompound activators/activated geopolymer pastes, PC-based SP is the most effective high-range water-reducing admixture with 46% increase of the relative slump, followed by N-based SP with increment of 8% while

M-based SP exhibited a decrease of relative slump by 3%. Though all the commercial SPs used were found to be chemically unstable in high basic media (NaOH + Na$_2$SiO$_3$), which reduce their plasticizing effect, PC-based SP showed a more prominent plasticizing effect as compared to N- and M-based SP owing to the existence of lateral ether chains in its structure, which resulted in steric repulsion in addition to the electrostatic repulsion effect possessed by N-, M-, and PC-based SP.

Being a silica–rich source material with over 90% of SiO$_2$ presents in the chemical composition, the alkali activation of RHA needed additional aluminum compound in order to achieve higher geopolymerization efficiency [86]. Rattanasak et al. [64] studied the effect of Al(OH)$_3$) substitution and curing temperature on the disintegration of RHA-based geopolymer paste in boiling water. RHA was replaced at 2.5, 5, 10, 20, and 30 wt% by Al(OH)$_3$. 10 M NaOH, Na$_2$SiO$_3$/NaOH = 1.5, and solid/total mixture = 0.6 was used throughout the experimental study. 1 wt% of boric acid (H$_3$BO$_3$) powder was added to each mixture to solve the problem of disintegration in water. Upon completion of compaction into steel molds, the geopolymer paste was wrapped with cling film and cured at 70, 85, 100, and 115°C for 48 h before being subjected to the boiling test. Al(OH)$_3$ is needed in RHA-based geopolymer because RHA contains mostly silica (SiO$_2$). The added Al(OH)$_3$ served as aluminum source for RHA geopolymer. Results showed that all the pastes cured at 70°C, 85°C, and 100°C disintegrated upon being subjected to boiling test while specimens with 2.5–30 wt% of Al(OH)$_3$ cured at 115°C maintained their structural integrity after immersion in boiling water. At curing temperature of 115°C, which is above the dehydration temperature of H$_3$BO$_3$ (110°C), H$_3$BO$_3$ starts to react with SiO$_2$ and other constituents in a high–silica aluminosilicate material and forms a more stable geopolymer matrix. The H$_3$BO$_3$ is relatively inert at temperature below 110°C in RHA geopolymer. Although the RHA geopolymer with 20 and 30 wt% of Al(OH)$_3$ does not disintegrate in boiling water, swelling of specimens was observed due to the excessive Al(OH)$_3$ content. Shorter setting time with the addition of aluminum compound was also observed by De Silva et al. [16].

Tables 11.5–11.7 summarized the effect of particle-size distribution of the binder phase and additives towards the mechanical, rheological, and physical properties of geopolymers derived from various industrial byproducts.

## 11.3.3 Microstructure of Geopolymer Matrix

Another little-known industrial byproduct, namely flue gas desulfurization gypsum (FGDG), has found its way into the fabrication process of

**Table 11.5** Effect of particle size distribution of binder phase on the compressive strength of geopolymer concrete

| Types of geopolymer | Fineness | Compressive strength | Primary findings |
|---|---|---|---|
| **FA-based** | | | |
| Nazari et al. [21] | FA: 75 and 3 µm RHBA: 90 and 7 µm | 15–45 MPa | Combination of FFA and RHBA give rise to highest strength |
| Chindaprasirt et al. [51] | FA with Blaine fineness of 2700, 3900, and 4500 cm²/g | 39–75 MPa | FA with highest Blaine fineness give rise to optimum strength |
| **RHA-based** | | | |
| He et al. [14] | RHA: 100% passes 150 µm sieve | 16 MPa | 37.43% strength increment as compared to unground samples |
| Detphan and Chindaprasirt [50] | RHA: 5%, 3%, and 1% retained on No. 325 sieve | 34.5–43.0 MPa | Finer RHA gives rise to highest compressive strength |
| **BA-based** | | | |
| Sata et al. [62] | BA: 15.7, 24.5, and 32.2 µm | 35–61.5 MPa | Finer BA gives rise to higher strength |

geopolymers. This waste gypsum originating from the coal-burning industry has the potential to enhance the geopolymerization process, e.g., by enhancing hydration of OPC concrete by pure gypsum [87,88]. However, the aforementioned waste materials have so far found limited research in the emerging geopolymer field [65]. In one of the few studies that utilized FGDG in the fabrication of geopolymers, Boonserm et al. [65] studied the microstructure characteristic of BA:FA geopolymer paste with different BA:FA ratios and also different replacement levels by FGDG. Sodium silicate/NaOH ratio and liquid/ash ratio were fixed at 1.0 and 0.6 respectively for all the geopolymer pastes. Freshly mixed pastes were subjected to 40°C electric oven curing for 48 h and were analyzed using SEM at

**Table 11.6** Effect of additives on the compressive strength and physical properties of geopolymer concrete

| Types of geopolymer | Types of additives | Compressive strength | Rheological and physical properties | Primary findings |
|---|---|---|---|---|
| **BA-based** | | | | |
| Boonserm et al. [65] | ASTM Class C FA (0, 25, 50, 75, 100%), FGDG (0, 5, 10, 15%) | 5–55 MPa | | 50% FA and 5% FGDG replacement level optimal |
| **FA-based** | | | | |
| Nematollahi and Sanjayan [67] | 1% addition by mass of binder of N, M, and PC based SP | 47–81.3 MPa | Relative slump: PC > M > N | PC–based SP showed highest plasticizing effect and least strength reduction |
| Puertas et al. [76] | 0.5–1.5% addition of vinyl copolymer and polyacrylate copolymer based SP | 30–35 MPa | | Insignificant changes in strength and workability with addition of both SPs |
| Phoo-ngernkham et al. [13] | 1–3% of nano-SiO$_2$ and nano-Al$_2$O$_3$ by binder weight | 20.2–56.4 MPa | Decrease of initial and final setting time | 2% nano-SiO$_2$ and 1% nano-Al$_2$O$_3$ optimal |
| Nath and Kumar [77] | 0–50% GBFS and GCS replacement by weight of binder | 8.5–93.4 MPa | | – 50% GBFS and GCS replacement optimal<br>– FA-GCS yielded higher strength than FA-GBFS |
| Chindaprasirt et al. [52] | Addition of 2.5–5 wt% of Al-rich waste calcined at 400, 600, 800, and 1000°C | 27.4–34.2 MPa | | Al-rich waste with 2.5 wt% and 1000°C calcined temperature optimal |
| **POFA-based** | | | | |
| Mijarsh et al. [11] | 15–25 wt% Ca(OH)$_2$, 5–10 wt% Al(OH)$_3$, 2.5–7.5 wt% SF | 15.67–44.74 MPa | | 20 wt% Ca(OH)$_2$, 5 wt% SF and 10 wt% Al(OH)$_3$ optimal |

Table 11.7 Effect of additives on the flexural strength and splitting tensile strength of geopolymer concrete

| Types of geopolymer | Types of additives | Flexural strength | Splitting tensile strength | Primary findings |
|---|---|---|---|---|
| **FA-based** | | | | |
| Chen et al. [68] | 1, 2, and 3 wt% addition of sweet sorghum fibers | 3.2–5.6 MPa | 2.2–3.4 MPa | Addition of 2 wt% of sweet sorghum fibers optimal |
| Alomayri et al. [83] | 0–8.3 wt% addition of horizontally and vertically oriented cotton fabric | 8–32 MPa | | 8.3 wt% and horizontally oriented cotton fabric optimal |
| Memon et al. [81] | 0–15 wt% addition of SF | 4.09–4.56 MPa | 4.14–4.67 MPa | 10 wt% of SF optimal |
| Chindaprasirt et al. [13] | 0–3% addition of nano–$SiO_2$ and nano–$Al_2O_3$ | 3.66–5.12 MPa | | 1 wt% $SiO_2$ and 2 wt% $Al_2O_3$ optimal |
| **Slag-based** | | | | |
| Zhang et al. [69] | 1–15 wt% addition of polymer resin | 4.8–8.6 MPa | | 41% enhancement in flexural strength by 1 wt% resin addition |

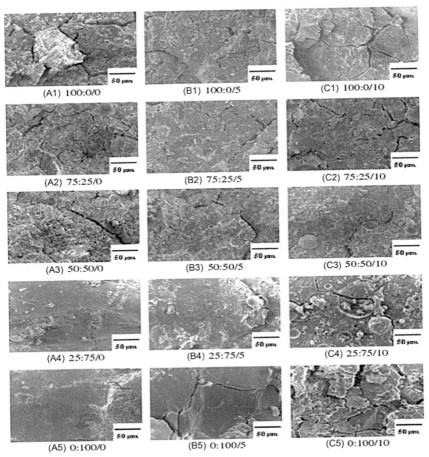

(A1) 100:0/0    (B1) 100:0/5    (C1) 100:0/10

(A2) 75:25/0    (B2) 75:25/5    (C2) 75:25/10

(A3) 50:50/0    (B3) 50:50/5    (C3) 50:50/10

(A4) 25:75/0    (B4) 25:75/5    (C4) 25:75/10

(A5) 0:100/0    (B5) 0:100/5    (C5) 0:100/10

**Figure 11.5** SEM micrograph of fractured BA and FA geopolymer pastes (BA:FA:FGDG) [65].

the age of 28 days. A SEM micrograph as per Fig. 11.5 shows that the geopolymerization products of BA:FA blended mortars consisted of well-connected structures, mostly glassy-phase structures with no definite grain boundary. Without FGDG, very dense geopolymer matrices were obtained with mixes containing high-FA content, i.e., 100% and 70% of FA as in Fig. 11.5A4,5. The matrices were less dense and less homogeneous with low-FA content as shown in Fig. 11.5A1–3. On the other hand, the addition of 5% and 10% of FGDG adversely affected the microstructure of geopolymer containing high-FA content (50%, 75%, and 100% FA;

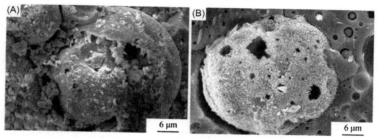

**Figure 11.6** SEM images of Al-rich waste added high-calcium FA geopolymer: (A) 2.5 wt% noncalcined Al-rich waste sample and (B) 2.5 wt% 1000°C calcined Al-rich waste sample [52].

Fig. 11.5C4,5) while improved microstructure was observed with mixes containing low-FA content (0% and 25% FA; Fig. 11.5A1–C1, A2–C2). The enhancement of the microstructure observed in low-FA content is attributed to the ability of sulfate ions in FGDG to dissolve $Al^{3+}$ ions in BA while the weakened matrices observed in high-FA mixes with the addition of FGDG is due to the presence of thenardite phase, which existed as an impurity in the geopolymer system, and which obstructed the geopolymerization reaction.

Addition of 2.5 wt% of Al-rich waste sludge calcined at 1000°C was found to produce a homogeneous matrix if compared to noncalcined Al-rich waste with the same dosage in the fabrication of high-calcium FA-based geopolymers, as can be seen in Fig. 11.6. The high calcination temperature transformed the inactive boehmite to active alumina ($\theta$-$Al_2O_3$), which in turn can be used to adjust the $SiO_2/Al_2O_3$ ratio in the high-calcium FA-based geopolymer and enhanced the geopolymeric gel formation in the resultant matrix [52].

## 11.3.4 FTIR Analysis

FTIR provides useful information about the vibrational transitions and rigidity of chemical bonds present in organic and inorganic materials [89,90]. In the emerging field of geopolymers, the changes in chemical bonds and spectrum upon the alkali–activation process can be identified and comparative study can be done efficiently using FTIR [52,65].

Boonserm et al. [65] analyzed the effect of FGDG on the IR spectra of BA and FA geopolymer pastes. The FGDG was added in 0%, 5%, 10%, and 15% to the BA and FA geopolymer pastes, respectively. Both BA and FA geopolymer pastes were activated using sodium silicate and NaOH, and were cured at 40°C in an electric oven for 48 h. The 28-day samples were then ground to particle size less than 75 μm and used for FTIR analysis at the range of 4000–400 cm$^{-1}$. FTIR results showed considerably broad bands at 3700–2200 cm$^{-1}$ and 1700–1600 cm$^{-1}$ for all geopolymer pastes which are assigned to O-H stretching and H-O-H bending were due to the weakly bound water molecules which were adsorbed on the surface or trapped in large pores between the rings of geopolymer products. $SO_4^{2-}$ bonding was detected at wave numbers of 1200 and 636 cm$^{-1}$ and this suggested that there is a reaction between $SO_4^{2-}$ ions and the alkaline solution that forms the $SO_4$ compound. With the exception of BA geopolymer paste with 5% FGDG, all other BA and FA geopolymer pastes exhibited distinctive peaks of $SO_4$ especially for the 10% and 15% FGDG-added FA geopolymer pastes. The large quantity of $SO_4$ compound detected was in line with the low strength exhibited by FA geopolymer mortars.

RHA is another industrial byproduct that possesses great potential to be utilized as construction and building materials as its high silica content (>80 wt%) renders it a potential source material for geopolymer fabrication. Due to its low aluminum content, an external aluminum source such as $Al(OH)_3$ is usually added to RHA-based geopolymer to enhance the geopolymerization process. Rattanasak et al. [64] performed FTIR analysis to determine the effect of $Al(OH)_3$ substitution to the RHA-based geopolymer. RHA was replaced at 2.5, 5, and 10 wt% by $Al(OH)_3$. For each mix proportion, 1 wt% of boric acid ($H_3BO_3$) was added to the mixture to overcome the problem of disintegration in water of the hardened geopolymer pastes. A mixture of NaOH and $Na_2SiO_3$ was used as the chemical activator solution for the geopolymer paste. 10 M of NaOH and $Na_2SiO3/NaOH$ of 1.5 was fixed for all the geopolymer mixes. The freshly casted geopolymer pastes were subjected to oven curing at 115°C for 48 h before being tested for FTIR analysis. Si-O bending and stretching peaks were observed at 470 and 1100 cm$^{-1}$ for the raw RHA sample. Upon incorporation of $Al(OH)_3$, reduction of Si-O bending peak at 470 cm$^{-1}$ and occurrence of a new peak at 780 cm$^{-1}$, which is assigned to Si-Al-O symmetric stretching, were observed. The reduction in Si-O bending peak is more pronounced with the increase in $Al(OH)_3$ content, suggesting that more silica has reacted with aluminum and formed geopolymer gels. The main spectra band, which was observed at 1100 cm$^{-1}$ of the raw RHA sample, was shifted to a lower frequency of 1040 cm$^{-1}$ (Si-Al bonding) for all the composites. The peak intensity at 1040 cm$^{-1}$ increased as the $Al(OH)_3$ content increased and this is associated with the increase in mean chain length of the aluminosilicate composite (ASC).

## 11.4 THE EFFECT OF AGGRESSIVE ENVIRONMENTAL EXPOSURE ON PROPERTIES OF GEOPOLYMERS

Similar with OPC concrete, geopolymers concrete will be exposed to severe environments, such as marine environments, where sulfate and acid attacks are predominant during their service life. Therefore, thorough understanding of the effect of geopolymer binders exposed to aggressive environment is imperative. Thus far, due to the total contrast of hydration reaction and the reaction H gels, geopolymer binders were reported to have superior resistance towards sulfate and acid attacks by various researchers [91–94]. The current literature reports the sulfate, acid, and chloride resistance of geopolymeric binders by using various methods and analytical techniques, such as the direct immersion in predetermined sulfate and acid solution followed by subsequent measurement of strength and mass loss [92], measurement of corroded depth [95], and accelerated laboratory electrochemical method [96].

In the following sections, the effect of aggressive environmental exposure on properties of geopolymers will be deliberated in terms of the influence of environmental exposure condition on the mechanical properties, physical properties, and microstructure changes.

### 11.4.1 Mechanical Properties

Apart from its environmental friendliness, the distinct advantage of geopolymer binders is their excellent acid–resistance properties if compared to OPC binders [3]. Ariffin et al. [92] exposed geopolymer concrete based on blended PFA and POFA to 2% solution of sulfuric acid up to 18 months. The ratio of PFA to POFA used was 70:30. The blended ash geopolymer (BAG) concrete was activated using commercial grade sodium hydroxide (NaOH) and sodium silicate ($Na_2SiO_3$) with alkaline-solution-to-blended-ash ratio of 0.4 and $NaOH:Na_2SiO_3$ of 2.5 by mass. An OPC concrete with water-to-cement ratio of 0.59 was fabricated and used as the control specimen. Upon casting, both the BAG and OPC concrete were subjected to room temperature (28°C) curing for 28 days before immersion in sulfuric acid. Compressive strength of both BAG and OPC concrete was examined before and after 1, 3, 6, 12, and 18 months of sulfuric acid exposure. Based on the compressive strength results,

BAG concrete exhibited superior resistance to sulfuric acid as compared to OPC concrete, with an average of 7.3% strength loss after a month, 1.6% strength loss for the subsequent months, and a total of 35% strength loss after 18 months of exposure. In contrast, the compressive strength of OPC concrete was severely affected with a total of 68% strength loss after 18 months of sulfuric acid exposure. The findings were in agreement with other similar studies where geopolymer concrete exhibited minimal strength loss upon prolonged acidic environment exposure, which justified its superior acid resistance in comparison with OPC concrete [62,97].

Ahmari and Zhang [94] measured the UCS of copper MTs-based geopolymer bricks after immersion in in pH 4 (nitric acid) and pH 7 solutions up to 4 months. NaOH concentration and curing temperature of 15 M and 90°C respectively were fixed throughout the experimental study. Based on the previous study [25], MT-based geopolymer bricks with initial water content/forming pressure of 12%/25 MPa and 16%/0.5 MPa were selected to study the effect after immersion in pH 4 and pH 7 solutions. Based on the compressive strength results, the authors carved out the following conclusions: (1) MT-based geopolymer bricks with water content/forming pressure of 12%/25 MPa exhibited UCS loss of 59.3% at pH 4 and 53.3% at pH 7. Meanwhile 16%/0.5 MPa MT-based geopolymer bricks exhibited UCS loss of 78.4% for pH = 4 and 75.2% for pH = 7, respectively, after immersion periods of 4 months; (2) the substantial strength loss by both 12%/25 MPa and 16%/0.5 MPa MT-based geopolymer bricks is caused by incomplete geopolymerization of MT, modification of chemical compositions of the geopolymer gels formed, high Si/Al ratio and also high degree of unreacted alkali in the specimens; (3) 12/25 specimens exhibited lower strength loss compared with 15/0.5 specimens due to the more prevalent effect of reduction in mix porosity due to the exerted compression pressure over the compact structure resulting from geopolymer gels towards acid-attack resistance. The aforementioned phenomenon is also related to the Na/Al ratio of 12/25 specimens, which is closer to unity than 16/0.5 specimens, and thus fewer geopolymer gels were dissolved in the acid solution.

Rattanasak et al. [64] studied the effect of sulfuric acid ($H_2SO_4$) and magnesium sulfate ($MgSO_4$) immersion on the compressive strength of RHA-based geopolymer mortars. RHA was replaced by $Al(OH)_3$ at 2.5, 5, and 10 wt%. 10 M NaOH and $Na_2SiO_3$ were used as chemical activator and the $Na_2SiO_3$/NaOH mass ratio was fixed at 1.5. Sand-to-powder-mass ratio of 2:1 was used for all the specimens. Upon being casted into 50-mm cubic molds, the fresh geopolymer mortars were wrapped with cling film to prevent moisture loss and were subjected to oven curing at 115°C for 48 h. After being cooled to room temperature, all the specimens were then immersed in 5 wt% $MgSO_4$ and 3 vol.% $H_2SO_4$ for 90 days, after which the compressive strength loss for each specimen was determined. For the unexposed specimens, the compressive strength was observed to increase as the $Al(OH)_3$ content increases with the highest compressive strength was recorded at 20.0 MPa. This was due to the lower Si/Al ratio as a result of $Al(OH)_3$ addition, which in turns promotes the formation of a more crosslinked aluminosilicate structure, thus leading to higher strength. After the immersion periods, specimens that were subjected to 5 wt% $MgSO_4$ immersion showed a higher rate of strength losses as compared to specimens that were immersed in 3 vol.% of $H_2SO_4$. This phenomenon was due to the formation of magnesium hydroxide $(Mg(OH)_2)$ resulting from the reaction between hydroxyl ions $(OH^-)$ of the ASC and the magnesium ions from $MgSO_4$ and caused the migration of hydroxide ions towards the surface of the specimen that produced the insoluble brucite. Furthermore, in a high-silica system, the $Mg(OH)_2$ reacted with the silica gels and formed the hydrated magnesium silicate, which possesses no binding capability and resulted in a sharp drop in compressive strength. The results justified that there is some distinct correlation between the mechanical, microstructure, and durability properties of RHA-based geopolymers [14,50,98].

An interesting phenomena has been observed by Bascarevic et al. [99] in assessing the sulfate resistance of two types of FA-based geopolymer mortars. While the lower-porosity FA-based geopolymer mortars only exhibited a noticeable reduction in compressive strength after 180 days of exposure period, the compressive strength of the more porous samples instead show a steady increase in strength along the exposure period up to 365 days, although the strength first drops during the 28-day exposure period. The authors concluded that the unusual observation was most probably attributed to the continuing alkali activation in the sulfate solution by the more porous geopolymer mortars. The aforementioned findings were in total contrast to the current body of knowledge, where a denser matrix or reduced porosity either for geopolymers or OPC concrete resulted in enhancement in the durability performance [64,91,100].

## 11.4.2 Microstructure Analysis of Geopolymer

Bhutta et al. [93] investigated the mineralogical phase changes of PFA-POFA based blended fuel ash geopolymer concrete (BFAGC) before and

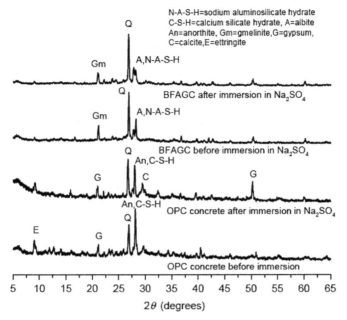

**Figure 11.7** XRD of BFAGC and OPC concrete before and after immersion in 5% Na₂SO₄ for 18 months [93].

after immersion in 5% sodium sulfate ($Na_2SO_4$) solution for a period of 18 months. As can be seen in Fig. 11.7, after 18 months of 5% $Na_2SO_4$ exposure, the semicrystalline aluminosilicates gel (N-A-S-H), which is present before the immersion, remains intact and showed little changes. On the other hand, the OPC specimen, which was used as the control specimen, showed significant changes in the diffraction pattern after 18 months of 5% $Na_2SO_4$ exposure due to the formation of gypsum and ettringite, which subsequently lead to expansion and spalling of surface layers.

The effect of 2% sulfuric acid immersion up to 18 months on the XRD pattern of BAG concrete based on PFA and POFA activated by a mixture of NaOH and $Na_2SiO_3$ solution was studied by Ariffin et al. [92]. OPC concrete was used as the control specimen. Before the acid exposure, the primary mineral phase detected in BAG concrete was a crystalline N-A-S-H phase, while a different crystalline phase of C-S-H was detected in the OPC concrete. After 18 months of sulfuric acid exposure, the main phases detected in BAG concrete, e.g., sodalite, gmelinite, natrolite, and N-A-S-H, were still intact and some traceable amount of gypsum was also detected as a result of reaction with atmospheric $CO_2$. As expected, no ettringite was detected in BAG concrete as the Al ions participated in the formation of N-A-S-H gels, thus making the available Al ions insufficient in the formation of ettringite as in the Portland cement binder system. In the other study, Bascarevic et al. [99] investigated the effect of sodium sulfate solution (50 g/L) exposure on the mineralogical phase changes of FA-based geopolymers up to 365 days. XRD analysis showed that no new phases were formed in the geopolymer sample even after 365 days of exposure period, implying the superior sulfate-resistance characteristic of the FA-based geopolymers.

XRD patterns of copper MTs-based geopolymer bricks subjected to immersion in pH 4 (nitric acid) and pH 7 solutions for 4 months showed increase in the crystalline peaks intensity after immersion. However, the crystalline phases before and after immersion showed very few differences. This is due to the effect of immersion in pH 4 and pH 7 solutions for a prolonged period of time, which caused the dissolution of geopolymer and subsequently led to the exposure of unreacted crystalline MT grains [94].

### 11.4.3 FTIR Analysis

Ariffin et al. [92] performed FTIR analysis on the BAG concrete based on PFA and POFA before and after immersion in sulfuric acid environment for up to 18 months. Mixture of NaOH and $Na_2SiO_3$ solution were used as chemical activator. OPC concrete was used as the control specimen. Major bands at approximately 3440, 1645, 1425, 1010 cm$^{-1}$ and 3465, 1645, 1425, 1040 cm$^{-1}$ were detected in unexposed OPC and BAG concrete, respectively. The stretching band of O-H, the bending of chemically bonded H-O-H and carbonate in the system were located at 3200–3700 cm$^{-1}$, 1645 cm$^{-1}$, and 1425 cm$^{-1}$, respectively. The main binder gel band for OPC, which is the asymmetric stretching mode of the C-S-H structure, was detected at 1010 cm$^{-1}$ while the N-A-S-H gels formed in the geopolymer binder system appeared at 1040 cm$^{-1}$. Upon exposure to 2% solution of sulfuric acid for 18 months, FTIR spectra of BAG concrete showed little or no difference from their unexposed counterpart. In contrast, marked decomposition of C-S-H gel and O-H phases was detected in the OPC concrete after exposure to acid environment. Water component was shifted from 3435 to 3405–3555 cm$^{-1}$ and the chemically bonded water molecules changed from 1625 to 1625–1690 cm$^{-1}$ due to the presence of gypsum. Meanwhile the presence of calcite shifted the C-S-H gel phase from 1010 to 1145 cm$^{-1}$ which indicates the decomposition of the

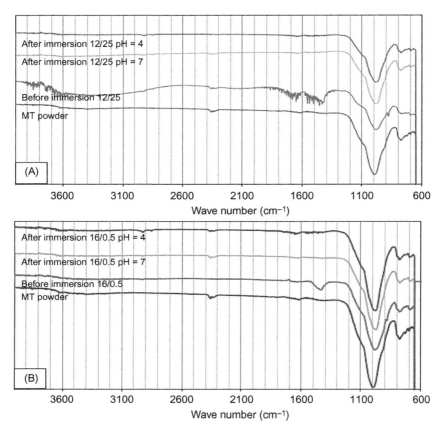

**Figure 11.8** IR spectra of MT powder. (A) 12/25 specimens and (B) 16/0.5 specimens before and after immersion in pH = 4 (nitric acid) and 7 solutions for 4 months [94].

main binder system in OPC concrete. The authors also concluded that the decomposition of the C-S-H gel has provided additional calcium for the formation of gypsum in the exposed OPC concrete.

Ahmari and Zhang [94] performed FTIR analysis on copper MTs-based geopolymer bricks after immersion in pH = 4 (nitric acid) and 7 solutions for a period of 4 months. Two batches of MT-based geopolymer bricks were fabricated, one with initial water content/forming pressure of 12%/25 MPa and another with ratios of 16%/0.5 MPa. The IR spectra of the MT powder, and geopolymer bricks before immersion and after immersion in different solutions are shown in Fig. 11.8. Significant difference was observed for geopolymer bricks before and after immersion, though there is not much of difference in the IR spectra between specimens immersed in pH 4 (nitric acid) and pH 7 solutions. For MT powder, after geopolymerization, the band around $1000\,cm^{-1}$, which is associated to the stretching vibrations of Si-O-T (T = Al or Si) bonds, shifted to a lower wave numbers while the weak shoulder at $1070\,cm^{-1}$ becomes stronger. Upon immersion in the aforementioned solutions for 4 months, the amorphous geopolymer gels were partially dissolved and the underlying crystalline phases of MT grains were exposed, leading to the Si-O-T bonds becoming sharper and shifting towards higher wave numbers. Also, the weak shoulder around $1070\,cm^{-1}$ becomes weaker as a consequence of geopolymer gels dissolutions. Carbonate compounds formed due to geopolymerization at around $1450\,cm^{-1}$ was untraceable after immersion in pH = 4 (nitric acid) and 7 solutions due to the fact that carbonates have been dissolved in the solutions.

## 11.4.4 Thermogravimetry Analysis

In order to study the effect of sulfuric acid exposure of BAG concrete based on PFA and POFA, Ariffin et al. [92] performed thermogravimetry analysis (TGA) on BAG concrete before and after 18 months of 2% solution of sulfuric acid exposure. OPC concrete was used as the control specimen. The ground BAG and OPC samples were held under isothermal condition for 60 min at 40°C and then heated to 900°C at 10°C/min in a nitrogen environment. Before exposure to acid environment, the mass loss in the measure period for OPC and BAG concrete was 18% and 10%, respectively. Fig. 11.9 showed there were four distinct peaks observed in the differential thermograms (DTGs) at various temperatures for both OPC and BAG concrete. The first two peaks, which occurred below 100°C, were attributed to the removal of free evaporable water trapped in the pores of binder's gel, i.e., C-S-H or N-A-S-H gel system. The dehydration of calcium-rich silicate gel was detected at peak just below 200°C while the mass loss at 264°C was attributed to the dehydration of gypsum (OPC) and gmelinite (BAG). Broad mass loss was detected at approximately 704°C for OPC concrete and was assigned to the decomposition of the carbonate minerals. Upon 18 months of 2% sulfuric acid immersion, the TGA and DTG diagram of BAG concrete showed little or no difference compared to their unexposed counterpart. In contrast, the mass loss of exposed OPC concrete was much higher at 18–22% in comparison with the unexposed OPC samples. The temperature by which the evaporation of free water occurred was higher for both OPC (140°C) and BA (122°C). Furthermore, the intensity of mass loss was so much higher

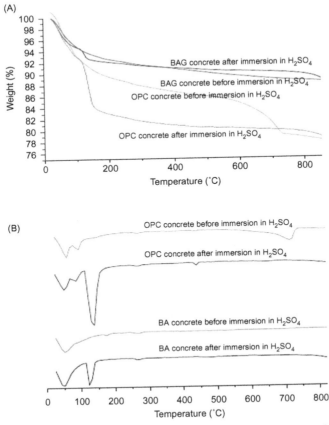

**Figure 11.9** Thermogravimetry (TGA–DTG) curves for OPC and BAG concretes. (A) Themogravimetric data (TGA). (B) DTGs [92].

in the exposed OPC samples. In the same study conducted, the authors concluded that gypsum is the dominant product in the samples exposed to sulfuric acid and the majority of the mass loss in the aforementioned intense peaks could be attributed to the decomposition of the resultant gypsum as gypsum was known to decompose from 110°C to 150°C. The results showed that the BAG concrete held the upper hand in terms of sulfuric acid resistance compared to the OPC concrete.

## 11.4.5 Physical Properties of Geopolymer

PFA–POFA based BFAGC showed superior resistance to prolonged sulfate environment exposure as compared to OPC concrete, as can be seen

OPC concrete specimen                    BFAGC specimen

**Figure 11.10** Visual appearance of OPC concrete and BFAGC after 18 months of 5% $Na_2SO_4$ immersion [93].

in Fig. 11.10. Also, the decrease in mass of OPC concrete after prolonged exposure to sulfate environment was as high as 20% as compared to a 4% decrease in mass of BFAGC. It was found that the superior difference in mass change of OPC concrete and BFAGC was due to the low–calcium content in BFAGC, which renders it more resistant to sulfate attack [93]. However, the mass change method in examining the sulfate and acid resistance of geopolymeric binders [93,101] has been cast into doubt as there are studies that claim that the results obtained from the mass change methods are not representative for assessing the sulfate and acid resistance of geopolymeric binders [95]. Lloyd et al. [95] claimed the utilization of corroded depth method over the conventional change in mass method as a more effective and representative testing method to determine the acid resistance of inorganic polymer binders. This is due to the tendency of formation of apparently intact, but physically weak and porous, reaction products on the sample surface upon attack by acid substances, a phenomenon that differs from the acid-attack mechanism for other type of binders, in which a complete disappearance of the binder phase was observed. The aforementioned statement needs further clarification in order to justify the methods currently being employed by various researchers in assessing that the acid resistance of geopolymeric binders is valid.

In another related study, the effect of 5% sodium sulfate and 3% sulfuric acid immersion over 360 days on the physical properties of BA geopolymers mortars was evaluated by Sata et al. [62]. PC mortar with the same binder-to-sand ratio was used as the control specimen and was subjected to the same environment as the BA geopolymer mortars specimens. BA geopolymer mortar exhibited excellent resistance after immersion in 5% sodium sulfate for 360 days where the length changes were only 65–121 microstrain. On the other hand were the PC mortars, which exhibited expansion of 7600 microstrain for the same duration of immersion in 5% sodium sulfate. Deterioration of PC mortar in the sulfate environment was due to the formation of gypsum and ettringite from the reaction of calcium hydroxide and calcium monosulfoaluminate, which leads to the expansion and cracking of the mortar's surface layers. Under 3% sulfuric acid solution, BA geopolymer mortar again exhibited superior resistance if compared to PC mortar, where the weight loss of BA geopolymer mortars after 360 days immersion in 3% sulfuric acid was only 1.4–3.6%, which is relatively insignificant as compared to PC mortar, which recorded a weight loss of 95.7%. The weight loss in acid environment is very much related to the calcium content in the system, where higher calcium content leads to higher amount of calcium hydroxide, which in turns reacts with the acid solution forming salt crystals. The salt crystals formed within the paste matrix induce internal tension stress, which causes the formation of a crack and scaling within the paste matrix.

Reddy et al. [96] attempted to simulate the exposure of FA-based geopolymer concrete to marine environment by utilizing an accelerated laboratory electrochemical method for the corrosion test, relative to OPC concrete. 150 × 150 × 525 mm geopolymer and OPC concrete beams with centrally reinforced 13 mm rebars were fabricated and tested for the corrosion test using the aforementioned method after curing for 28 days. Results showed that geopolymer concrete possessed superior resistance towards salt attack relatively with OPC concrete. OPC beams start to crack after 60 h of accelerated corrosion test while geopolymer concrete showed no sign of cracking even after the end of the accelerated corrosion test. The superior durability performance of geopolymer concrete was further verified where no mass loss were recorded for the rebars upon the completion of accelerated corrosion test where for OPC concrete, the maximum mass loss of rebars was recorded as 71.2% after the completion of the test.

## 11.5 THE EFFECT OF WATER CONTENT AND FORMING PRESSURE ON THE PROPERTIES OF GEOPOLYMERS

Water content and forming pressure has a significant effect on the mechanical strength and sorptivity performance of geopolymers, particularly in the fabrication of geopolymer pressed block [25]. This is because both water content and forming pressure have a direct influence on the total porosity of the geopolymer matrix, similar to conventional OPC concrete mixture design. Generally, higher water content will result in increased total porosity [102]. On the other hand, higher forming pressure will reduce the total porosity of the geopolymer matrix. Also, the utilization of pressed forming methods in fabricating geopolymers allows significant reduction in water requirement in comparison with vibratory forming methods where order-suitable workability of fresh mixtures must be achieved for proper compaction [103]. In this section, the effect of water content and forming pressure on the properties of geopolymers is deliberated in depth.

## 11.5.1 Mechanical Properties

In the alkali-activation process, the geopolymerization reaction primarily involves the chemical reaction between the dissolved species of silicates and aluminates in the presence of a highly alkaline environment. The presence of water in the geopolymer system merely acts as a transport medium between the dissolved silicate and aluminate ions. Besides, mixing water also provides workability to the freshly mixed geopolymer mortars, as it does not participate directly in the geopolymeric reaction [104]. However, Komljenovic et al. [31] reported there are some effects, if not significant ones, of the water/FA ratio on the strength of geopolymer mortars, depending on the type of activators used. Generally, for NaOH, $Na_2SiO_3$ and $Ca(OH)_2$ activated FA geopolymer mortars, the compressive strength increases with the decrease in water/FA ratio. For KOH-activated geopolymer mortars, the geopolymer mortars exhibited low compressive strength even at low water/FA ratio. This is indicative of the low activation potential of KOH for FA activation as compared with other alkali activators.

The importance of water content in geopolymer mix design was further highlighted in a study where the liquid medium is reported to be crucial for the diffusion of the dissolved alumina and silica species. Ahmari and Zhang [25] studied the effect of initial water content on the UCS of copper MTs-based geopolymer bricks. A number of geopolymer bricks were fabricated based on six initial water contents, i.e., 8, 10, 12, 14, 16, and 18%, and tested on the 7th day of curing duration. The initial water content was referred to the mass ratio between the water in the activating solution (NaOH) and the solid content of the mixture. The fresh geopolymer pastes were put inside a steel mold and compressed to an extent whereby they reached saturation state. The results indicated that the UCS of MT-based geopolymer bricks increased with higher initial water content. The aforementioned observation was attributed to the role of water as a liquid medium during geopolymerization. Besides, it is also related to the availability of a sufficient amount of NaOH in liquid phase during geopolymerization. The amount of NaOH, in turn, is strongly dependent on the Na/Al and Na/Si ratios as shown by other related studies [16,60,105]. UCS as high as 33.7 MPa was obtained at 18% initial water content and 0.2 MPa forming pressure. In another experimental work, the compressed autoclaved bricks based on low silicon tailings and alkali-activated cementing materials (slag and FA) attained the optimum compressive strength of 16.0 MPa at the water content range of between 6.5–8.0% and forming pressure of 20 MPa. Any values in excess of the aforementioned range of water content and forming pressure had resulted in a decrease in compressive strength [102].

In the fabrication of compressed building bricks, forming pressure plays an important role in achieving optimum densification of intra- or interparticle packing by pushing the entrapped air out of the binder matrix. Thus, bricks with the lowest possible level of porosity and higher strength can be fabricated using the pressure-forming method [102,103]. An experimental study was conducted by Ahmari and Zhang [25] to study the effect of forming pressure on the UCS of copper MTs-based geopolymer bricks. The MT-based geopolymer bricks were activated with 15 M of NaOH and curing temperature of 90°C based on the optimization results from the same research work. Various forming pressures were applied and the geopolymer bricks were tested after 7 days of oven curing. From the UCS results, it can be concluded that (1) all the MT-based geopolymer bricks showed an increasing trend of UCS up to a certain level of forming pressure; (2) the decline in UCS for high forming pressure was due to the loss of NaOH solution that was squeezed out during the forming process at high forming pressure, hence reducing the degree of geopolymerization; (3) forming pressure is related to the initial water content of geopolymer mix, i.e., higher forming pressure may result in lower initial water content.

Zhao et al. [102] studied the effect of forming pressure on the compressive strength of autoclaved bricks made from low-silicon tailings. Alkali-activated cementitious materials based on slag and FA were incorporated into the mixture in an attempt to produce high-strength low-silicon autoclaved bricks with load-bearing capacity. Water content in materials of 7.5%, tailings to cementing material mass ratio of 85:15, autoclaved curing regime of 2-8-2 (temperature rising–holding–dropping stages), and autoclaved steam pressure of 1.0 MPa was fixed throughout the study. Results showed that the compressive strength of bricks increased with the forming pressure. However, the magnitude of strength increase in forming pressure exceeding 20 MPa was deemed insignificant. Thus, the optimum forming pressure should range between 18 and 20 MPa in consideration of producing bricks with adequate mechanical strength with minimal energy usage. The compressive strength of autoclaved bricks obtained using the aforementioned forming pressure range exceeds the target strength of 15.0 MPa for load–bearing brick.

## 11.5.2 Water Absorption

Water absorption is a very important parameter for the fabrication of geopolymer bricks as it indicates the permeability and the degree of reaction for geopolymer bricks. Generally a higher degree of geopolymerization gives rise to a less porous and permeable geopolymer matrix. Ahmari and Zhang [25] evaluated the effect of forming pressure on the water absorption of copper MTs-based geopolymer bricks. NaOH concentration, initial water content, and curing temperature were made constant at 15 M, 16%, and 90°C, respectively. The freshly mixed MT-based geopolymer paste was compressed at five different forming pressures, namely 0.5, 1.5, 3, 5, and 15 MPa to form the geopolymer bricks. The authors found that the water absorption after 4 days of soaking varies from 2.26% to 4.73%, corresponding to the forming pressure from 0.5 to 15 MPa. The increase in water absorption with the increasing forming pressure was attributed to a higher amount of NaOH solution being squeezed out at higher forming pressure. Under such circumstances, the geopolymerization reaction is hindered, hence, fewer geopolymer gels were formed, subsequently resulting in higher porosity of the geopolymer matrix. On the other hand, Freidin et al. [106] reported water absorption less than 10% for FA-based geopolymer bricks upon the addition of hydrophobic additives. All the water absorption values of the MT-based geopolymer bricks were well below the maximum water absorption value allowable for different kind of bricks as in accordance to various ASTM standards namely ASTM C34-03, C62-10, C126-99, C216-07a, and C902-07.

## 11.6 BLENDED GEOPOLYMER

Blended geopolymers is a new category of geopolymeric binder, which is produced by selective hybridization of two or more industrial waste ashes followed by subsequent stabilization and solidification using chemical activators. The dual advantages of waste utilization and more importantly the alteration in Si/Al and Ca/Si in the geopolymer system prompted a sudden rush in the amount of research in the field of blended geopolymer during recent years [21,23,24,103,107,108]. Ever since it was known that C-S-H gel can coexist with geopolymeric gel in a single system and that it contributes to the overall strength gain [40], various researchers have utilized high-calcium waste material such as GGBFS and ASTM class C FA to blend with ASTM class F FA in order to achieve a higher early strength gain and shorter setting time, which is beneficial particularly for application in the precast industry [32,77]. Canfield et al. [109] investigated the role of calcium in FA-based geopolymers by blending high- and low-calcium FA. It was found that calcium played two major roles during the geopolymerization of the blended FA geopolymer specimen: (1) calcium was found to aid the dissolution of silica and alumina species from the FA particles, yielding higher tetrahedral silicate and aluminate monomer concentration as shown in FTIR, XRD, and TGA/DSC results; (2) calcium also functions as a counterbalancing cation when incorporated into the geopolymer pore structure. In a separate study, the workability of fresh geopolymer concrete consisting of GGBFS and FA showed a decreasing trend with an increase in slag content and decrease of SS/SH ratio due to the enhancement in reactivity in the presence of GGBFS in a FA-based geopolymer system [107]. Other waste materials such as RHA and POFA have also found considerable interest among geopolymer researchers in the blended geopolymer field [11,38,50,64].

The following section covers the mechanical, durability, and microstructure properties of a number of emerging blended geopolymers.

## 11.6.1 Mechanical Properties

He et al. [14] attempted to incorporate silica-rich RHA into silica-deficient RM to produce a new class of blended geopolymer. The 60-days compressive strength of geopolymer pastes derived from RHA and RM with varying RHA/RM ratios (0.3, 0.4, 0.5, and 0.6) was examined. All the geopolymer pastes were cured under room temperature and atmospheric pressure until the specified testing age. NaOH was used as the chemical activator and the concentration and liquid-to-solid weight ratio were fixed at 4M and 1.2 throughout the experimental study. The aforementioned RHA/RM ratios of 0.3, 0.4, 0.5, and 0.6 employed in this study were corresponding to Si/Al ratios of 1.68, 2.24, 2.80, and 3.35. Results showed compressive strength increased up to RHA/RM ratio of 0.5 before decreased for the paste specimens with RHA/RM ratio of 0.6. The compressive strength obtained ranged from 3.2 to 20.5 MPa. The enhancement in strength was due to the increasing amount of reactive silica and higher specific surface area of RHA. Meanwhile the deterioration in strength observed in specimens with RHA/RM ratio of 0.6 was due to the high amount of unreacted RHA particles in the mixture. Also, the higher concentration of soluble Si ion in that particular mixture that hinders the restructuring of Si and Al geopolymer network, subsequently results in the formation of a weaker geopolymer matrix. In another study involving RHA, Detphan and Chindaprasirt [50] studied the feasibility of producing geopolymers based on the hybridization of RHA and FA. The hybridization ratios of RHA/FA used to fabricate the geopolymer mortars were 0/100, 20/80, 40/60, and 60/40. The effect of sodium–silicate-to-sodium-hydroxide ratio, the $SiO_2/Al_2O_3$ ratio, and curing regime (delay time, curing temperature, and curing period) on the compressive strength development of the RHA/FA geopolymer mortars were investigated. The following conclusions were made as a result of the laboratory investigation: (1) Compressive strength range from 13 to 42 MPa can be achieved using sodium–silicate-to-sodium-hydroxide ratio from 1.5 to 5.9, with the optimum ratio at 4.0 regardless of the RHA/FA hybridization ratios; (2) a reduction in compressive strength was observed with the increasing $SiO_2/Al_2O_3$ ratio mainly due to the cellular structure of RHA, which resulted in high water uptake of the corresponding mortar mixes: (3) a delay time of 1 h, curing period of 48 h, and curing temperature of 60°C was found to be the most ideal curing regime to achieve high strength, with the effect of curing temperature being more prevalent to mortar mixes that contain high FA dosage.

RHBA is generated in biomass electricity power plants where rice husks and bark are burned at a temperature around 400°C, yielding silica-rich ash with similar chemical composition to RHA, with the exception of slightly lower silica and higher calcium compounds. Songpiriyakij et al. [49] investigated the effect of $SiO_2/Al_2O_3$ ratios towards the compressive strength development of RHBA/FA blended geopolymers. Due to the silica-rich RHBA, a wide range of $SiO_2/Al_2O_3$ ratios, i.e., 4.03–1035, were obtained by hybridizing the two aforementioned base materials. Results shown that the optimum compressive strength at 28 days of curing obtained is 51.0 MPa for $SiO_2/Al_2O_3$ ratio of 15.9 due to the formation of stronger Si-O-Si bonds contributed by the addition of silica-rich RHBA. However, the authors observed expansion and cracking on specimens with $SiO_2/Al_2O_3$ ratios greater than 15.9 over time. Also, specimen failure mode transition from crushing to deformation was also observed for $SiO_2/Al_2O_3$ greater than 15.9. The authors also concluded that other than the reactivity of the source materials, the quality of the matrix phase developed is also an essential factor that contributes to the compressive strength development of the RHBA/FA blended geopolymer paste. The strength obtained was in agreement with another related study [21].

Due to the emergence of fluidized bed combustion (FBC) as a promising clean-coal technology compared to the traditional pulverized coal combustion (PCC) method due to the reduction of $SO_2$ and $NO_x$ gasses emitted in flue gas, Chindaprasirt et al. [110] utilized FA obtained from FBC as a source material for geopolymer fabrication. Geopolymer mortars consisted of FBC-FA and PCC-FA with mass ratios 0:100, 20:80, 40:60, 60:40, 80:20, 100:0 were activated by 10M NaOH and sodium silicate with $Na_2SiO_3$ to NaOH ratio of 1.5 by weight. Liquid content for each mix was adjusted accordingly to achieve a workable mix due to the irregular shape of FBC-FA. Sand-to-ash ratio of 2 was maintained for all the mixes. Freshly mixed geopolymer was poured into a 25-mm-diameter and 25-mm-height mold, oven cured at 65°C for 48 h, and cured continuously at controlled room temperature of 25°C until the testing ages. The compressive strength results at 7-days specimen age were indicative that the compressive strength of geopolymer mortar decreased as the amount of FBC-FA increased. This is due to the lower reactivity of FBC-FA as compared to primary PCC-FA. The low amorphous phase and the porous nature of FBC-FA is the governing reason for the aforementioned observation. PCC-FA to FBC-FA ratio of 60:40, which exhibited 7 days compressive strength of 30.0 MPa, was recommended as the mix proportion for the fabrication of geopolymer due to the relatively high strength and also the significant amount of FBC-FA used.

While the addition of waste-paper sludge from the paper-recycling industries in OPC concrete brings about adverse effect to various properties [111,112], Yan and Sagoe-Crentsil [37] attempted to incorporate dry waste-paper sludge to FA-based geopolymer mortars on the mechanical properties. The dry waste-paper sludge was added to the geopolymer system as a sand-replacement material in the range of 0–10 wt%. Laboratory-grade sodium silicate solution and sodium hydroxide pellets were used as activating solutions and the $SiO_2/Na_2O$ molar ratio and $H_2O/Na_2O$ ratio were fixed at 1.5 and 11, respectively. Also, the other constant parameters in the experimental study are the sand/FA ratio of 3 and the liquid/ solids ratio of 0.2. The freshly casted specimens were initially cured in a steam chamber at 60°C for 8 h before being demolded, followed moist curing (sealed in plastic bags at room temperature) until the testing ages. The authors concluded that the decrease in compressive strength of the geopolymer mortars up to 10 wt% of dry waste–paper sludge was due to the presence of surfactants (dissolved lignin residues) in dry waste–paper sludge, which act as air-entraining admixtures and alter the porosity and pore-size distribution of the geopolymer matrix, resulting in a lower-density geopolymer mortar as the percentage of dry waste–paper sludge increases. The average 91-days compressive strengths of geopolymer mortars containing 2.5 and 10 wt% of dry waste–paper sludge were 55.7 and 31.2 MPa, respectively, retaining 92% and 52% of the control mortars' strength, which attained 60.6 MPa.

GGBFS is the one of the most popular source materials to be blended with other geopolymer source materials as the addition of GGBFS increases the amount of amorphous silica and alumina and also the CaO in the resultant geopolymer system, which in turn will greatly improve the mechanical performance of the blended geopolymers. For instance, different grades of GGBFS, i.e., 80, 100, and 120 were incorporated into FA-based geopolymers in a fixed FA/GGBFS weight ratio of 5/3. Results suggested that without the incorporation of external amorphous silica source, higher-grade GGBFS is only beneficial for the early strength development of FA geopolymer. The reactivity of the higher-grade GGBFS can be exploited by the enhancement in $SiO_2/Al_2O_3$ and $SiO_2/Cao$ ratios by the incorporation of additional amorphous silica source [108]. GGBFS/FA geopolymer concrete with 28-days compressive strength as high as 51 MPa could be achieved with the GGBFS/FA hybridization ratio of 20/80, 40% of activator liquid and SS/SH ratio of 1.5 when cured at ambient temperature [107].

Deb et al. [107] investigated the splitting tensile strength of GGBFS/FA-based geopolymer concrete cured at ambient temperature. The tensile strength increased with increasing slag content and decreasing SS/SH ratio, providing a strong correlation with the corresponding compressive strength development. GGBFS/FA geopolymer concrete with 20% of GGBS content and SS/SH ratio of 1.5 exhibited 55% higher 28-days tensile strength than the geopolymer concrete mixture with 10% GGBFS and SS/SH ratio of 2.5. The ranges of tensile strength obtained were also consistently higher than the OPC concrete specimens.

The number of researches utilizing POFA in the fabrication of geopolymers or blended geopolymers has risen in recent years due to the abundance of POFA wastes especially in Southeast Asian countries, e.g., Malaysia and Thailand [113–117]. Approximately 3 million tons and 0.1 million tons of POFA were produced annually in Malaysia and Thailand, respectively [4,118]. The aluminosilicate waste material was obtained from the palm oil industry and has usually undergone a pretreatment process involving calcination, grinding, and sieving before it can be suitably used as geopolymer feedstocks [11]. As of now, researchers are focusing on the optimization of blended geopolymers involving GGBFS/POFA [23,115], FA/POFA [24,46,92,93]. A ternary blended geopolymer system of GGBFS/POFA/RHA was also proposed [117]. Various factors such as $H_2O/Na_2O$ ratio [114], $M_s$ [38], curing regime, and chemical activator dosage [116] were investigated on a GGBFS/POFA blended geopolymer system and excellent mechanical properties were obtained from the optimum mixes from each of the factors studied. The excellent mechanical properties were due to the formation of a dense geopolymer matrix and the coexistence of C-S-H gels and geopolymeric gels (NASH/CASH) with the inclusion of GGBFS in the blended system [114]. Yusuf et al. [115] recommended the partial replacement of POFA by GGBFS to be at 20% in order to maintain an excellent mechanical properties profile. POFA, when used as a replacement material in FA-based geopolymers, reduces the early strength and delayed the geopolymerization process [24]. However, POFA/FA geopolymers exhibited a gain in strength even upon being subjected to elevated temperature as high as 500°C. The authors also reported the addition of POFA into FA-based geopolymer mortars reduced the compressive strength and density of the mixtures [46]. Excellent compressive and flexural strength was observed in a ternary blended geopolymer system comprising POFA, FA, and RHA [117].

## 11.6.2 Microstructure of Geopolymer Matrix

The microstructure of FBC-FA and PCC-FA blended geopolymer paste was studied by Chindaprasirt et al. [110]. The geopolymer paste was activated using the NaOH to $Na_2SiO_3$ ratio of 1.5 and was oven cured at 65°C for 48 h, followed by subsequent curing at a controlled temperature of 25°C. The morphology of the blended geopolymer paste was examined at the age of 7 days. Fig. 11.11A and B show the morphology of 60:40 PCC-FA: FBC-FA blend and its geopolymer paste counterpart.

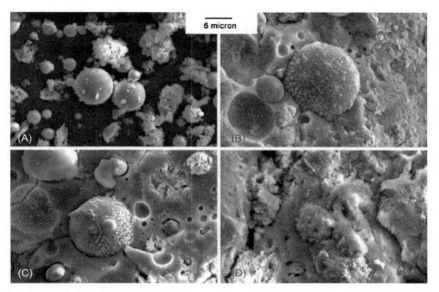

**Figure 11.11** Morphology of 7-days specimens: (A) 60/40 blend PCC-FA: FBC-FA; (B) 60/40 PCC-FA: FBC-FA blended geopolymer; (C) PCC-FA geopolymer paste; (D) FBC-FA geopolymer paste [110].

The spherical shape of PCC-FA induced the ball–bearing effect and improved the workability of the resultant paste as compared with FBC-FA, which consisted of irregular and porous particles. The morphology of the 60:40 blended geopolymer paste, PCC-FA geopolymer paste (Fig. 11.11C), and FBC-geopolymer (Fig. 11.11D) paste show continuous mass of aluminosilicate, indicating a relatively well-developed geopolymer network. However, the unreacted/partially reacted grains of irregular FBC-FA are much more porous than PCC-FA, which culminated in a lower strength of blended geopolymer paste with higher FBC-FA content. In the same study, XRD results indicated that at the age of 7 days, blended geopolymer pastes with a high amount of PCC-FA (60%, 80%, and 100% PCC-FA) showed a high amount of amorphous phases and trace amount of crystalline products. Meanwhile geopolymer pastes with a high amount of FBC-FA (60%, 80%, and 100% FBC-FA) exhibited intense peaks of crystalline phases with a reduced amount of amorphous gel. Calcium silicate similar to the hydration product of Portland cement was detected in all the blended hardened geopolymer pastes.

The pore size of POFA/FA geopolymer mortars significantly increased upon being subjected to elevated temperature beyond 800°C. High POFA content in the blended geopolymer mortar mixes deformed at 800°C while FA-based geopolymer mortars maintain their structural integrity up to a temperature of 1000°C, suggesting lower thermal stability upon addition of POFA in FA-based geopolymer mortars [46]. In a separate study, the addition of POFA in a FA geopolymer system increases the porosity in the resultant blended geopolymer matrix [24]. This is due to the unreacted POFA particles having the tendency to trap air because of their inherent crumbled shape.

### 11.6.3 Dimensional Stability

Yusuf et al. [113] investigated various factors influencing the shrinkage behavior of GGBFS/POFA blended-geopolymer mortar. The factors studied included the effect of GGBFS addition, effect of SS/SH ratio, and also the effect of pore sizes and volume of base materials. The following conclusions can be drawn from the study: (1) Internal microcracks and enhancement in pore filling and pore refinement, which resulted in significant reduced shrinkage, can be achieved with the addition of GGBFS up to 60%; (2) SS/SH ratio of 2.5 was recommended for a GGBFS/POFA blended-geopolymer system, where enhancement in product glassy phase, reduction in carbonation, and tendency of C-A-S-H gel formation resulted in significantly reduced shrinkage.

The drying shrinkage behavior of dry waste-paper sludge added-FA geopolymer mortars was monitored up to 91 days by Yan and Sagoe-Crentsil [37]. The dry waste-paper sludge was used in the range of 0–10 wt% as sand-replacement material. Results showed that the addition of dry waste-paper sludge up to 10 wt% reduces the drying shrinkage of the resultant geopolymer mortars compared with the reference mortars, which contain only FA as the binder phase material. The drying shrinkage of 91 days cured mortars specimens of 10 wt% dry waste-paper sludge is 492 µε, which corresponds to a remarkable 64% reduction if compared with the reference mortars (1346 µε). The reduction in drying shrinkage with the increasing dry waste-paper sludge was due to the presence of cellulose fibers, which existed predominantly in dry waste-paper sludge. The expansion of cellulose fibers in the presence of moisture compensated the actual drying shrinkage of geopolymer mortars specimens incorporating dry wastepaper sludge, which is expected to exhibit a higher degree of drying shrinkage based on the results of moisture loss analysis.

## 11.7  SUMMARY OF THE CURRENT BODY OF KNOWLEDGE AND DISCUSSIONS

Following the reviews performed on the essential factors influencing the properties of geopolymers derived from industrial byproducts, the following summaries can be made:

1. Generally the inclusions of 2–14 M of NaOH as chemical activators into the geopolymer matrix increased the compressive strength of the hardened geopolymer concrete.

2. The multicompound chemical activators consisting of $Na_2SiO_3$/ NaOH were found to be the most effective chemical activators to be added into the geopolymer matrix for the purpose of mechanical strength enhancement.

3. Coupled with the addition of chemical activators, the application of heat curing in terms of duration and temperature is essential in accelerating the early age strength development of geopolymer concrete. A maximum duration of 24 h and a heat curing temperature range of 50–90°C was found to be beneficial towards the short- and long-term strength development and stability of geopolymer concrete.

4. Higher liquid alkaline/ash ratio generally resulted in higher workability of the fresh geopolymer mixture. Higher $Na_2SiO_3$/NaOH ratio generally reduced the workability of the fresh mixtures due to the higher viscosity of $Na_2SiO_3$, while a higher concentration of NaOH was found to increase the setting time of the resultant geopolymer mixture.

5. Generally the shrinkage of geopolymer concrete is governed by the corresponding strength. The increase in linear shrinkage is associated with the low strength development of the geopolymers, and vice versa.

6. In calcium-added or high-calcium geopolymeric systems, C-S-H gels were found to coexist with the geopolymeric gels and enhance the microstructural and mechanical strength development of the resulting geopolymer concrete.

7. High-strength (>60 MPa) geopolymer concrete can be fabricated with the higher fineness of the binder-phase materials.

8. A small amount (≤3% by binder weight) of nanosized particle additions to the geopolymer matrix resulted in significant enhancement in mechanical and microstructural properties of the geopolymer concrete and caused reduction in both initial and final setting time of the geopolymer paste.

9. Generally, naphthalene-based SPs are the most effective high-range water-reducing admixtures for geopolymer systems activated by NaOH only. For geopolymer systems activated by both $Na_2SiO_3$ and NaOH, polycarboxylate-based plasticizers exhibited the best plasticizing effect amongst other commercial SPs (naphthalene-based and melamine-based superplasticizers).

10. Hybridization of different sources of binder materials with the optimized proportion of chemical activators and optimum curing method depending on the suitability of source materials can lead to the production of geopolymer concrete with superior mechanical, physical, durability, and microstructural properties.

11. Geopolymer concrete exhibited superior resistance upon exposure to aggressive environments such as sulfuric acid, magnesium sulfate, nitric acid immersion, etc., with an average of 12–40% of compressive strength loss compared to OPC concrete, which generally exhibited 40–65% compressive strength loss within the same exposure period.

12. The effect of water content in geopolymer concrete on the strength development was generally governed by the type of chemical activators used. For example, for FA-based geopolymer, the strength increased with the decreased water/FA ratio for the NaOH, $Na_2SiO_3$, and $Ca(OH)_2$ activated FA-based geopolymer. On the other hand, the reversed trend was observed for the FA-based geopolymer activated by KOH.

13. In the fabrication of press-formed geopolymer bricks, an appropriate forming pressure that minimized the interparticle spacing in the ITZ and also the leaching out of the alkaline pore solution will yield geopolymer bricks with excellent compressive strength ($\geq 30\,MPa$) and low water absorption (2–5%).

OPC concrete has received wide criticism over the past decades with increasing awareness on its inherently high-carbon footprint and embodied energy of production, which has prompted the rise of geopolymer technology. Geopolymer was seen as a solution to both the construction industry's problems and also the waste management issues suffered by various industries such as the coal-burning industry, palm oil, rice milling industry, etc. However, despite the intense studies performed over the years, geopolymer technology is still far from achieving its ultimate goal, which is to replace OPC in actual industrial practices. Based on the reviews performed, the most influencing factors that governed the mechanical, physical, durability, and microstructure performance of geopolymers are the alkaline activator, curing regime, and also the physical properties and chemical composition of the raw materials.

Several researchers have examined the environmental impact of geopolymer concrete in comparison with OPC concrete [119–122]. For instance, Habert et al. [119] evaluated the environmental impact of geopolymer concrete against OPC concrete using the life-cycle assessment (LCA) approach. The authors found that the production of a typical geopolymer concrete exhibited lower global warming potential value than OPC concrete with identical strength properties, with the former exhibiting 168.5 kg $CO_2$ eq. while the latter with global warming potential value of 305.9 kg $CO_2$ eq. However, geopolymer concrete exhibited higher values in three other impact categories examined, namely, abiotic depletion, marine ecotoxicity, and acidification, as compared with OPC concrete, mainly due to the effects originating from the production of sodium silicate solution. The aforementioned findings were supported by another research work by Turner and Collins [120], who concluded that the reduction in carbon dioxide equivalent emissions ($CO_2$-e) of geopolymer concrete in comparison with OPC concrete is a mere 9%. This is in contrast to the widespread claim that the production of geopolymer binders reduced the carbon footprint of OPC binders by as high as 60% [123]. The main reasons that culminated in the higher-than-expected figure are the enormous amount of energy incurred during the mining, treatment, and transport of raw materials to manufacture the chemical activator. Besides, the energy-intensive manufacturing process, which includes the mixing and melting of sodium carbonate and sand for the production of sodium silicate solution, is also the primary contributor to the embodied energy of the geopolymer binder system.

Based on the reviewed literature, in order to achieve similar or better properties in comparison with OPC concrete, most of the geopolymers require either the utilization of high dosage of alkaline activator [13] or heat treatment [1], or in many cases the application of both [4,19,46]. It is believed that the aforementioned factors proved to be the stumbling blocks in the transition of geopolymers technology from research basis towards industrial application. This is because the addition of a high dosage of chemical activators in the geopolymer mix design will result in a spike increase in the manufacturing cost, twice as high as OPC concrete [122]. Moreover, the necessity of elevated temperature treatment rendered mass industrial fabrication of geopolymer concrete impractical. Coupled with the environmental impact assessment discussed in the previous paragraph, future research in the field of geopolymers should focus on minimizing or eliminating the usage of chemical activators and heat treatment. This is necessary in order to justify the feasibility of geopolymer concrete to replace OPC concrete from both the economic and carbon footprint perspective.

Although geopolymeric binders present the highest potential to emerge as the new low-carbon footprint binder to replace OPC, one of the foreseeable challenges that warrants a proper solution is the inconsistency in properties and performance shown by various geopolymer source materials. It is well known that the performance of geopolymers is very much governed by the physical and chemical properties of the aluminosilicate source material. Popular geopolymer source materials originating from industry wastes, such as FA, GGBFS, RHA, and POFA, have their own unique chemical composition and physical properties, and thus require distinctly different alkaline-activator dosage and processing methods in order to achieve similar performance. Furthermore, the properties of the same source materials, but from different locations, possessed different characteristics in term of chemical composition and physical properties. The aforementioned variations will definitely pose problems when transferring geopolymer knowledge to the industrial practitioners.

In light of the various uncertainties in terms of the actual environmental impact and challenges faced in the field of geopolymer concrete, the following strategies/recommendations are proposed to significantly reduce the carbon footprint of geopolymer concrete and at the same time facilitate the implementation of geopolymer technology in the concrete production industry:

- Future research should focus on fabricating geopolymer concrete with minimal alkaline-activator dosage and elevated temperature treatment in order to produce a sustainable product with low embodied energy and low carbon footprint, which is low in production cost and safe for site handling. For example, geopolymer concrete made from FA and GGBFS hybrid aluminosilicate raw material have a lower environmental impact as it required less sodium silicate in order to be activated as compared to geopolymer concrete made with pure MK [119].
- Hybridization of various industrial waste materials such as FA, GGBFS, RHA, POFA, and any other aluminosilicate-rich raw materials is necessary in order to strike a balance in the eventual Si/Al molar ratio for optimum geopolymerization reaction to occur.
- Leveraging the well-established particle technology in OPC to improve on the granular distribution of geopolymer materials to achieve higher packing density thus reducing the amount of alkaline activators and active binders required [124].

- Development of new activation and curing methods such as utilizing waste materials, e.g., sodic waste or other additives that possessed similar properties to the commercial alkaline activators.
- Establishment of standard specification and testing methods designed specifically for geopolymer concrete or mortar might be one of the few steps in convincing the widespread acceptance of geopolymers technology in replacing OPC concrete.
- A complete elucidation and modeling of geopolymerization reaction kinetics and chemistry based on different source materials should be created to serve as a general guideline for the geopolymers researchers and engineers. This is essential in identifying the crucial parameters and factors to be considered during the design and fabrication stage of geopolymer concrete material.
- Utilization of advanced analytical methods such as the nuclear magnetic resonance technique to elucidate the structural unit of the amorphous geopolymer products formed by single or hybridized source materials that cannot be derived quantitatively using other analytical methods such as XRD.
- LCA of each geopolymer concrete developed should be derived in order to truly justify the environmental and economic benefits offered by geopolymeric binders over OPC concrete. This is because besides the energy incurred during the production of geopolymer source materials and other additives other factors such as the location of source materials, the energy source, and mode of transportation are equally important considerations. These factors determine the actual environmental impact of geopolymer concrete when it is in actual industrial applications, as pointed out by McLellan et al. [122]. With reference to a typical Australian material supply chain, the variability of the aforementioned factors could lead to a very wide range of greenhouse gases emissions by geopolymer concrete as compared to an equivalent OPC concrete. Values could be ranging from a reduction of 97% up to an increment of 14% in terms of total carbon footprint.

## 11.8 CONCLUSIONS

This chapter reviews and summarizes the essential factors that have considerable effect on the properties of geopolymers derived from source materials originated from industrial waste streams. The issues regarding the environmental impact of geopolymer concrete in comparison with OPC concrete were also deliberated. Based upon the review work done, numerous challenges and issues faced by the practitioners of geopolymer technology and the steps needed to overcome the barriers were highlighted and discussed as well.

The present literature provides a detailed elucidation of various factors that influence the properties of geopolymer concrete. With reference to the current body of knowledge in geopolymer technology, a rigorous amount of study has been performed to cover the various aspects of established geopolymers such as FA- and GGBFS-based geopolymers. However, there is still a significant gap of knowledge related to the geopolymerization reaction kinetics, material properties, and rheological behavior of a number of emerging geopolymers such as blended geopolymers and biomass ash geopolymers. Hence, detailed studies such as those related to the derivation and modeling of reaction kinetics under various treatments and fabrication conditions of the emerging class of geopolymer source materials such as POFA, RHA, and blended geopolymers are required. Besides, contradicting findings in terms of embodied energy and carbon footprint and embodied energy of geopolymers in comparison with conventional OPC concrete must be addressed. Hence, the development of new geopolymer material design, fabrication, and postfabrication treatment technology, which are oriented towards minimizing the production cost, embodied energy, and carbon footprint is essential. These can be achieved when the LCA of the geopolymer concrete is taken as a primary consideration during the design stage. All of this is a certain necessity to ensure the sustainability and effective implementation of geopolymer technology as low-carbon concrete materials.

Successful utilization of geopolymers derived from industrial byproducts in actual industrial application will bring about numerous benefits to the construction industry and solve various industrial waste management issues. It is a promising sign that the research in geopolymers, especially in the utilization of industrial waste materials, has been intensified and it is certainly a step forward in achieving the ultimate goal of geopolymer use as a complete replacement of OPC as the primary construction material with a significantly low carbon footprint.

## REFERENCES

[1] Ahmari S, Ren X, Toufigh V, Zhang L. Production of geopolymeric binder from blended waste concrete powder and fly ash. Constr Build Mater 2012;35:718–29.
[2] Schneider M, Romer M, Tschudin M, Bolio H. Sustainable cement production—present and future. Cem Concr Res 2011;41(7):642–50.

[3] Davidovits J. Global warming impact on the cement and aggregate industries. World Res Rev 1994;6:263–78.

[4] Yusuf MO, Johari MAM, Ahmad ZA, Maslehuddin M. Strength and microstructure of alkali-activated binary blended binder containing palm oil fuel ash and ground blast-furnace slag. Constr Build Mater 2014;52:504–10.

[5] Damtoft JS, Lukasik J, Herfort D, Sorrentino D, Gartner EM. Sustainable development and climate change initiatives. Cem Concr Res 2008;38(2):115–27.

[6] Cheah CB, Ramli M. Mechanical strength, durability and drying shrinkage of structural mortar containing HCWA as partial replacement of cement. Constr Build Mater 2012;30:320–9.

[7] Kroehong W, Sinsiri T, Jaturapitakkul C, Chindaprasirt P. Effect of palm oil fuel ash fineness on the microstructure of blended cement paste. Constr Build Mater 2011;25(11):4095–104.

[8] Nath P, Sarker P. Effect of fly ash on the durability properties of high strength concrete. Proc Eng 2011;14:1149–56.

[9] Cheah CB, Ramli M. The engineering properties of high performance concrete with HCWA-DSF supplementary binder. Constr Build Mater 2013;40:93–103.

[10] Cheah CB, Ramli M. The fluid transport properties of HCWA-DSF hybrid supplementary binder mortar. Composites Part B 2014;56:681–90.

[11] Mijarsh MJA, Megat Johari MA, Ahmad ZA. Synthesis of geopolymer from large amounts of treated palm oil fuel ash: application of the Taguchi method in investigating the main parameters affecting compressive strength. Constr Build Mater 2014;52(0):473–81.

[12] Hanjitsuwan S, Hunpratub S, Thongbai P, Maensiri S, Sata V, Chindaprasirt P. Effects of NaOH concentrations on physical and electrical properties of high calcium fly ash geopolymer paste. Cem Concr Compos 2014;45(0):9–14.

[13] Phoo-ngernkham T, Chindaprasirt P, Sata V, Hanjitsuwan S, Hatanaka S. The effect of adding nano-$SiO_2$ and nano-$Al_2O_3$ on properties of high calcium fly ash geopolymer cured at ambient temperature. Mater Des 2014;55:58–65.

[14] He J, Jie Y, Zhang J, Yu Y, Zhang G. Synthesis and characterization of red mud and rice husk ash-based geopolymer composites. Cem Concr Compos 2013;37(0):108–18.

[15] Cheah CB, Part WK, Ramli M. The hybridizations of coal fly ash and wood ash for the fabrication of low alkalinity geopolymer load bearing block cured at ambient temperature. Constr Build Mater 2015;88(0):41–55.

[16] Silva PD, Sagoe-Crenstil K, Sirivivatnanon V. Kinetics of geopolymerization: role of $Al_2O_3$ and $SiO_2$. Cem Concr Res 2007;37(4):512–8.

[17] Davidovits J. Properties of geopolymer cements, alkaline cements and concretes. Kiev, Ukraine: 1994.

[18] Joseph B, Mathew G. Influence of aggregate content on the behaviour of fly ash based geopolymer concrete. Sci Iranica 2012;19(5):1188–94.

[19] Gorhan G, Kurklu G. The influence of the NaOH solution on the properties of the fly ash-based geopolymer mortar cured at different temperatures. Composites Part B 2013;58:371–7.

[20] Giasuddin HM, Sanjayan JG, Ranjith PG. Strength of geopolymer cured in saline water in ambient conditions. Fuel 2013;107:34–9.

[21] Nazari A, Bagheri A, Riahi S. Properties of geopolymer with seeded fly ash and rice husk bark ash. Mater Sci Eng A 2011;528(24):7395–401.

[22] Aydin S, Baradan B. Mechanical and microstructural properties of heat cured alkali-activated slag mortars. Mater Des 2012;35(0):374–83.

[23] Islam A, Johnson Alengaram U, Jumaat MZ, Bashar II. The development of compressive strength of ground granulated blast furnace slag-palm oil fuel ash-fly ash based geopolymer mortar. Mater Des 2014;56:833–41.

[24] Ranjbar N, Mehrali M, Behnia A, Johnson Alengaram U, Jumaat MZ. Compressive strength and microstructural analysis of fly ash/palm oil fuel ash based geopolymer mortar. Mater Des 2014;59:532–9.

[25] Ahmari S, Zhang L. Production of eco-friendly bricks from copper mine tailings through geopolymerization. Constr Build Mater 2012;29(0):323–31.

[26] Chen Y, Zhang Y, Chen T, Zhao Y, Bao S. Preparation of eco-friendly construction bricks from hematite tailings. Constr Build Mater 2011;25(4):2107–11.

[27] de Vargas AS, Dal Molin DCC, Vilela ACF, Silva FJD, Pavao B, Veit H. The effects of $Na_2O/SiO_2$ molar ratio, curing temperature and age on compressive strength, morphology and microstructure of alkali-activated fly ash-based geopolymers. Cem Concr Compos 2011;33(6):653–60.

[28] Hu M, Zhu X, Long F. Alkali-activated fly ash-based geopolymers with zeolite or bentonite as additives. Cem Concr Compos 2009;31(10):762–8.

[29] Panias D, Giannopoulou IP, Perraki T. Effect of synthesis parameters on the mechanical properties of fly ash-based geopolymers. Colloids Surf A 2007;301(1–3):246–54.

[30] Somna K, Jaturapitakkul C, Kajitvichyanukul P, Chindaprasirt P. NaOH-activated ground fly ash geopolymer cured at ambient temperature. Fuel 2011;90(6):2118–24.

[31] Komljenovic M, Bascarevic Z, Bradic V. Mechanical and microstructural properties of alkali-activated fly ash geopolymers. J Hazard Mater 2010;181(1–3):35–42.

[32] Sathonsaowaphak A, Chindaprasirt P, Pimraksa K. Workability and strength of lignite bottom ash geopolymer mortar. J Hazard Mater 2009;168(1):44–50.

[33] North MR, Swaddle TW. Kinetics of silicate exchange in alkaline aluminosilicate solutions. Inorg Chem 2000;39(12):2661–5.

[34] Sukmak P, Horpibulsuk S, Shen S-L. Strength development in clay-fly ash geopolymer. Constr Build Mater 2013;40(0):566–74.

[35] Ridtirud C, Chindaprasirt P, Pimraksa K. Factors affecting the shrinkage of fly ash geopolymers. Int J Miner Metall Mater 2011;18(1):100–4.

[36] Salih MA, Abang Ali AA, Farzadnia N. Characterization of mechanical and microstructural properties of palm oil fuel ash geopolymer cement paste. Constr Build Mater 2014;65(0):592–603.

[37] Yan S, Sagoe-Crentsil K. Properties of wastepaper sludge in geopolymer mortars for masonry applications. J Environ Manage 2012;112(0):27–32.

[38] Yusuf M, Megat Johari M, Ahmad Z, Maslehuddin M. Impacts of silica modulus on the early strength of alkaline activated ground slag/ultrafine palm oil fuel ash based concrete. Mater Struct 2014:1–9.

[39] Law D, Adam A, Molyneaux T, Patnaikuni I, Wardhono A. Long term durability properties of class F fly ash geopolymer concrete. Mater Struct 2014:1–11.

[40] Guo X, Shi H, Dick WA. Compressive strength and microstructural characteristics of class C fly ash geopolymer. Cem Concr Compos 2010;32(2):142–7.

[41] Topark-Ngarm P, Chindaprasirt P, Sata V. Setting time, strength and bond of high calcium fly ash geopolymer concrete. J Mater Civ Eng 2014;27(7).

[42] Khandelwal M, Ranjith PG, Pan Z, Sanjayan JG. Effect of strain rate on strength properties of low-calcium fly-ash-based geopolymer mortar under dry condition. Arabian J Geosci 2013;6(7):2383–9.

[43] Olivia M, Nikraz H. Properties of fly ash geopolymer concrete designed by Taguchi method. Mater Des 2012;36(0):191–8.

[44] Sofi M, van Deventer JSJ, Mendis PA, Lukey GC. Engineering properties of inorganic polymer concretes (IPCs). Cem Concr Res 2007;37(2):251–7.

[45] Ryu GS, Lee YB, Koh KT, Chung YS. The mechanical properties of fly ash-based geopolymer concrete with alkaline activators. Constr Build Mater 2013;47(0):409–18.

[46] Ranjbar N, Mehrali M, Alengaram UJ, Metselaar HSC, Jumaat MZ. Compressive strength and microstructural analysis of fly ash/palm oil fuel ash based geopolymer mortar under elevated temperatures. Constr Build Mater 2014;65(0):114–21.

[47] Khater HM. Effect of calcium on geopolymerization of aluminosilicate wastes. J Mater Civil Eng 2011;24(1):92–101.

[48] Chindaprasirt P, Rattanasak U, Taebuanhuad S. Resistance to acid and sulfate solution of microwave-assisted high calcium fly ash geopolymer. Mater Struct 2013:375–81.

[49] Songpiriyakij S, Kubprasit T, Jaturapitakkul C, Chindaprasirt P. Compressive strength and degree of reaction of biomass- and fly ash-based geopolymer. Constr Build Mater 2010;24(3):236–40.

[50] Detphan S, Chindaprasirt P. Preparation of fly ash and rice husk ash geopolymer. Int J Miner Metall Mater 2009;16(6):720–6.

[51] Chindaprasirt P, Chareerat T, Hatanaka S, Cao T. High strength geopolymer using fine high calcium fly ash. J Mater Civil Eng 2010;23(3):264–70.

[52] Chindaprasirt P, Rattanasak U, Vongvoradit P, Jenjirapanya S. Thermal treatment and utilization of Al-rich waste in high calcium fly ash geopolymeric materials. Int J Miner Metall Mater 2012;19(9):872–8.

[53] Bakharev T. Geopolymeric materials prepared using Class F fly ash and elevated temperature curing. Cem Concr Res 2005;35(6):1224–32.

[54] Perera DS, Uchida O, Vance ER, Finnie KS. Influence of curing schedule on the integrity of geopolymers. J Mater Sci 2007;42(9):3099–106.

[55] Puertas F, Fernandez-Jimenez A, Blanco-Varela MT. Pore solution in alkali-activated slag cement pastes. Relation to the composition and structure of calcium silicate hydrate. Cem Concr Res 2004;34(1):139–48.

[56] Mishra A, Choudhary D, Jain N, Kumar M, Sharda N, Dutt D. Effect of concentration of alkaline liquid and curing time on strength and water absorption of geopolymer concrete. ARPN J Eng Appl Sci 2008;3(1):14–18.

[57] Jeyasehar CA, Saravanan G, Ramakrishnan AK, Kandasamy S. Strength and durability studies on fly ash based geopolymer bricks. Asian J Civil Eng 2013;14(6):797–808.

[58] Rattanasak U, Pankhet K, Chindaprasirt P. Effect of chemical admixtures on properties of high-calcium fly ash geopolymer. Int J Miner Metall Mater 2011;18(3):364–9.

[59] Alonso S, Palomo A. Calorimetric study of alkaline activation of calcium hydroxide-metakaolin solid mixtures. Cem Concr Res 2001;31(1):25–30.

[60] Rattanasak U, Chindaprasirt P. Influence of NaOH solution on the synthesis of fly ash geopolymer. Miner Eng 2009;22(12):1073–8.

[61] Chindaprasirt P, De Silva P, Sagoe-Crentsil K, Hanjitsuwan S. Effect of SiO2 and Al2O3 on the setting and hardening of high calcium fly ash-based geopolymer systems. J Mater Sci 2012;47(12):4876–83.

[62] Sata V, Sathonsaowaphak A, Chindaprasirt P. Resistance of lignite bottom ash geopolymer mortar to sulfate and sulfuric acid attack. Cem Concr Compos 2012;34(5):700–8.

[63] Rashad AM. A comprehensive overview about the influence of different admixtures and additives on the properties of alkali-activated fly ash. Mater Des 2014;53(0):1005–25.

[64] Rattanasak U, Chindaprasirt P, Suwanvitaya P. Development of high volume rice husk ash alumino silicate composites. Int J Miner Metall Mater 2010;17(5):654–9.

[65] Boonserm K, Sata V, Pimraksa K, Chindaprasirt P. Improved geopolymerization of bottom ash by incorporating fly ash and using waste gypsum as additive. Cem Concr Compos 2012;34(7):819–24.

[66] Alomayri T, Shaikh FUA, Low IM. Synthesis and mechanical properties of cotton fabric reinforced geopolymer composites. Composites Part B 2014;60(0):36–42.

[67] Nematollahi B, Sanjayan J. Effect of different superplasticizers and activator combinations on workability and strength of fly ash based geopolymer. Mater Des 2014; 57:667–72.

[68] Chen R, Ahmari S, Zhang L. Utilization of sweet sorghum fiber to reinforce fly ash-based geopolymer. J Mater Sci 2014;49:2548–58.

[69] Zhang YJ, Wang YC, Xu DL, Li S. Mechanical performance and hydration mechanism of geopolymer composite reinforced by resin. Mater Sci Eng A 2010;527(24–25):6574–80.

[70] Chindaprasirt P, Jaturapitakkul C, Chalee W, Rattanasak U. Comparative study on the characteristics of fly ash and bottom ash geopolymers. Waste Manage 2009;29(2):539–43.

[71] Jaturapitakkul C, Cheerarot R. Development of bottom ash as pozzolanic material. J Mater Civil Eng 2003;15:48–53.

[72] Li G. Properties of high-volume fly ash concrete incorporating nano-SiO$_2$. Cem Concr Res 2004;34(6):1043–9.

[73] Kong DLY, Sanjayan JG. Effect of elevated temperatures on geopolymer paste, mortar and concrete. Cem Concr Res 2010;40(2):334–9.

[74] Palacios M, Puertas F. Effect of superplasticizer and shrinkage-reducing admixtures on alkali-activated slag pastes and mortars. Cem Concr Res 2005;35(7):1358–67.

[75] Palacios M, Houst YF, Bowen P, Puertas F. Adsorption of superplasticizer admixtures on alkali-activated slag pastes. Cem Concr Res 2009;39(8):670–7.

[76] Puertas F, Palomo A, Fernandez-Jimenez A, Izquierdo M, Granizo M. Effect of superplasticisers on the behaviour and properties of alkaline cements. Adv Cem Res 2003;15(1):23–8.

[77] Nath SK, Kumar S. Influence of iron making slags on strength and microstructure of fly ash geopolymer. Constr Build Mater 2013;38(0):924–30.

[78] Sun P, Wu H-C. Transition from brittle to ductile behavior of fly ash using PVA fibers. Cem Concr Compos 2008;30(1):29–36.

[79] He P, Jia D, Lin T, Wang M, Zhou Y. Effects of high-temperature heat treatment on the mechanical properties of unidirectional carbon fiber reinforced geopolymer composites. Ceram Int 2010;36(4):1447–53.

[80] Zhao Q, Nair B, Rahimian T, Balaguru P. Novel geopolymer based composites with enhanced ductility. J Mater Sci 2007;42(9):3131–7.

[81] Memon FA, Nuruddin MF, Shafiq N. Effect of silica fume on the fresh and hardened properties of fly ash-based self-compacting geopolymer concrete. Int J Miner Metall Mater 2013;20(2):205–13.

[82] Ramakrishna G, Sundararajan T. Impact strength of a few natural fibre reinforced cement mortar slabs: a comparative study. Cem Concr Compos 2005;27(5):547–53.

[83] Alomayri T, Shaikh FUA, Low IM. Effect of fabric orientation on mechanical properties of cotton fabric reinforced geopolymer composites. Mater Des 2014;57(0):360–5.

[84] Alomayri T, Low IM. Synthesis and characterization of mechanical properties in cotton fiber-reinforced geopolymer composites. J Asian Ceram Soc 2013;1(1):30–4.

[85] Zhang YJ, Li S, Wang YC, Xu DL. Microstructural and strength evolutions of geopolymer composite reinforced by resin exposed to elevated temperature. J Non-Cryst Solids 2012;358(3):620–4.

[86] Fletcher RA, MacKenzie KJD, Nicholson CL, Shimada S. The composition range of aluminosilicate geopolymers. J Eur Ceram Soc 2005;25:1471.

[87] Gutti CS, Roy A, Metcalf JB, Seals RK. The influence of admixtures on the strength and linear expansion of cement-stabilized phosphogypsum. Cem Concr Res 1996;26(7):1083–94.

[88] Talero R. Comparative XRD analysis ettringite originating from pozzolan and from Portland cement. Cem Concr Res 1996;26(8):1277–83.

[89] Socrates G. Infrared and Raman characteristic group frequencies, third ed. England: John Wiley & Sons; 2001.

[90] Komnitsas K, Zaharaki D. Geopolymerisation: a review and prospects for the minerals industry. Miner Eng 2007;20(14):1261–77.

[91] Bakharev T. Durability of geopolymer materials in sodium and magnesium sulfate solutions. Cem Concr Res 2005;35(6):1233–46.

[92] Ariffin MAM, Bhutta MAR, Hussin MW, Mohd Tahir M, Aziah N. Sulfuric acid resistance of blended ash geopolymer concrete. Constr Build Mater 2013;43(0):80–6.

[93] Bhutta M, Hussin M, Ariffin M, Tahir M. Sulphate resistance of geopolymer concrete prepared from blended waste fuel ash. J Mater Civ Eng 2014;26(11).

[94] Ahmari S, Zhang L. Durability and leaching behavior of mine tailings-based geopolymer bricks. Constr Build Mater 2013;44:743–50.

[95] lloyd RR, Provis JL, van Deventer JSJ. Acid resistance of inorganic binders. 1. Corrosion rate. Mater Struct 2012;45:1–14.

[96] Reddy DV, Edouard JB, Sobhan K. Durability of fly-ash based geopolymer structural concrete in the marine environment. J Mater Civil Eng 2012;25(6):781–7.

[97] Bakharev T. Resistance of geopolymer materials to acid attack. Cem Concr Res 2005;35(4):658–70.

[98] Chindaprasirt P, Rukzon S. Strength, porosity and corrosion resistance of ternary blend Portland cement, rice husk ash and fly ash mortar. Constr Build Mater 2008;22(8):1601–6.

[99] Bascarevic Z, Komljenovic M, Miladinovic Z, Nikolic V, Marjanovic N, Petrovic R. Impact of sodium sulfate solution on mechanical properties and structure of fly ash based geopolymers. Mater Struct 2014:1–15.

[100] Maes M, De Belie N. Resistance of concrete and mortar against combined attack of chloride and sodium sulphate. Cem Concr Compos 2014;53(0):59–72.

[101] Rangan BV, Wallah SE. Low calcium fly ash based geopolymer concrete: long term properties. Research Report GC2. Perth: Faculty of Engineering, Curtin University of Technology; 2006.

[102] Zhao F-Q, Zhao J, Liu H-J. Autoclaved brick from low-silicon tailings. Constr Build Mater 2009;23(1):538–41.

[103] Cheah CB, Noor Shazea AN, Part WK, Ramli M, Kwan WH. The high volume reuse of hybrid biomass ash as a primary binder in cementless mortar block. Am J Appl Sci 2014;11(8):1369–78.

[104] Chindaprasirt P, Chareerat T, Sirivivatnanon V. Workability and strength of coarse high calcium fly ash geopolymer. Cem Concr Compos 2007;29:224–9.

[105] Davidovits J. Mineral polymers and methods of making them, in US Patent 4349386. 1982.

[106] Freidin C. Cementless pressed blocks from waste products of coal-firing power station. Constr Build Mater 2007;21(1):12–18.

[107] Deb PS, Nath P, Sarker PK. The effects of ground granulated blast-furnace slag blending with fly ash and activator content on the workability and strength properties of geopolymer concrete cured at ambient temperature. Mater Des 2014;62(0):32–9.

[108] Xu H, Gong W, Syltebo L, Izzo K, Lutze W, Pegg IL. Effect of blast furnace slag grades on fly ash based geopolymer waste forms. Fuel 2014;133(0):332–40.

[109] Canfield GM, Eichler J, Griffith K, Hearn JD. The role of calcium in blended fly ash geopolymers. J Mater Sci 2014;49:5922–33.

[110] Chindaprasirt P, Rattanasak U, Jaturapitakkul C. Utilization of fly ash blends from pulverized coal and fluidized bed combustions in geopolymeric materials. Cem Concr Compos 2011;33:55–60.

[111] Yan S, Sagoe-Crentsil K, Shapiro G. Reuse of de-inking sludge from wastepaper recycling in cement mortar products. J Environ Manage 2011;92(8):2085–90.

[112] Naik T. Greener concrete using recycled materials. Concr Int 2002;7:45–9.

[113] Yusuf MO, Megat Johari MA, Ahmad ZA, Maslehuddin M. Shrinkage and strength of alkaline activated ground steel slag/ultrafine palm oil fuel ash pastes and mortars. Mater Des 2014;63(0):710–8.

[114] Yusuf MO, Megat Johari MA, Ahmad ZA, Maslehuddin M. Effects of H2O/Na2O molar ratio on the strength of alkaline activated ground blast furnace slag-ultrafine palm oil fuel ash based concrete. Mater Des 2014;56(0):158–64.

[115] Yusuf MO, Megat Johari MA, Ahmad ZA, Maslehuddin M. Evolution of alkaline activated ground blast furnace slag-ultrafine palm oil fuel ash based concrete. Mater Des 2014;55(0):387–93.

[116] Yusuf MO, Megat Johari MA, Ahmad ZA, Maslehuddin M. Influence of curing methods and concentration of NaOH on strength of the synthesized alkaline activated ground slag-ultrafine palm oil fuel ash mortar/concrete. Constr Build Mater 2014;66(0):541–8.

[117] Karim MR, Zain MFM, Jamil M, Lai FC. Fabrication of a non-cement binder using slag, palm oil fuel ash and rice husk ash with sodium hydroxide. Constr Build Mater 2013;49(0):894–902.

[118] Chindaprasirt P, Homwuttiwong S, Jaturapitakkul C. Strength and water permeability of concrete containing palm oil fuel ash and rice husk bark ash. Constr Build Mater 2007;21(7):1492–9.

[119] Habert G, d'Espinose de Lacaillerie JB, Roussel N. An environmental evaluation of geopolymer based concrete production: reviewing current research trends. J Clean Prod 2011;19(11):1229–38.

[120] Turner LK, Collins FG. Carbon dioxide equivalent (CO2-e) emissions: A comparison between geopolymer and OPC cement concrete. Constr Build Mater 2013;43(0):125–30.

[121] Jamieson E, McLellan B, van Riessen A, Nikraz H. Comparison of embodied energies of ordinary Portland cement with Bayer-derived geopolymer products. J Clean Prod 2015;99:112–8.

[122] McLellan BC, Williams RP, Lay J, van Riessen A, Corder GD. Costs and carbon emissions for geopolymer pastes in comparison to ordinary Portland cement. J Clean Prod 2011;19(9–10):1080–90.

[123] Duxson P, Provis JL, Lukey GC, van Deventer JSJ. The role of inorganic polymer technology in the development of 'green concrete'. Cem Concr Res 2007;37(12):1590–7.

[124] Provis JL, Duxson P, van Deventer JSJ. The role of particle technology in developing sustainable construction materials. Adv Powder Technol 2010;21(1):2–7.

# CHAPTER 12

# Performance on an Alkali-Activated Cement-Based Binder (AACB) for Coating of an OPC Infrastructure Exposed to Chemical Attack: A Case Study

W. Tahri[1], Z. Abdollahnejad[2], F. Pacheco-Torgal[2,3], and J. Aguiar[2]
[1]University of Sfax, Sfax, Tunisia
[2]University of Minho, Guimarães, Portugal
[3]University of Sungkyunkwan, Suwon, Republic of Korea

## 12.1 INTRODUCTION

Premature degradation of ordinary Portland cement (OPC) concrete infrastructure is a current and serious problem related to the fact that OPC concrete presents a higher permeability that allows water and other aggressive elements to enter, leading to carbonation and chloride-ion attack, resulting in corrosion problems [1].

Pacheco-Torgal et al. [2] mentioned the case of a tunnel in Dubai, which had been concluded in 1975 and needed to be completely repaired after just 11 years, a case of pile foundations that had disintegrated after just 12 years, and also a study on Norway OPC concrete bridges that indicated that several presented corrosion problems 24 years after they were built. As a consequence, worldwide concrete infrastructure rehabilitation costs are staggering. For example in the United States, where about 27% of all highway bridges are in need of repair or replacement, the needs are estimated to be over US$1.6 trillion by 2021, and the corrosion deterioration cost due to deicing and sea salt effects is estimated at over US$150 billion. In the European Union, nearly 84,000 reinforced and prestressed concrete bridges require maintenance, repair, and strengthening with an annual budget of £215M, and that estimate does not include traffic management costs [3].

Many of the degraded concrete structures were built decades ago when little attention was given to durability issues. Concrete durability means above all minimizing the possibility of aggressive elements to enter the

*Handbook of Low Carbon Concrete.*
DOI: http://dx.doi.org/10.1016/B978-0-12-804524-4.00012-9

concrete, under certain environmental conditions for any of the following transport mechanisms: permeability, diffusion, or capillarity. The use of concrete surface treatments with waterproofing materials (also known as sealers) to prevent the access of aggressive substances is an important way of contributing to concrete durability. Almusallam et al. [4] studied several concrete coatings concluding that epoxy and polyurethane coatings performed better than acrylic, polymer, and chlorinated rubber coatings.

Other authors [5,6] showed that although some waterproof materials are effective for a particular transport mechanism (diffusion, capillarity, permeability), they may not be for another. They compared the waterproofing capacity of concrete with three polymeric resins (epoxy, silicone, acrylic) and mentioned that the silicone-based one is more effective (99.2%) in reducing water absorption by capillarity than the epoxy resin (93.6%), but in terms of chloride diffusion the epoxy resin is 100% effective, while the silicone varnish does not go beyond 67.5%. Epoxy coatings exhibited excellent durability under the laboratory and field-test conditions and are recommended for protecting concrete in cooling tower basins against sulfur-oxidizing or other acid-producing bacteria [7].

Medeiros and Helene [8] used a water-repellent material based on silane-siloxane, noticing that although it is effective to reduce the water absorption by capillarity of concrete (reduced from two to seven times), it only managed to achieve a reduction of the chloride diffusion from 11% to 17% and also failed to prevent the access of water by permeability.

Pacheco-Torgal and Jalali [9] confirm that the surface treatment of concrete with a water-repellent material is effective, but above all more cost effective when compared with the alternative of using a polymer additive in the composition of concrete.

In 2013, Brenna et al. [10] studied the efficiency of four commercial concrete coatings (a polymer-modified cementitious mortar and three elastomeric coatings) against chloride-induced corrosion, concluding that the polymer-containing mortar shows the best effect on delaying chloride penetration in concrete. In summary, the most common surface treatments use polymeric resins based on epoxy, silicone (siloxane), acrylics, chlorinated rubber, polyurethanes, or polymethacrylate.

Bijen [11] mentioned that the epoxy resins have low resistance to ultraviolet radiation and polyurethanes are sensitive to high-alkalinity environments. Polyurethane is obtained from isocyanates, known worldwide for their tragic association with the Bhopal disaster. As for chlorinated rubber it derives from reacting butyl rubber with chlorine; it is important to remember that chlorine is associated with the production of dioxins and furans, which are extremely toxic and also bioaccumulative. Several scientist groups already suggest that chlorine-based industrial products should be prohibited [12].

Besides, the European Union has approved Regulation (EU) 305/2011, related to construction products regulation, which will replace the current Directive 89/106/CEE, already amended by Directive 1993/68/EEC, known as the Construction Products Directive. A crucial aspect of the new regulation relates to the information regarding hazardous substances [13].

Recent investigations on the geopolymer field [14] reveal a third category of mortars with high potential to enhance the durability of concrete structures. Investigations in the field of geopolymers have exponentially increased after the research results of Davidovits [15], who developed and patented binders obtained from the alkali activation of metakaolin, coining the term "geopolymer" in 1978. The technology of alkali activation, however, predates this terminology by several decades [16].

For the chemical designation of the geopolymer, Davidovits suggested the name "polysialates," in which sialate is an abbreviation for aluminosilicate oxide. The sialate network is composed of tetrahedral anions $[SiO_4]^{4-}$ and $[AlO_4]^{5-}$ sharing the oxygen, which needs positive ions such as ($Na^+$, $K^+$, $Li^+$, $Ca^{++}$, $Na^+$, $Ba^{++}$, $NH_4^+$, $H_3O^+$) to compensate for the electric charge of $Al^{3+}$ in tetrahedral coordination (after dehydroxilation the aluminum changes from coordination 6 (octahedral) to coordination 4 (tetrahedral). However, Provis and Van Deventer [17] mentioned that the sialate nomenclature "implies certain aspects of the geopolymer gel structure which do not correspond to reality."

In 2014, Provis presented a rigorous a useful definition of these materials: "alkali-activated materials are produced through the reaction of an aluminosilicate—normally supplied in powder form as an industrial by-product or other inexpensive material—with an alkaline activator, which is usually a concentrated aqueous solution of alkali hydroxide, silicate, carbonate or sulfate" [18].

In the last decade several authors have reported research in a large number of aspects related to geopolymers.

However, very few studies [19–21] have addressed the use of geopolymers for enhancement of concrete structures' durability. Since geopolymer performance concerns the resistance to acid attack, is far better than that of Portland cement [16], which means that these materials could be an alternative low-toxicity coating material.

This paper presents results of an experimental investigation on the resistance to chemical attack (with sulfuric, hydrochloric, and nitric acid) of several materials: OPC concrete, high-performance concrete (HPC), epoxy resin, acrylic painting, and a fly ash–based geopolymeric mortar.

## 12.2 EXPERIMENTAL WORK

### 12.2.1 Materials, Mix Design, Mortar and Concrete Mixing, and Concrete Coating

The characteristics of the aggregates (coarse and sand) used are shown in Table 12.1 and in Fig. 12.1. The fly ash used in the geopolymeric mortars was supplied by Sines-EDP and according to the NP EN 450-1 it belongs to the B-class and has an N-class fineness modulus. Geopolymeric mortars were a mixture of aggregates, fly ash, calcium hydroxide, and alkaline silicate solution. The mass ratio for aggregates/fly ash and activator was 2/1/0.6. A 10% percentage substitution of fly ash by calcium hydroxide in the mixture was also used. This is because the use of minor

**Table 12.1** Characteristics of the aggregates

|  | Max dimension | Fine content | Density (kg/m³) | Water absorption |
|---|---|---|---|---|
| Sand | 4.0 | $\leq 3$ | 2660 | 0.2 |
| Coarse aggregates | 8.0 | $\leq 1.5$ | 2620 | 0.6 |

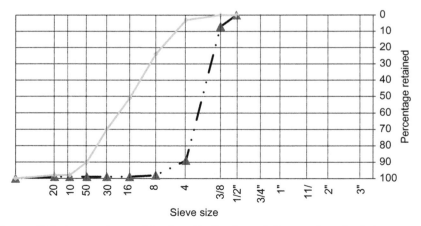

**Figure 12.1** Aggregate particle-size distribution of the sand and of the coarse aggregate.

calcium hydroxide percentages is pivotal for the strength and durability of geopolymers [22,23]. The alkaline activator was prepared prior to use. An activator with sodium hydroxide and sodium silicate solution ($Na_2O$ = 13.5%, $SiO_2$ = 58.7%, and water = 45.2%) was used with a mass ratio of 1:2.5. Previous investigations showed that this ratio lead to the highest compressive strength results in geopolymeric mortars [14]. The sand, fly ash, and calcium hydroxide were dry mixed before being added to the activator. Three different sodium hydroxide concentrations (10, 14, and 18 M) were used. The fresh mortar was cast and allowed to set at room temperature for 24 h before being removed from the molds and kept at room temperature (20°C) until tested in compression and flexural strength. An OPC (CEM I 42,5 N) was used to prepare the concrete mixtures. Two concrete mixes (normal and HPC) were designed using the Faury concrete mix design method (Table 12.2). The concrete mixing starts with the introduction of the coarse aggregates in the mixer, followed by the sand for 2 min; then OPC is introduced and mixed to the aggregates for 2 more minutes. Then, 70% of the water is introduced in the mixer and all the ingredients are mixed for 2 min. Finally, the remaining water is added for 2 min and everything is mixed for 2 more minutes.

The concrete specimens were conditioned at a temperature equal to 21 ± 2°C cured in a moist chamber until they have reached 28 days. An epoxy resin often used as concrete coating protection against acid attack with a commercial reference Sikagard 62 PT was used for coating of the two concrete mixtures. The epoxy adhesive is a two-component system (resin and hardener) with a bulk density of $1.35\,kg/dm^3$. After mixing the two components, the mixture remains workable for 20 min at 20°C or just 0 min at 30°C. An acrylic paint often used as concrete coating protection to prevent the access of aggressive substances with a commercial reference Sikagard 660 ES was also used for coating of the two concrete mixtures. This material has a bulk density of $1.30\,kg/dm^3$ and is provided by the manufacturer as ready to be used.

**Table 12.2** Concrete mix proportions per cubic meter of concrete

|  | Cement (kg) | Sand (kg) | Coarse aggregates (kg) | Water | W/C |
|---|---|---|---|---|---|
| NC | 270 | 1135 | 732 | 182 | 0.65 |
| BED | 442 | 876 | 782 | 205 | 0.45 |

## 12.3 EXPERIMENTAL PROCEDURES

### 12.3.1 Compressive Strength

The compressive strength was performed under NP EN 206-1. Tests were performed on $100 \times 100 \times 100 \, mm^3$ concrete specimens. The compressive and flexural strength data of geopolymeric mortars was obtained using $160 \times 40 \times 40 \, mm^3$ specimens according to EN 1015-11. Compressive strength for each mixture was obtained from an average of three cubic specimens determined at the age of 28 days of curing.

### 12.3.2 Water Absorption by Immersion

Tests were performed on $40 \times 400 \times 80 \, mm^3$ specimens. Specimens were tested with 28 days curing. The specimens were immersed in water at room temperature for 24 h. First, the weight of the specimens while suspended by a thin wire and completely submerged in water is recorded as $W_{im}$ (immersed weight). After that, the specimens were removed from water, and placed for 1 min on a wire mesh allowing water to drain; then visible surface water is removed with a damp cloth and weight is recorded as $W_{sat}$ (saturated weight). All specimens were placed in a ventilated oven at 105°C for not less than 24 h and allowing that two successive weighings at intervals of 2 h show an increment of loss not greater than 0.1% of the last previously determined weight of the specimen. The weight of the dried specimens is recorded as $W_{dry}$ (oven-dry weight). The absorption coefficient is determined as following equation:

$$A(\%) = \frac{W_{sat} - W_{dry}}{W_{sat} - W_{im}} \times 100 \tag{12.1}$$

### 12.3.3 Capillary Water Absorption

Capillary water absorption was carried out using $40 \times 400 \times 80 \, mm^3$ specimens in the case of geopolymeric mortars and $100 \times 100 \times 100 \, mm^3$ specimens for concrete. After 28 days in a moist chamber the specimens were placed in a 105°C oven for 24 h. The test consists of placing the specimens in a container with enough water so that one side of the sample will remain immersed. This test is carried out according to Standard LNEC E393. Water absorption has been measured after 5, 10, 20, 30, 60, 90, 120, 180, 240, 300, 360, 420, and 480 min. Capillarity water absorption was obtained from an average of three specimens.

## 12.3.4 Resistance to Chemical Attack

The resistance to chemical attack followed a variation of the ASTM C-267 (Standard test methods for chemical resistance of mortars, grouts, and monolithic surfacing's and polymer concretes). The test used in the present investigation consists of the immersion of $100 \times 100 \times 100 \, mm^3$ concrete (NC, HPC, coated concrete specimens) and fly ash geopolymeric mortar specimens with 28 days curing in acid solution. Three different acids were used (sulfuric, hydrochloric, and nitric). Three acid concentrations were used (10%, 20%, and 30%) to simulate long-term exposure at lower concentrations. Other authors used 5% $Na_2SO_4$ concentrations and immersion for 12 months [24]. The resistance to acid attack was assessed by the differences in weight of dry specimens before and after acid attack at 1, 7, 14, 28, and 56 days. The chemical resistance was assessed by the differences in weight of dry specimens before and after acid attack, since compressive strength of specimens immersed in acid media could not be evaluated. The fly ash–based geopolymeric mortar used in the resistance to acid attack was the one associated with the highest compressive strength and low water absorption.

## 12.4  RESULTS AND DISCUSSION

### 12.4.1 Compressive Strength

Fig. 12.2 shows the results of the compressive strength of the fly ash–based geopolymeric mortars after 28 days curing as well as of the two concrete mixtures. The results show that the compressive strength of geopolymeric mortars is very dependent on the molarity of the sodium hydroxide. Increasing the molarity from 10 to 14 M leads to a relevant compressive strength loss.

However, further increase from 14 to 18 M shows no noticeable effects. Previous investigations [25] have shown that although a high alkali content favors the dissolution of Al and Si species of fly ash it can also negatively affect its strength. Pacheco-Torgal et al. [26], who studied the geopolymerization of mine wastes, noticed the opposite phenomenon. Other authors [27] mentioned that when $OH^-$ concentration was high enough, dissolution of fly ash was accelerated, but polycondensation was hindered. Normal concrete (NC) has a compressive strength around 30 MPa while HPC compressive strength slightly exceeds 45 MPa. The standard deviation was low and the coefficient of variation does not exceed 12% meaning that the results were statistical relevant.

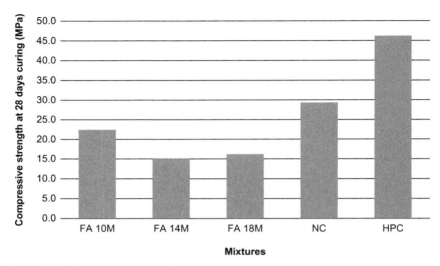

**Figure 12.2** Compressive strength.

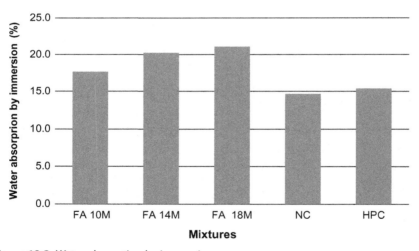

**Figure 12.3** Water absorption by immersion.

## 12.4.2  Water Absorption by Immersion

The results of water absorption by immersion are showed in Fig. 12.3. These results are aligned with compressive strength performance. The fly ash geopolymeric mortar with the least water absorption by immersion is the one with the highest compressive strength.

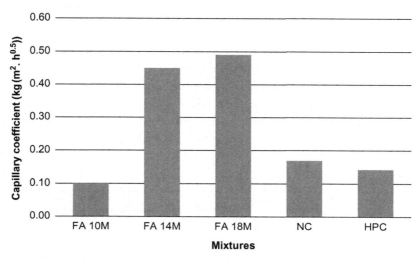

**Figure 12.4** Water absorption capillary coefficients.

The geopolymeric mortars with a sodium hydroxide molarity of 14 and 18 M show a water absorption around 20%. This means that compressive strength is directly influenced by open porosity. However, previous investigations [28] on the field of geopolymers showed that low porosity does not always mean high compressive strength; being that compressive strength is more influenced by NaOH concentration than it is from porosity. Both NC and HPC show a water absorption around 15%. This falls in the current water absorption by immersion range of current OPC concretes used by the construction industry (compressive strength at 28 days curing between 25 and 45 MPa), of 12–16%.

### 12.4.3 Capillary Water Absorption

Fig. 12.4 shows the capillary water absorption coefficients. While the fly ash geopolymeric mortars with a sodium hydroxide molarity of 14 and 18 M show a capillary water absorption around $0.45\,\text{kg/m}^2.\text{h}^{0.5}$ the geopolymeric mortar with the lowest open porosity and the highest compressive strength has a $0.1\,\text{kg/m}^2.\text{h}^{0.5}$ capillary water absorption coefficient.

The capillary water absorption of the two concrete mixes used in this investigation is very low, around $0.15\,\text{kg/m}^2.\text{h}^{0.5}$. As a comparison, a plain C30/37-strength class concrete has a capillarity coefficient of $0.251\,\text{kg/m}^2.\text{h}^{0.5}$ for 28 days curing [29], while a plain C20/25-strength class concrete (the most used strength class in Europe [30]) has capillarity coefficients between 0.85 and 2.6 kg/m$^2$.h$^{0.5}$ [31].

## 12.4.4 Resistance to Chemical Attack

### 12.4.4.1 Resistance to Sulfuric Acid Attack

Fig. 12.5 shows the weight loss after sulfuric acid attack for the different acid concentrations. NC coated with epoxy resin shows the most stable performance for all three acid concentrations confirming previous investigations.

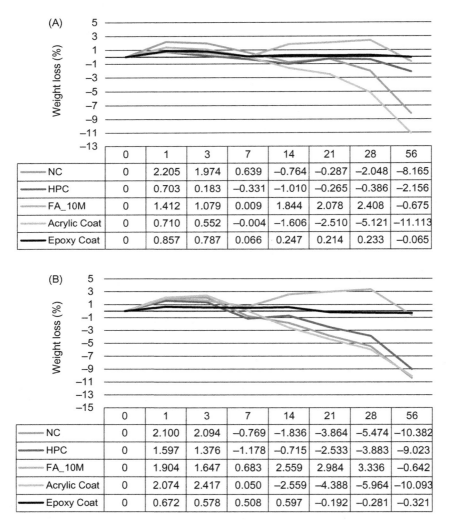

| (A) | 0 | 1 | 3 | 7 | 14 | 21 | 28 | 56 |
|---|---|---|---|---|---|---|---|---|
| NC | 0 | 2.205 | 1.974 | 0.639 | −0.764 | −0.287 | −2.048 | −8.165 |
| HPC | 0 | 0.703 | 0.183 | −0.331 | −1.010 | −0.265 | −0.386 | −2.156 |
| FA_10M | 0 | 1.412 | 1.079 | 0.009 | 1.844 | 2.078 | 2.408 | −0.675 |
| Acrylic Coat | 0 | 0.710 | 0.552 | −0.004 | −1.606 | −2.510 | −5.121 | −11.113 |
| Epoxy Coat | 0 | 0.857 | 0.787 | 0.066 | 0.247 | 0.214 | 0.233 | −0.065 |

| (B) | 0 | 1 | 3 | 7 | 14 | 21 | 28 | 56 |
|---|---|---|---|---|---|---|---|---|
| NC | 0 | 2.100 | 2.094 | −0.769 | −1.836 | −3.864 | −5.474 | −10.382 |
| HPC | 0 | 1.597 | 1.376 | −1.178 | −0.715 | −2.533 | −3.883 | −9.023 |
| FA_10M | 0 | 1.904 | 1.647 | 0.683 | 2.559 | 2.984 | 3.336 | −0.642 |
| Acrylic Coat | 0 | 2.074 | 2.417 | 0.050 | −2.559 | −4.388 | −5.964 | −10.093 |
| Epoxy Coat | 0 | 0.672 | 0.578 | 0.508 | 0.597 | −0.192 | −0.281 | −0.321 |

**Figure 12.5** Weight loss due to sulfuric acid attack: (A) 10% acid concentration, (B) 20% acid concentration, and (C) 30% acid concentration.

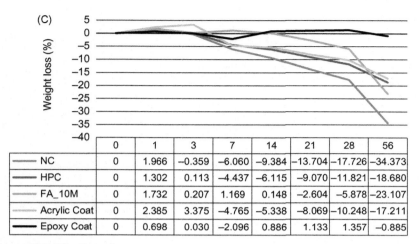

| (C) | 0 | 1 | 3 | 7 | 14 | 21 | 28 | 56 |
|---|---|---|---|---|---|---|---|---|
| —— NC | 0 | 1.966 | −0.359 | −6.060 | −9.384 | −13.704 | −17.726 | −34.373 |
| —— HPC | 0 | 1.302 | 0.113 | −4.437 | −6.115 | −9.070 | −11.821 | −18.680 |
| ---- FA_10M | 0 | 1.732 | 0.207 | 1.169 | 0.148 | −2.604 | −5.878 | −23.107 |
| ···· Acrylic Coat | 0 | 2.385 | 3.375 | −4.765 | −5.338 | −8.069 | −10.248 | −17.211 |
| —— Epoxy Coat | 0 | 0.698 | 0.030 | −2.096 | 0.886 | 1.133 | 1.357 | −0.885 |

**Figure 12.5** (Continued)

The fly ash geopolymeric mortar shows a good performance for both 10% and 20% sulfuric acid concentration. Fig. 12.6 shows photos of the different specimens after immersion in a 20% sulfuric acid concentration. Even for a 30% sulfuric acid concentration this mortar shows a good acid resistance for immersion until 14 days. HPC specimens show the third-best performance. It shows a minor weight loss after 56 days in a 10% sulfuric acid concentration. For a 20% sulfuric acid concentration the weight loss is clear beyond 14 days, reaching a maximum of 9%. When the concentration increases to 30%, the weight loss starts after 7 days immersion and reaches a maximum of 20% after 56 days. Specimens of NC coated with acrylic paint show the same performance of uncoated concrete specimens for both 10% and 20% sulfuric acid concentration. Only for the 30% acid concentration and long-term immersion can this coat be of some use.

Since NC and HPC have almost similar capillary water absorption, the differences in acid resistance lie in the leaching of calcium hydroxide $(Ca(OH)_2)$ from the pore solution and decalcification of CSH, which must be lower in the latter case due to a much higher Portland cement content. In the sulfuric acid attack, sulfate ions react with calcium hydroxide, forming calcium sulfate dihydrate-gypsum (Fig. 12.2), and with aluminate hydrates, forming ettringite (Fig. 12.3).

$$H_2SO_4 + Ca(OH)_2 \rightarrow CaSO_4 \tag{12.2}$$

$$3CaSO_4 + 3CaO \cdot Al_2O_3 \cdot 6H_2O + 25H_2O \rightarrow 3CaO \cdot Al_2O_3 \cdot 3CaSO_4 \cdot 31H_2O \tag{12.3}$$

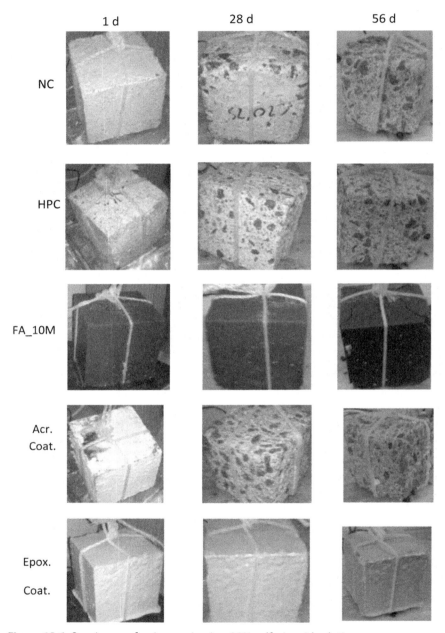

**Figure 12.6** Specimens after immersion in a 20% sulfuric acid solution.

## 12.4.4.2 Resistance to Nitric Acid Attack

Weight loss after nitric acid attack is shown in Fig. 12.7. Again, NC coated with epoxy resin shows the most stable performance for all three acid concentrations. Nitric acid attack at 10% concentration is especially destructive for NC even after just 7 days immersion. Nitric acid reacts with calcium compounds, forming calcium nitrate, which has a solubility of 56%. All the other mixtures show a weight loss not exceeding 2% even after 56 days immersion. The behavior for a 20% nitric acid concentration

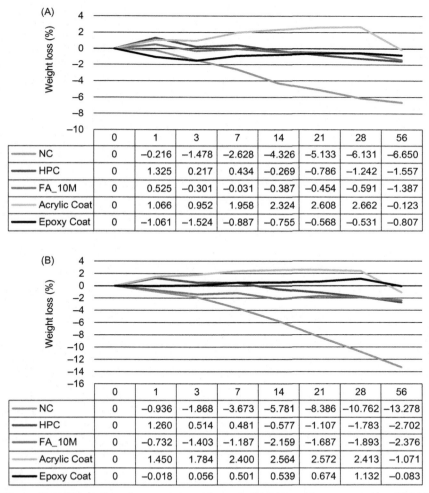

**(A)**

| | 0 | 1 | 3 | 7 | 14 | 21 | 28 | 56 |
|---|---|---|---|---|---|---|---|---|
| —— NC | 0 | −0.216 | −1.478 | −2.628 | −4.326 | −5.133 | −6.131 | −6.650 |
| —— HPC | 0 | 1.325 | 0.217 | 0.434 | −0.269 | −0.786 | −1.242 | −1.557 |
| —— FA_10M | 0 | 0.525 | −0.301 | −0.031 | −0.387 | −0.454 | −0.591 | −1.387 |
| —— Acrylic Coat | 0 | 1.066 | 0.952 | 1.958 | 2.324 | 2.608 | 2.662 | −0.123 |
| —— Epoxy Coat | 0 | −1.061 | −1.524 | −0.887 | −0.755 | −0.568 | −0.531 | −0.807 |

**(B)**

| | 0 | 1 | 3 | 7 | 14 | 21 | 28 | 56 |
|---|---|---|---|---|---|---|---|---|
| —— NC | 0 | −0.936 | −1.868 | −3.673 | −5.781 | −8.386 | −10.762 | −13.278 |
| —— HPC | 0 | 1.260 | 0.514 | 0.481 | −0.577 | −1.107 | −1.783 | −2.702 |
| —— FA_10M | 0 | −0.732 | −1.403 | −1.187 | −2.159 | −1.687 | −1.893 | −2.376 |
| —— Acrylic Coat | 0 | 1.450 | 1.784 | 2.400 | 2.564 | 2.572 | 2.413 | −1.071 |
| —— Epoxy Coat | 0 | −0.018 | 0.056 | 0.501 | 0.539 | 0.674 | 1.132 | −0.083 |

**Figure 12.7** Weight loss due to nitric acid attack: (A) 10% acid concentration, (B) 20% acid concentration, and (C) 30% acid concentration.

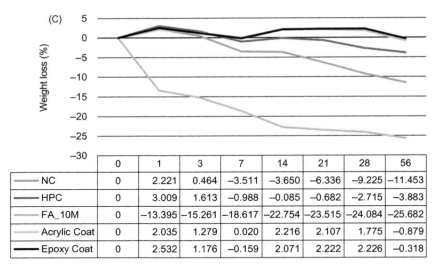

| (C) | 0 | 1 | 3 | 7 | 14 | 21 | 28 | 56 |
|---|---|---|---|---|---|---|---|---|
| NC | 0 | 2.221 | 0.464 | −3.511 | −3.650 | −6.336 | −9.225 | −11.453 |
| HPC | 0 | 3.009 | 1.613 | −0.988 | −0.085 | −0.682 | −2.715 | −3.883 |
| FA_10M | 0 | −13.395 | −15.261 | −18.617 | −22.754 | −23.515 | −24.084 | −25.682 |
| Acrylic Coat | 0 | 2.035 | 1.279 | 0.020 | 2.216 | 2.107 | 1.775 | −0.879 |
| Epoxy Coat | 0 | 2.532 | 1.176 | −0.159 | 2.071 | 2.222 | 2.226 | −0.318 |

**Figure 12.7** (Continued)

is almost the same, the difference being that NC shows a higher weight loss. When the acid concentration is increased to 30% NC does not show an increase in the weight loss. For this very high acid concentration the geopolymeric mortar shows a disappointing performance. Allahverdi and Škvára [32,33] suggested that the electrophilic attack of nitric acid protons results in the ejection of tetrahedral aluminum from the aluminosilicate framework and in the formation of an imperfect highly siliceous framework. Other authors [34] also suggested this aluminosilicate depolymerization.

### 12.4.4.3 Resistance to Hydrochloric Acid Attack

Fig. 12.8 shows the weight loss after hydrochloric acid attack for the different acid concentrations. The results are every similar to those of the nitric acid attack. A 10% hydrochloric acid concentration is responsible for a relevant NC weight loss even after just 7 days immersion. This type of acid reacts with calcium compounds, leading to the formation of calcium chloride, which has extremely high solubility (46.1 wt%) [35]. The behavior for a 20% nitric acid concentration is almost the same.

The difference being that NC shows a higher weight loss. All the other mixtures show a weight loss not exceeding 2% even after 56 days immersion. When the hydrochloric acid concentration is increased to 30%, NC does not show a relevant increase in the weight loss. However,

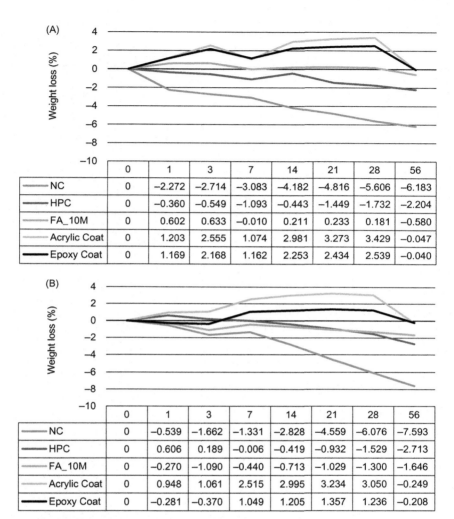

| (A) | 0 | 1 | 3 | 7 | 14 | 21 | 28 | 56 |
|---|---|---|---|---|---|---|---|---|
| NC | 0 | −2.272 | −2.714 | −3.083 | −4.182 | −4.816 | −5.606 | −6.183 |
| HPC | 0 | −0.360 | −0.549 | −1.093 | −0.443 | −1.449 | −1.732 | −2.204 |
| FA_10M | 0 | 0.602 | 0.633 | −0.010 | 0.211 | 0.233 | 0.181 | −0.580 |
| Acrylic Coat | 0 | 1.203 | 2.555 | 1.074 | 2.981 | 3.273 | 3.429 | −0.047 |
| Epoxy Coat | 0 | 1.169 | 2.168 | 1.162 | 2.253 | 2.434 | 2.539 | −0.040 |

| (B) | 0 | 1 | 3 | 7 | 14 | 21 | 28 | 56 |
|---|---|---|---|---|---|---|---|---|
| NC | 0 | −0.539 | −1.662 | −1.331 | −2.828 | −4.559 | −6.076 | −7.593 |
| HPC | 0 | 0.606 | 0.189 | −0.006 | −0.419 | −0.932 | −1.529 | −2.713 |
| FA_10M | 0 | −0.270 | −1.090 | −0.440 | −0.713 | −1.029 | −1.300 | −1.646 |
| Acrylic Coat | 0 | 0.948 | 1.061 | 2.515 | 2.995 | 3.234 | 3.050 | −0.249 |
| Epoxy Coat | 0 | −0.281 | −0.370 | 1.049 | 1.205 | 1.357 | 1.236 | −0.208 |

**Figure 12.8** Weight loss due to hydrochloric acid attack: (A) 10% acid concentration, (B) 20% acid concentration, and (C) 30% acid concentration.

the geopolymeric mortar shows a high weight loss. Davidovits et al. [36] reported a 78% weight loss for OPC concrete specimens immersed for 4 weeks in a 5% hydrochloric acid solution, which is much higher than the weight loss of NC after immersion for 56 days in a 30% hydrochloric acid solution, which was lower than 10%. This difference is so high that it cannot be explained by the specimen's geometry or OPC concrete composition. A possible explanation could be related to the periodic replacement

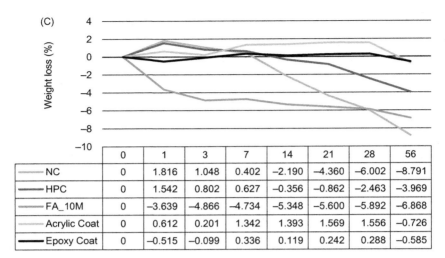

| (C) | 0 | 1 | 3 | 7 | 14 | 21 | 28 | 56 |
|---|---|---|---|---|---|---|---|---|
| NC | 0 | 1.816 | 1.048 | 0.402 | −2.190 | −4.360 | −6.002 | −8.791 |
| HPC | 0 | 1.542 | 0.802 | 0.627 | −0.356 | −0.862 | −2.463 | −3.969 |
| FA_10M | 0 | −3.639 | −4.866 | −4.734 | −5.348 | −5.600 | −5.892 | −6.868 |
| Acrylic Coat | 0 | 0.612 | 0.201 | 1.342 | 1.393 | 1.569 | 1.556 | −0.726 |
| Epoxy Coat | 0 | −0.515 | −0.099 | 0.336 | 0.119 | 0.242 | 0.288 | −0.585 |

**Figure 12.8** (Continued)

of the acid solution by the Davidovits study. Just because the pH is raised with time, for instance, a solution of sulfuric acid at 5% concentration evolves from pH = 1.05–6.95 after 28 days [37].

## 12.5  COST ANALYSIS

In order to evaluate the economic efficiency of several structural solutions, comparisons between the costs of materials were made. The cost calculations were related to $1\,m^2$ of concrete pavement with $0.3\,m$ thickness. Two noncoated solutions (NC, HPC), one with $0.275\,m$ NC thickness coated with $0.025\,m$ fly ash geopolymer and two coated with acrylic paint and epoxy resin were analyzed. Fig. 12.9 shows the costs of the different solutions. The concrete pavement coated by epoxy resin is by far the most costly solution. Epoxy coating costs exceed the NC solution costs by as much as 100%. Fig. 12.10 shows the cost to remaining mass (after acid attack) ratio according to acid concentration. The results show that for 10% and even 20% acid concentrations NC shows the best cost efficiency. The cost efficiency of the HPC-based solution is similar to the fly ash–based geopolymeric mortar except for a 30% acid concentration.

The results also show that no matter how well epoxy resin performs under acid attack its economic efficiency is the worst between all five solutions, being 70% above the cost efficiency of the fly ash–based geopolymeric mortar. Only for a 30% acid concentration does the

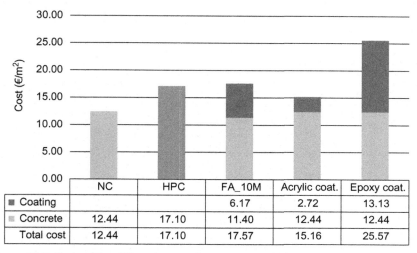

| | NC | HPC | FA_10M | Acrylic coat. | Epoxy coat. |
|---|---|---|---|---|---|
| ■ Coating | | | 6.17 | 2.72 | 13.13 |
| ▨ Concrete | 12.44 | 17.10 | 11.40 | 12.44 | 12.44 |
| Total cost | 12.44 | 17.10 | 17.57 | 15.16 | 25.57 |

**Figure 12.9** Costs of the different concrete pavement solutions.

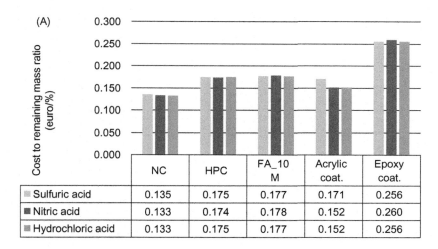

(A)

| | NC | HPC | FA_10 M | Acrylic coat. | Epoxy coat. |
|---|---|---|---|---|---|
| ▨ Sulfuric acid | 0.135 | 0.175 | 0.177 | 0.171 | 0.256 |
| ■ Nitric acid | 0.133 | 0.174 | 0.178 | 0.152 | 0.260 |
| ▨ Hydrochloric acid | 0.133 | 0.175 | 0.177 | 0.152 | 0.256 |

(B)

| | NC | HPC | FA_10 M | Acrylic coat. | Epoxy coat. |
|---|---|---|---|---|---|
| ▨ Sulfuric acid | 0.139 | 0.188 | 0.177 | 0.169 | 0.256 |
| ■ Nitric acid | 0.143 | 0.176 | 0.180 | 0.153 | 0.256 |
| ▨ Hydrochloric acid | 0.135 | 0.176 | 0.179 | 0.152 | 0.257 |

**Figure 12.10** Cost to remaining mass ratio (euro/%): (A) 10% acid concentration, (B) 20% acid concentration, and (C) 30% acid concentration.

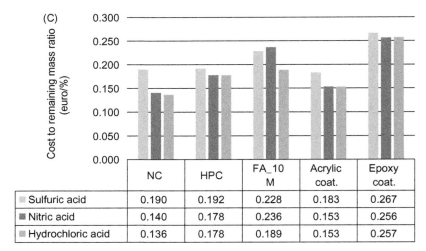

| (C) | NC | HPC | FA_10 M | Acrylic coat. | Epoxy coat. |
|---|---|---|---|---|---|
| Sulfuric acid | 0.190 | 0.192 | 0.228 | 0.183 | 0.267 |
| Nitric acid | 0.140 | 0.178 | 0.236 | 0.153 | 0.256 |
| Hydrochloric acid | 0.136 | 0.178 | 0.189 | 0.153 | 0.257 |

**Figure 12.10** (Continued)

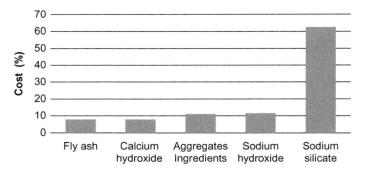

**Figure 12.11** Cost percentage of fly ash geopolymeric mortar ingredients.

epoxy-based solution gain some interest. It is important to remember that the cost of the fly ash–based geopolymeric mortar is very dependent on the cost of sodium silicate (Fig. 12.11). Fig. 12.12 shows a simulation of the cost-to-remaining-mass (after acid attack) ratio according to acid concentration when the sodium silicate cost is around 30% of its current cost. This means that current investigations aiming to replace sodium silicate with low-cost waste glass [38] will increase the cost efficiency of the fly ash–based geopolymeric mortar as a coating material for OPC concrete infrastructures exposed to harsh chemical environments. Furthermore, the future use of waste glass as sodium silicate replacement fits the European zero-waste program COM 398 [39].

(A)

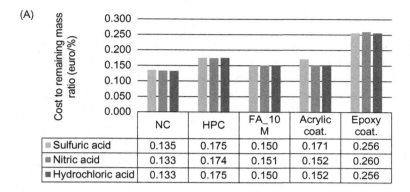

|  | NC | HPC | FA_10 M | Acrylic coat. | Epoxy coat. |
|---|---|---|---|---|---|
| Sulfuric acid | 0.135 | 0.175 | 0.150 | 0.171 | 0.256 |
| Nitric acid | 0.133 | 0.174 | 0.151 | 0.152 | 0.260 |
| Hydrochloric acid | 0.133 | 0.175 | 0.150 | 0.152 | 0.256 |

(B)

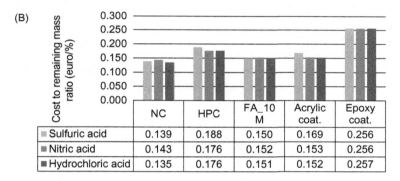

|  | NC | HPC | FA_10 M | Acrylic coat. | Epoxy coat. |
|---|---|---|---|---|---|
| Sulfuric acid | 0.139 | 0.188 | 0.150 | 0.169 | 0.256 |
| Nitric acid | 0.143 | 0.176 | 0.152 | 0.153 | 0.256 |
| Hydrochloric acid | 0.135 | 0.176 | 0.151 | 0.152 | 0.257 |

(C)

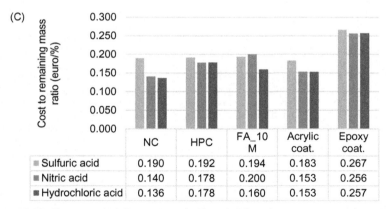

|  | NC | HPC | FA_10 M | Acrylic coat. | Epoxy coat. |
|---|---|---|---|---|---|
| Sulfuric acid | 0.190 | 0.192 | 0.194 | 0.183 | 0.267 |
| Nitric acid | 0.140 | 0.178 | 0.200 | 0.153 | 0.256 |
| Hydrochloric acid | 0.136 | 0.178 | 0.160 | 0.153 | 0.257 |

**Figure 12.12** Cost to remaining mass ratio (euro/%) for a low-cost waste glass sodium silicate replacement simulation: (A) 10% acid concentration, (B) 20% acid concentration, and (C) 30% acid concentration.

## 12.6 CONCLUSIONS

Worldwide infrastructure rehabilitation costs are staggering. Premature degradation of OPC concrete infrastructure is a current and serious problem related to the fact that OPC concrete presents a higher permeability that allows water and other aggressive elements to enter, leading to carbonation and chloride-ion attack, resulting in corrosion problems. This article presents results of an experimental investigation on the resistance to chemical attack of several materials. NC coated with epoxy resin shows the most stable performance for all three acid types and acid concentrations. For a very high nitric acid concentration the geopolymeric mortar shows a disappointing performance that could be due to the ejection of tetrahedral aluminum from the aluminosilicate framework and in the formation of an imperfect highly siliceous framework. The results show that no matter how well epoxy resin performs under acid attack its economic efficiency is the worst between all the five solutions, being 70% above the cost efficiency of the fly ash–based geopolymeric mortar. Current investigations aiming to replace sodium silicate with low-cost waste glass will increase the cost efficiency of the fly ash–based geopolymeric mortar as coating material for OPC concrete infrastructure exposed to harsh chemical environments.

## REFERENCES

[1] Glasser F, Marchand J, Samson E. Durability of concrete. Degradation phenomena involving detrimental chemical reactions. Cem Concr Res 2008;38:226–46. http://dx.doi.org/10.1016/j.cemconres.2007.09.015.

[2] Pacheco-Torgal F, Gomes JP, Jalali S. Alkali—activated binders: a review. Part 1 Historical background, terminology, reaction mechanisms and hydration products. Constr Build Mater 2008;22:1305–14. http://dx.doi.org/10.1016/j.conbuildmat.2007.10.015.

[3] Pacheco-Torgal F, Abdollahnejad Z, Miraldo S, Baklouti S, Ding Y. An overview on the potential of geopolymers for concrete infrastructure rehabilitation. Constr Build Mater 2012;36:1053–8. http://dx.doi.org/10.1016/j.conbuildmat.2012.07.003.

[4] Almusallam A, Khan F, Dulaijan S, Al-Amoudi O. Effectiveness of surface coatings in improving concrete durability. Cem Concr Compos 2003;25:473–81. http://dx.doi.org/10.1016/S0958-9465(02)00087-2.

[5] Aguiar JB, Camões A, Moreira PM. Coatings for concrete protection against aggressive environments. J Adv Concr Technol 2008;6(1):243–50.

[6] Moreira P. Using polymeric coatings to improve the durability of concrete exposed to aggressive media. Master thesis. University of Minho; 2006.

[7] Berndt M. Evaluation of coatings, mortars and mix design for protection of concrete against sulphur oxidising bacteria. Constr Build Mater 2011;25:3893–902. http://dx.doi.org/10.1016/j.conbuildmat.2011.04.014.

[8] Medeiros M, Helene P. Efficacy of surface hydrophobic agents in reducing water and chloride ion penetration in concrete. Mater Struct 2008;41:59–71. http://dx.doi.org/10.1617/s11527-006-9218-5.

[9]  Pacheco-Torgal F, Jalali S. Sulphuric acid resistance of plain, polymer modified, and fly ash cement concretes. Constr Build Mater 2009;23:3485–91. http://dx.doi.org/10.1016/j.conbuildmat.2009.08.001.

[10]  Brenna A, Bolzoni F, Beretta S, Ormellese M. Long-term chloride-induced corrosion monitoring of reinforced concrete coated with commercial polymer-modified mortar and polymeric coatings. Constr Build Mater 2013;48:734–44. http://dx.doi.org/10.1016/j.conbuildmat.2013.07.099.

[11]  Bijen J. Durability of engineering structures. Design, repair and maintenance. Abington Hall, Cambridge: Woodhead Publishing Limited; 2000.

[12]  Pacheco Torgal F, Jalali S. Toxicity of building materials. A key issue in sustainable construction. Int J Sustainable Eng 2011;4:281–7. http://dx.doi.org/10.1080/19397038.2011.569583.

[13]  Pacheco-Torgal F, Jalali S, Fucic A. Toxicity of building materials. Cambridge: Woodhead Publishing Limited; 2012. 480p.

[14]  Pacheco-Torgal F, Gomes J, Jalali S. Adhesion characterization of tungsten mine waste geopolymeric binder. Influence of OPC concrete substrate surface treatment. Constr Build Mater 2008;22:154–61. http://dx.doi.org/10.1016/j.conbuildmat.2006.10.005.

[15]  Davidovits J. Synthesis of new high temperature geo-polymers for reinforced plastics/composites. Brookfield Center: SPE PACTEC 79 Society of Plastic Engineers; 1979;151–4

[16]  Pacheco-Torgal F, Labrincha JA, Leonelli C, Palomo A, Chindaprasirt P. Handbook of alkali-activated cements, mortars and concretes, ed. 1 Abington Hall, Cambridge, UK: Woodhead Publishing Limited—Elsevier Science and Technology; 2014.

[17]  Provis JL, Van Deventer JSJ, (editors.) Geopolymers: structure, processing, properties and industrial applications. Cambridge, UK: Woodhead Publishing; 2009.

[18]  Provis J. Geopolymers and other alkali activated materials: why, how, and what? J Mater Struct 2014;47:11–25.

[19]  Papakonstantinou CG, Balaguru PN. Geopolymer protective coatings for concrete. International SAMPE symposium and exhibition (Proceedings) (2007) 52.

[20]  Zhang Z, Yao X, Zhu H. Potential applications of geopolymers as protection coatings for marine concrete I. Basic properties. Appl Clay Sci 2010;49:1–6. http://dx.doi.org/10.1016/j.clay.2010.01.014.

[21]  Zhang Z, Yao X, Zhu H. Potential application of geopolymers as protection coatings for marine concrete II. microstructure and anticorrosion mechanism. Appl Clay Sci 2010;49:7–12. http://dx.doi.org/10.1016/j.clay.2010.04.024.

[22]  Yip C, Lukey G, Deventer SJS. The coexistence of geopolymeric gel and calcium silicate hydrate gel at the early stage of alkaline activation. Cem Concr Res 2005;35:1688–97. http://dx.doi.org/10.1016/j.cemconres.2004.10.042.

[23]  Van Deventer JSJ, Provis J, Duxson P. Technical and commercial progress in the adoption of geopolymer cement. Miner Eng 2012;29:89–104. http://dx.doi.org/10.1016/j.mineng.2011.09.009.

[24]  Bakharev T, Sanjayan JG, Cheng Y-B. Sulfate attack on alkali-activated slag concrete. Cem Concr Res 2002;32:211–6. http://dx.doi.org/10.1016/j.conbuildmat.2007.07.015.

[25]  Lee WKW, Van Deventer JSJ. The effect of ionic contaminants on the early-age properties of alcali-activated fly ash-based cements. Cem Concr Res 2002;32:577–84. doi:10.1016/S0008-8846(01)00724-4.

[26]  Pacheco-Torgal F, Gomes JP, Jalali S. Investigations on mix design of tungsten mine waste geopolymeric binders. Constr Build Mater 2008;22:1939–49.

[27]  Somna K, Jaturapitakkul C, Kajitvichyanukul P, Chindaprasirt P. NaOH-activated ground fly ash geopolymer cured at ambient temperature. Fuel 2011;90:2118–24.

[28]  Granizo ML, Blanco-Varela MT, Martinez-Ramirez S. Alkali activation of metakaolins: parameters affecting mechanical, structural and microstructural properties. J Mater Sci 2007;42(9):2934–43. http://dx.doi.org/10.1007/s10853-006-0565-y.

[29] Ferreira R. 246p Evaluation of durability test. Master Thesis. Guimaraes, Portugal: University of Minho; 2000.
[30] ERMCO. European ready-mixed concrete industry statistics—2013, 2014.
[31] Pacheco-Torgal F, Castro-Gomes JP. Influence of physical and geometrical properties of granite and limestone aggregates on the durability of a C20/25 strength class concrete. Constr Build Mater 2006;20:1079–88. http://dx.doi.org/10.1016/j.conbuildmat.2005.01.063.
[32] Allahverdi A, Škvára F. Nitric acid attack on hardened paste of geopolymeric cements, Part 1. Ceram – Silik 2001;45(3):81–8.
[33] Allahverdi A, Škvára F. Nitric acid attack on hardened paste of geopolymeric cements, Part 2. Ceram – Silik 2001;45(4):143–9.
[34] Fernandez-Jimenez A, García-Lodeiro I, Palomo A. Durability of alkali-activated fly ash cementitious materials. J Mater Sci 2007;42:3055–65. http://dx.doi.org/10.1007/s10853-006-0584-8.
[35] Zivica V, Bazja A. Acid attack of cement based materials—a review. Part 1, Principle of acid attack. Cem Concr Res 2001;15:331–40. http://www.ceramics-silikaty.cz/2005/pdf/2005_04_225.pdf.
[36] Davidovits J, Comrie DC, Paterson JH, Ritcey DJ. Geopolymeric concretes for environmental protection. ACI Concr Int 1990;12:30–40.
[37] Roy DM, Arjunan P, Silsbee MR. Effect of silica fume, metakaolin, and low-calcium fly ash on chemical resistance of concrete. Cem Concr Res 2001;31:1809–18013. doi:10.1016/S0008-8846(01)00548-8.
[38] Puertas F, Torres-Carrasco M, Alonso M. Reuse of urban and industrial waste glass as novel activator for alkali-activated slag cement pastes: a case study Pacheco-Torgal F, Labrincha JA, Leonelli C, Palomo A, Chindaprasirt P, editors. Handbook of alkali-activated cements, mortars and concretes (ed. 1). Abington Hall, Cambridge, UK: Woodhead Publishing Limited—Elsevier Science; 2014.
[39] COM (2014) 398 of July 2. http://ec.europa.eu/transparency/regdoc/rep/1/2014/EN/1-2014-398-EN-F1-1.Pdf.

# CHAPTER 13

# Alkali-Activated Cement (AAC) From Fly Ash and High-Magnesium Nickel Slag

Z. Zhang[1], T. Yang[2] and H. Wang[1]
[1]University of Southern Queensland, Toowoomba, Australia
[2]China Oilfield Services Limited, Yanjiao, China

## 13.1 INTRODUCTION

Alkali–activation technology has shown the potential of converting solid aluminosilicate materials into green cement, also known as alkali–activated cement (AAC) and geopolymer, in terms of fewer $CO_2$ emissions and lower energy requirements compared to ordinary Portland cement (OPC) [1]. The aluminosilicate precursors are usually clays, calcined clays (metakaolin), and fly ash, which contains $SiO_2$ and $Al_2O_3$ as the main composition. AAC has been widely applied to a large variety of alkali-activated materials in the past 30 years, including those derived from calcium, magnesium, and iron-bearing solid materials. The common feature of these industrial wastes is that they all contain high concentrations of silica and alumina and certain amounts of iron, magnesium, and/or calcium. The simple process of converting the industrial wastes into value-added materials is so attractive for many industrial sectors, particularly the metallurgical industries, which generate large amounts of solid industrial wastes annually.

Metal nickel production from high-magnesium nickel oxide ore follows a pyrometallurgical process including prereduction, smelting in a blast furnace, and nickel-enrichment refining. At the smelting stage, high-magnesium nickel slag (HMNS) is formed; it consists of a range of oxides and the remaining magnesium. The HMNS is recycled back into the process until its nickel concentration is low enough, and then after quenching and grinding it is disposed of in piles onsite. A large amount of HMNS is produced every year worldwide from metal nickel production using high-magnesium nickel oxide ore. In China, the annual generation of nickel slag is more than 800 kt, while less than 8% is utilized by the construction and building industries (mainly by the cement industry)

*Handbook of Low Carbon Concrete.*
DOI: http://dx.doi.org/10.1016/B978-0-12-804524-4.00013-0

357

[2,3]. The problem with large amounts of utilization is the high-Mg concentration in HMNS, which may lead to stability problems in concrete structures. The reason is that HMNS contains a relatively high concentration of magnesium, which can lead to unpredictable consequences for the long-term engineering of properties, especially the volume stability. Most of the rest of HMNS, more than 90%, is disposed in an open environment, which has hazardous effects on ground and underground water due to the ambient leaching under rain or surface-water conditions. However, HMNS is often considered to be a hazard that does not comply with environmental standards for safe disposal, although the amount of residual nickel in it is too small to require immobilization. Therefore, the sustainable production model for waste management to implement the new strict environmental regulations requires low-cost technologies to utilize the slag for the production of added-value products. The alkali-activated technique is one of the most promising candidates. A number of slags, such as primary lead slag [4], Cu-Ni slag [5], and ferronickel slag [6,7], have been used in the manufacture of AACs.

In terms of its environmental footprint, this type of material, i.e., AACs, can be much greener than OPC. This is mainly because of the noncalcination process of applying alkali-activation technology. Although the preparation of metakaolin requires a heating process if clay is used as a raw material, the heating temperature, usually between 500°C and 800°C, is much lower than the clinkering temperature (1450°C) for OPC, and the heating period is also shorter. The industrial scale application uses fly ash as the major raw material, which does not require any heating at all. The alkali activator used can be a great contributor to the input energy and relative $CO_2$ emissions. Some carbon accountings regarding metakaolin and fly ash–based AAC binders and concretes have been reported recently [8–10]. It is generally accepted that the environmental footprint of AAC manufacturing is significantly affected by the use of alkali activators, the curing temperature, and the cost of sourcing the raw materials.

This chapter analyzes the feasibility of using HMNS as a raw material for AAC manufacturing. Fly ash is also used for the sake of supplementing alumina deficiency. The porosity, compressive strength, and drying shrinkage are major parameters of the deriving AAC, and are investigated to optimize the slag-blending ratio and to examine the high-magnesium content on the volume stability of products. X-ray diffractometry (XRD), Fourier transform infrared (FTIR) spectroscopy and scanning electron microscopy (SEM) are adopted to analyze the reaction products to understand the reaction mechanisms of HMNS. The sustainability of the deriving AACs, in terms of $CO_2$ emissions and energy consumption, is analyzed by taking into account some realistic conditions.

## 13.2 MANUFACTURE OF AACs

### 13.2.1 Materials

Fly ash obtained from Xuzhou, Jiangsu Province, China, which is classified as type F according to ASTM-618 [11], is used as the main solid material. HMNS is a waste product disposed by a Nanjing (Jiangsu Province, China) nickel production company, at an open environment (Fig. 13.1). It is required to be treated safely for the purposes of environmental protection and land use. To be used in AAC manufacturing, the HMNS was dried at 105°C for 24h and then ground by a ball mill for 1h at 250 rpm. The milling process makes a suitable particle size (1–100 μm) to mix with fly ash, as shown in the scanning electron microscope (SEM) images (Fig. 13.2). The SEM analysis was done on a ZEISS EVO MA18 machine, at 20 kV with samples coated with gold. The fly ash and the ground HMNS were also measured by a laser particle-size analyzer, and their surface areas were 1.4 and 1.3 $m^2$/g (Fig. 13.3).

The compositions of the HMNS and fly ash were determined by X-ray fluorescence method. In Table 13.1 it shows the HMNS contains $SiO_2$ and MgO as dominant compositions, while the fly ash contains more $Al_2O_3$. It is noted that the HMNS contains higher contents of Cr, Ni, and Zn than the fly ash, and their concentrations are also higher than typical blast furnace slag. The Cr and other harmful elements in HMNS are expected to be immobilized in the alkali-activated products

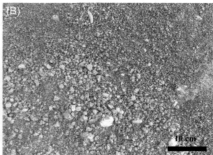

**Figure 13.1** HMNS disposed on an open field (A) and a closer look at the granules (B).

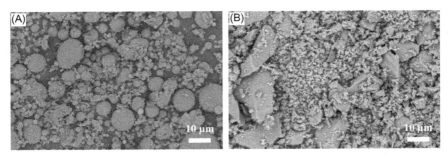

**Figure 13.2** Back-scattered electron (BSE) images of fly ash particles (A) and ground HMNS particles (B).

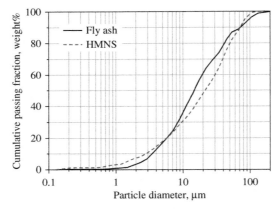

**Figure 13.3** Particle-size distributions of the raw materials as measured by laser particle-size analyzer.

**Table 13.1** Chemical compositions of fly ash and HMNS

| Content (wt%) | HMNS | Fly ash |
|---|---|---|
| $Al_2O_3$ | 6.2 | 30.6 |
| $SiO_2$ | 52.3 | 53.0 |
| $MgO$ | 26.9 | 1.2 |
| $CaO$ | 8.8 | 4.8 |
| $Fe_2O_3$ | 4.2 | 3.8 |
| $K_2O$ | 0.2 | 1.4 |
| $TiO_2$ | 0.1 | 1.1 |
| $Na_2O$ | 0.1 | 0.5 |
| Cr | 0.4 | n.d. |
| Ni | <0.1 | n.d. |
| Zn | <0.1 | n.d. |
| LOI | 0.5 | 2.3 |

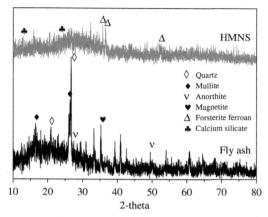

**Figure 13.4** XRD patterns of fly ash and HMNS. Quartz ($SiO_2$), PDF #83-539; mullite ($Al_{2.4}Si_{0.6}O_{4.8}$), PDF #73-1389; anorthite ($CaAl_2Si_2O_8$), PDF #89-1470; magnetite ($Fe_{2.939}O_4$), PDF #86-1356; calcium silicate ($Ca_2SiO_4$), PDF #36-0642; forsterite ferroan ($Mg_{1.824}Fe_{0.176}SiO_4$), PDF #88-1991.

[12,13]. The mineral components in the fly ash and HMNS were investigated using XRD, as shown in Fig. 13.4. The XRD data were recorded using a Thermo ARL'tra machine with Cu Kα radiation at a scanning rate of 10°/min from 10° to 80° 2θ. The majority in fly ash is amorphous, which is reflected by the broad hump among the range of $2\theta = 15–30°$. A small amount of crystalline phases, including quartz ($SiO_2$), mullite ($Al_{2.4}Si_{0.6}O_{4.8}$), anorthite ($CaAl_2Si_2O_8$), and magnetite ($Fe_{2.939}O_4$) are detectable. In the HMNS the majority is also amorphous phase, and trace amounts of magnesium silicate hydrate ($3MgO·2SiO_2·2H_2O$) and crystalline forsterite ferroan ($Mg_{1.824}Fe_{0.176}SiO_4$) are detected.

The alkaline activators used were four sodium silicate solutions with varying modulus ($M_s$), i.e., $SiO_2/Na_2O = 1.2, 1.4, 1.6,$ and 2.0. They were prepared by mixing NaOH, water and a $Na_2O·2.44SiO_2$ waterglass to solid concentration of 30%. All activators were prepared and allowed to cool down to room temperature before using.

## 13.2.2  AAC Manufacture and Characterization

The fly ash and HMNS were dry-mixed by hand at blending ratios of 100/0, 80/20, 60/40, 20/80, 0/100 and then mixed with alkali silicate activators ($M_s = 1.2, 1.4, 1.6,$ and 2.0) in a cement paste mixer to form the AAC binders. The liquid/binder ratio was constant at 0.42. Samples were cast to be Ø25.4 × 25.4 mm and 20 × 20 × 80 mm for compressive strength and drying shrinkage tests, respectively, using stainless steel molds cured at 20°C and 50% RH in an air-conditioned room. After 3 days the samples with 0–60% slag were strong enough to be demolded; unfortunately, the samples mixed with 80% and 100% slag did not harden and were excluded for further testing.

The compressive strength testing of the cylinder samples was conducted on a WHY-200 auto-test compression machine. Drying shrinkage was tested using the cuboid specimens. A length comparator was used to measure the linear variation of the sample along the longitudinal axis. Linear drying shrinkage of AACs were calculated through Eq. (13.1):

$$\text{Linear drying shrinkage } \delta = (L_{in} - L_{fi})/75 \times 100\% \qquad (13.1)$$

in which $L_{in}$ (mm) is initial length of each sample at demold time, $L_{fi}$ (mm) is the measured length of each sample at different ages, 75 (mm) is the effective length of each sample as cast.

Mercury intrusion porosimetry was performed using a Poremaster GT-60 (Quantachrome, United States). Before testing, the 60-day aged samples were crushed into suitable-sized pieces and stored in ethanol for 24 h to extract free water, and then dried at 60°C for 4 h. This relatively low temperature is believed to be harmless to the pore structure of AACs. The samples were also used for SEM analysis. Ground AAC powders were used for XRD analysis.

## 13.3 PROPERTIES OF AACs

### 13.3.1 Compressive Strength

Fig. 13.5 shows the compressive strength of alkali-activated fly ash/HMNS binders as a function of the blending content of slag at varying modulus of sodium silicate solution. The compressive strength increases as the modulus increases from 1.2 to 1.4 and starts to decrease at higher modulus in the study's range. This is consistent with the study of Maragkos et al. [14], where ferronickel slag was used for the generation of AAC. The decrease of molar ratios promotes the dissolution of solid precursors in the highly alkaline aqueous phase; as a consequence, the quantity of gel binder is increased. The gel phase is the binding material in the structure, making a great contribution to the increase of compressive strength [14]. However, an excessive amount of NaOH in the lowest modulus ($M_s = 1.0$) seems to inhibit the formation of polymeric binder, which is also reported in alkali activation of ferronickel slag [14]. This is probably due to the quick

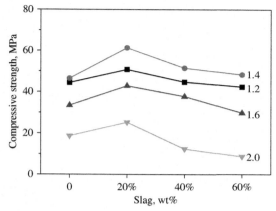

**Figure 13.5** Compressive strength of AAC pastes.

setting and hardening at low-modulus conditions, which in fact hinder the further dissolution of raw materials.

The blending of 20% HMNS leads to an increase of compressive strength by 20–40% at the optimal modulus conditions with $M_s = 1.4$. More HMNS (40% and 60%) results in decreased strength. It means that the extent of alkali activation of raw materials is not the only factor determining the mechanical properties of AAC. The benefits of blending 20% HMNS should be considered from two perspectives. In geometry, the irregular polyhedron HMNS particles that do not react completely will work as microaggregates in the AAC matrix, maintaining or increasing the strength. Also, in terms of pore structure, a more refined structure due to the presence of HMNS can lead to a more uniform stress distribution under loading, and increases the compressive strength of the samples [15]. The two perspectives will be discussed in more detail in the following sections. It was observed that the samples with 80% and 100% HMNS were not able to set within 24 h and were too weak to demold after 1 week of curing. The reason is due to the extremely low reactivity of HMNS, even though it is highly amorphous in nature.

The chemical compositions in HMNS are dominated by $SiO_2$ and MgO, while $Al_2O_3$ and CaO are not as high as in normal slag. The lack of alumina and calcium has some impact on the formation of aluminosilicate gels and calcium silicate hydrates, which are two major contributors of strength development [16]. It cannot exclude the possibility of forming Mg-containing gels during alkali activation; however, it seems that Mg is more likely to form hydrotalcite under alkali–activation conditions [17,18].

In order to utilize HMNS more effectively and produce fly ash–HMNS AAC with high compressive strength, further research is recommended to mix HMNS with other aluminosilicate solid precursors with relatively higher calcium and alumina contents, such as high-calcium fly ash and high-alumina fly ash. The high-calcium fly ash will lead to higher dissolution of precursors, and usually higher compressive strength, than will low-calcium fly ash [19,20].

## 13.3.2 Microstructure of AACs

The BSE images of the AAC pastes are presented in Fig. 13.6. The residue fly ash particles and angular HMNS particles, about 20 μm in diameter, are embedded in and bound with the gel phase. Along the residual particles there are many microcracks, which are probably due to the shrinkage of gel during drying. No large crystalline phase or ordered structure is observed. It seems that the AAC with 20% HMNS possesses the most compact microstructure in the four mixes.

The Si/Al ratios in gel phases are determined using the SEM–energy-dispersive X-ray spectroscopy (EDS) technique (Table 13.2). The average value increases from 1.34 to 3.50 as the HMNS content increases from

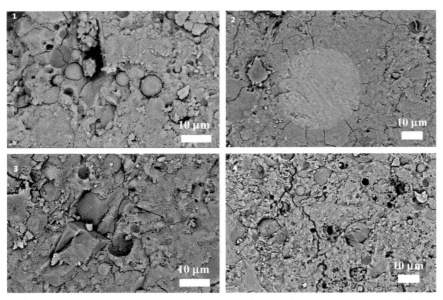

**Figure 13.6** SEM image for fracture section of geopolymer binders with different HMNS contents: (1–4) 0%, 20%, 40%, and 60%.

**Table 13.2** Si/Al molar ratios of gel phase in the binders

| HMNS | Areas | Si | Al | Si/Al | Average value |
|------|-------|-------|-------|-------|---------------|
| 0 | 1 | 17.36 | 19.93 | 0.87 | 1.34 |
|   | 2 | 26.53 | 14.53 | 1.81 |   |
| 20% | 1 | 25.02 | 11.37 | 2.20 | 2.07 |
|   | 2 | 24.56 | 12.66 | 1.94 |   |
| 40% | 1 | 22.37 | 7.82 | 2.86 | 3.34 |
|   | 2 | 26.94 | 7.04 | 3.81 |   |
| 60% | 1 | 15.72 | 4.21 | 3.73 | 3.50 |
|   | 2 | 22.76 | 6.96 | 3.27 |   |

Two areas of size $3 \times 3\,\mu m$ are detected for an average value in each binder.

0% to 60%. Under the attack of alkali, Al species in raw materials are more readily dissolved than Si species [21]; however, due to the lack of Al in HMNS, the rate of Al dissolution will be suppressed. Furthermore, the reactive Al deficiency will influence the polymerization stage whereby the aluminate and silicate species are polymerized to form the aluminosilicate gels. Thus, the evolution of alkali activation and degree of polymerization of the binders could thus be decreased. This is consistent with the compressive strength-testing result. The composition analysis shown here indicates that the gels containing Si/Al= 2.07 in the system with 20% HMNS endow the matrix with the highest compressive strength, which is consistent with the previous works [22–24].

## 13.3.3 Pore-Size Distribution

Fig. 13.7 shows the pore-size distribution and porosity of AAC pastes with different HMNS blending ratios. The curve of 100% fly ash binder presents dominant pore diameters at 10–100 nm. In this binder the maximal distribution at diameters at around 27 nm and the pores in the mesopores interval, 10–50 nm, account for 80% frequency of total porosity. This curve is different from the bimodal profile of the pore-size distributions as observed in alkali-activated fly ash cured for 28 days, where pores are respectively located at 100 and 1000 nm [22,25]. It indicates that the large capillary pores (around 100–1000 nm) are space-filled by the aluminosilicate gels, and thus there is only one broad peak in the binders with a denser pore structure developed.

All of the curves of the alkali-activated fly ash/HMNS blends present wide pore-size distributions over a range of 10–300 nm. The diameters of distribution-maximal pores increase from 24 to 64 nm as

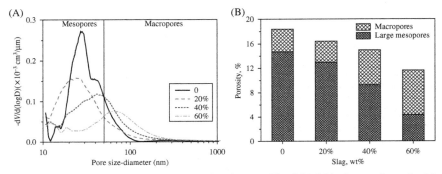

**Figure 13.7** Pore size (A) and porosity distribution (B) of AAC binders activated with activator $M_s$ = 1.4 and different HMNS contents at 60 days.

HMNS content increases; meanwhile the total porosity decreases. This means that when more HMNS is used, the total volume of pores in the AAC binders turns out to be lower, while more macropores (>50 nm) are formed. This is different from the effects of granulated blast furnace slag (GBFS) on the pore structure of fly ash–based AAC. Provis et al. [26] found that the addition of GBFS could not only provide an overall porosity reduction but also a pore-refinement effect. The presence of more bound water within the C–(A)–S–H gel predominantly formed in the slag-rich AAC systems would lead to more pore-filling in the binder matrix. However, in this work, HMNS particles will supply silica during the dissolution stage. The silicate ingredients appear to be helpful for forming more aluminosilicate gels with lower-bond water content in the AACs. The connectivity degree of the binders is supposed to be improved [20], and some mesopores will be filled with these reaction products, resulting in a decrease of porosity for the binders. Nevertheless, the reaction will provide the pore solution eventually bonded with aluminosilicate gels. It is interesting to note that the maximum amount of pores and total porosity of 20% HMNS geopolymer binders are a bit lower than those of the 100% fly ash binders. HMNS particles work as a microaggregate in the fly ash–based system, resulting in a refinement for the pore structures of the samples and the positive influences on their mechanical properties. In addition to the increased Si/Al ratios, the presence of more mesopores in the pastes with 40% and 60% HMNS probably also contributes to their lower strengths.

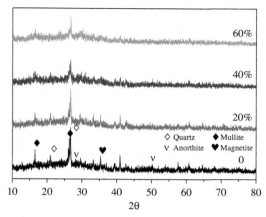

**Figure 13.8** XRD patterns of AAC binders activated with activator $M_s = 1.4$ and different HMNS contents at 60 days.

## 13.3.4 XRD Analysis

Fig. 13.8 plots the XRD patterns of 60-day aged AAC binders activated with activator $M_s = 1.4$. In the 100% fly ash binder, the major X-ray crystalline phases are mullite and quartz contributed by the unreacted fly ash. Minor calcite is also detected, due to the carbonation on the surface of the products under atmospheric conditions. The broad peak from 20° to 35° $2\theta$ in fly ash is broadened up to 40° after reaction, with the intensity shifting to higher degrees. This is well known as an indication due to the formation of N–A–S–H gels [27,28].

The intensity of the peaks for forsterite ferroan, which is contributed by the unreacted HMNS, present in the fly ash/HMNS AACs, increases as the HMNS blending ratio increases. It indicates that the crystalline phase is relatively stable during the alkali-activation process, unlike the iron-containing phases in ferronickel slags [29]. Meanwhile, the intensity of the peaks related to mullite and quartz due to uncreated fly ash decreases with the incorporation of HMNS. The peaks of trace calcium silicate identified in HMNS disappear after reaction. This suggests the dissolution of these crystalline phases under highly alkaline conditions. Semicrystalline to crystalline zeolitic phases, which are detectable in the alkali activation of low-calcium ferronickel slags [6,29], have not been found in the FA/HMNS binders. This is probably due to the room-temperature curing scheme, unlike hydrothermal curing [19]. The major products formed in alkali-activated fly ash/HMNS are a type of amorphous gel phase, also inferred by the broad peak of 20–40° $2\theta$. In conjunction with the SEM–EDS analysis, the gel phase is sodium aluminosilicate gel containing a certain amount of magnesium (N–(M)–A–S), which is different from the products formed in alkali-activated a single fly ash [30,31], or single GBFS [15,32,33]. To identify the influence of the addition of HMNS on the specific molecular structures of geopolymers, mainly the Si-O-T bonds that reflect the extent of polymerization, further infrared spectroscopy study and nuclear magnetic resonance are required [29]. In addition, Mg(OH)$_2$ crystalline products, which could cause undesired expansion to the structure [34], is not detected in the binders.

### 13.3.5 Drying Shrinkage

Fig. 13.9 plots the linear drying shrinkage of AAC binders cured at 20°C and 90% RH. The drying shrinkage takes place mainly in the early age (before 28 days). The increase of SiO$_2$/Na$_2$O ratio from 1.2 to 2.0 in the

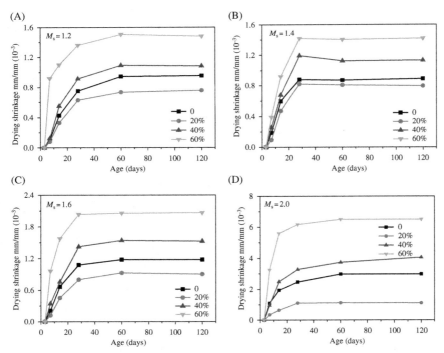

**Figure 13.9** Drying shrinkage of AACs: (A–D) present the effects of slag contents on drying shrinkage of the samples prepared using activators with $M_s$ = 1.2, 1.4, 1.6, and 2.0.

activators results in a drying shrinkage that is two to four times higher, due to the large shrinkage nature of silica gels of the sodium silicate solutions.

The linear drying shrinkage of 120-day cured geopolymer binders activated with $M_s = 1.2$ is in the range of $0.7–1.5 \times 10^{-3}$ mm/mm, which is much lower than $1–9 \times 10^{-3}$ mm/mm of alkali-activated GBFS binders under comparable reaction conditions [15,35]. It is probably attributed to two reasons: (1) lower degree of polymerization for high–calcium slag [36]; and (2) the difference between the shrinkage natures of AAC products for the two systems. The alkali-activated GBFS is well known to produce C–S–H gels [15,32,33], while the products here are probably N–(M)–A–S–H gels with frame structure.

It is interesting to note that only 20% HMNS addition leads to a decrease in drying shrinkage for the fly ash–based AAC binders prepared with the four different activators. This is in consistent with the variation tendency in pore structures of the AAC binders activated with $M_s = 1.4$. The drying shrinkage of AAC takes place and relates to the evaporation of free water from macropores and mesopores [35]. The 20% HMNS geopolymer binders possess a refiner porous structure and lower total porosity, compared with the matrix of 100% fly ash AAC. When the 20% HMNS samples are exposed to a relatively dry environment (pH = 90% is relatively high though), the evaporation rate of water from their mesopores and macropores is slower than the others. This is helpful for keeping a relative higher humidity of the pores in the matrix, and lower capillary tension and contraction force in the pores [35]. In the 40% and 60% HMNS binders, a large number of macropores are formed. The evaporation of free water from numerous macropores causes an increase in drying shrinkage, although the total volume of pores in the binders is low. The effects of pore structure on the shrinkage behavior of alkali-activated fly ash/HMNS blends show a consistent trend with those in an activated GBFS cement system [15,35]. Despite the role of the pore structure, the influence of the addition of HMNS on the shrinkage behavior of AACs can be explained from other perspectives. The HMNS has processed 1400°C calcination, so the large amount of magnesium in HMNS cannot work as a MgO-based expansive agent. This is confirmed by the XRD and SEM analysis. In addition, the residual HMNS particles work as microaggregates to fill into the macropores, and the shrinkage of AAC binders will be restricted with HMNS substitution. However, when more than 20% HMNS is used, the shrinking nature of high Si/Al gels overcomes the beneficial influence of the microaggregate effect.

## 13.4 SUSTAINABILITY OF AACs

To assess the sustainability of the alkali-activated fly ash/HMNS AACs, several factors are assumed as following:

- The embodied energies for dry sodium hydroxide and sodium silicate solution are 14.9 and 5.4 GJ/t, respectively. The embodied energy of the sodium silicate solution is calculated based on the analysis of Fawer et al. [37].
- The transportation distances of the fly ash and HMNS are supposed to be equivalent to that of cement, varying from 50 to 300 km, which is a realistic distance in many countries.
- The grinding process for HMNS requires 60 kWh/t to the fineness used in this study.

According to the above assumptions, the database for raw materials is summarized in Table 13.3. The embodied energy and effective emissions of OPC are also given for comparison purposes. The sustainability of AAC cement, in terms of embodied energy (EE) and effective $CO_2$ emissions (CE), can be calculated from the following equations:

$$EE = \sum E_{materials}/(1 + R_{NaOH} + R_{ss} + R_{H_2O}) \qquad (13.2)$$

$$CE = \sum C_{materials}/(1 + R_{NaOH} + R_{ss} + R_{H_2O}) \qquad (13.3)$$

in Eq. (13.2), EE is the embodied energy per ton paste, GJ/t; $\sum E_{materials}$ is the sum of energy consumptions for each reaction-ready component, including production and transportation of fly ash, HMNS, NaOH, and production of liquid sodium silicate; $R_{NaOH}$, $R_{ss}$, and $R_{H_2O}$ are the mass ratios of NaOH, liquid sodium silicate, and water to fly ash (or the blend of fly ash and HMNS). In Eq. (13.3), CE is the $CO_2$ emission per ton of paste, t $CO_2$/t; $\sum C_{materials}$ is the sum of $CO_2$ emissions for all of the reaction components; $R_{NaOH}$, $R_{ss}$, and $R_{H_2O}$ have the same meanings as in Eq. (13.2).

Because of the grinding process, HMNS possesses a much higher embodied energy than fly ash (Table 13.3). Using HMNS to partially substitute for fly ash will increase the embodied energy and $CO_2$ emissions of the alkali-activated blend system. The effect of blending HMNS on sustainability is shown in Fig. 13.10. The embodied energy and $CO_2$ emissions of the AAC paste are at the optimal modulus of 1.4. The embodied energy is only slightly increased from 1.20 to 1.25 GJ/t when HMNS substitution increases from 0% to 60%, at the assumed transportation distance

**Table 13.3** Embodied energy and associated $CO_2$ emissions (without transportation contribution) of raw materials for AAC production

| Raw materials | Embodied energy (GJ/t) | Effective emissions (t $CO_2$/t) |
|---|---|---|
| Fly ash | 0.087 | 0.027 |
| HMNS | 0.22 | 0.045 |
| Sodium hydroxide | 14.90 | 2.86 |
| Sodium silicate | 5.37 | 0.87 |
| Portland cement | 6.60 | 0.82 |

*Note:* The $CO_2$ emission data for fly ash and Portland cement are from the Ash Development Association of Australia, Fly Ash Technical Note 11 (2012). The embodied energy for fly ash is calculated based on its effective $CO_2$ emissions, which are assumed due to the electricity required for collection and classification. The embodied energy data for OPC is from the article *Cement and Concrete Environmental Considerations* (www.buildinggreen.com) (2011-09-11).

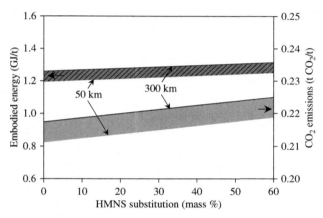

**Figure 13.10** Embodied energy and $CO_2$ emissions of the AAC pastes activated with activator $M_s = 1.4$.

of 50 km. The corresponding $CO_2$ emissions increase from 0.21 to 0.22 t $CO_2$/t. The increase of emission rate is less than 5%. The analysis again confirms that the alkali activator is the major contributor to the environmental footprint of AAC manufacturing [1]. These data also suggest that using other types of slag with suitable compositions to partially replace fly ash will not affect the "greener potential" of AAC paste. In comparison, the Portland cement paste at a w/c of 0.42 possesses embodied energy of 4.7 GJ/t and an emission rate of 0.58 t $CO_2$/t. The reductions of embodied energy and carbon emissions of the AACs manufactured in this study are up to 74% and 64%.

While this study focuses on the evaluation of the suitability and environmental impacts of HMNS addition in AAC manufacture, two other aspects need to be considered for future research:

1. Durability of resulting AAC concretes. The good volume stability of the HMNS is shown here; nevertheless, other properties regarding the durability are unclear, such as water resistance, carbonation rate, and sulfate resistance.

2. Leaching behavior of the AACs with varying slag compositions. The purpose of the proposed model in this study is to safely treat the industrial wastes through alkali-activation technology. Due to the activation process, some metals may not present as stale oxides in the glassy phases in slag. Attention should be paid on their leaching rates, although it is expected that the leaching of heavy metals may be reduced under high-pH conditions. The two aspects obviously deserve more research.

## CONCLUSIONS

An industrial solid-waste HMNS generated by pyrometallurgical production of nickel is attempted in the manufacture of fly ash–based AACs through an alkali activation technique. The major products formed in the binders are a class of magnesium containing sodium aluminosilicate gels. The blending of HMNS can provide a large number of silicate ingredients for the fly ash–based AAC, and thus the Si/Al ratio of the gel phase increases with HMNS content. HMNS particles can work as microaggregates in AAC binders. The total volume of pores in the AAC binders turns to be lower and more macropores are formed as the content of HMNS increases. When the molar ratio of activator solutions is in the range of 1.2–2.0, the optimal dosage of HMNS is 20%, resulting in the highest compressive strength and the lowest linear drying shrinkage. These performances correlate well with the most refined pore sizes and more compact microstructure in hardened binders compared to non- and more HMNS systems. Because of the high temperature of pyrometallurgical process, the HMNS is much less active than a normal MgO-based expansive agent, and thus will not trigger the volume stability problem of fly ash–based AAC. The embodied energy of AAC pastes is 1.20–1.25 GJ/t, and the $CO_2$ emissions is 0.21–0.22 t $CO_2$/t when the blending ratio increases from 0% to 60%. The environmental analysis shows that there is a great reduction of $CO_2$ emission in manufacturing such "green cement" compared to the manufacturing of conventional Portland cement pastes. This study shows a practical model of converting local industrial wastes into greener cement or cement-like products. The future work of developing this type of work should be focused on (1) the long-term properties of the deriving materials; (2) the match between raw materials and the activator and process, which requires a database for various materials; and (3) the recommended manufacturing and application based on the database.

## REFERENCES

[1] McLelellan BC, Williams RP, Lay J, van Riessen A, Corder GP. Costs and carbon emissions for geopolymer pastes in comparison to ordinary Portland cement. J Clean Prod 2011;19:1080–90.

[2] Shan C, Wang J, Zheng J, Yu Y. Study on application of nickel slag in cement concrete. Bull Chin Ceram Soc 2012;31:1263–8. (in Chinese with English abstract).

[3] Sheng G, Zhai J. Making metallurgical slag from nickel industry a resource. Metal Mine 2005;10:68–71. (in Chinese with English abstract).

[4] Onisei S, Pontikes Y, van Gerven T, Angelopoulos GN, Velea T, Predica V, et al. Synthesis of geopolymers using fly ash and primary lead slag. J Harzard Mater 2012;205-206:101–10.

[5] Kalinkin AM, Kumar S, Gurevich BI, Alex TC, Kalinkina EV, Tyukavkina VV, et al. Geopolymerization behavior of Cu–Ni slag mechanically activated in air and in $CO_2$ atmosphere. Int J Miner Process 2012;112-113:101–6.

[6] Maragkos I, Giannopoulou IP, Panias D. Synthesis of ferronickel slag-based geopolymers. Miner Eng 2009;22:196–203.

[7] Komnitsas K, Zaharaki D, Perdikatsis V. Geopolymerisation of low calcium ferronickel slags. J Mater Sci 2007;42:3073–82.

[8] Damtoft JS, Lukasik J, Herfort D, Sorrentino D, Gartner EM. Sustainable development and climate change initiatives. Cem Concr Res 2008;38:115–27.

[9] Yang KH, Song JK, Song KI. Assessment of $CO_2$ reduction of alkali-activated concrete. J Clean Prod 2013;39:265–72.

[10] Heath A, Paine K, McManus M. Minimising the global warming potential of clay based geopolymers. J Clean Prod 2014;78:75–83.

[11] ASTMC618-01. Standard specification for coal fly ash and raw or calcined natural pozzolan for use as a mineral admixture in concrete. In: Concrete and aggregates, ASTM Book of Standards, October 2008.

[12] Ogundiran MB, Nugteren HW, Witkamp GJ. Immobilisation of lead smelting slag within spent aluminate—fly ash based geopolymers. J Hazard Mater 2013;248–249: 29–36.

[13] Zhang J, Provis JL, Feng D, van Deventer JSJ. Geopolymers for immobilization of $Cr^{6+}$, $Cd^{2+}$, and $Pb^{2+}$. J Hazard Mater 2008;57:587–98.

[14] Maragkos I, Giannopoulou IP, Panias D. Synthesis of ferronickel slag-based geopolymers. Miner Eng 2009;22:196–203.

[15] Neto AAM, Cincotto MA, Repette W. Mechanical properties, drying and autogenous shrinkage of blast furnace slag activated with hydrated lime and gypsum. Cem Concr Compos 2010;32:312–8.

[16] Duxson P, Fernández-Jiménez A, Provis JL, Lukey GC, Palomo A, van Deventer JSJ. Geopolymer technology: the current state of the art. J Mater Sci 2007;42:2917–33.

[17] Ismail I, Bernal SA, Provis JL, San Nicolas R, Hamdan S, van Deventer JSJ. Modification of phase evolution in alkali-activated blast furnace slag by the incorporation of fly ash. Cem Concr Compos 2014;45:125–35.

[18] Puertas F, Fernández-Jiménez A. Mineralogical and microstructural characterisation of alkali-activated fly ash/slag pastes. Cem Concr Compos 2003;25:287–92.

[19] Duxson P, Provis JL. Designing precursors for geopolymer cements. J Am Ceram Soc 2008;91:3864–9.

[20] Diaz EI, Allouche EN, Eklund S. Factors affecting the suitability of fly ash as source material for geopolymers. Fuel 2010;89:992–6.

[21] Fernández-Jiménez A, Palomo A, Sobrados I, Sanz J. The role played by the reactive alumina content in the alkaline activation of fly ashes. Microporous Mesoporous Mater 2006;91:111–9.

[22] Ma Y, Hu J, Ye G. The pore structure and permeability of alkali activated fly ash. Fuel 2013;104:771–80.

[23] Khale D, Chaudhary R. Mechanism of geopolymerization and factors influencing its development: a review. J Mater Sci 2007;42:729–46.

[24] Duxson P, Provis JL, Lukey GC, Mallicoat SW, Kriven WM, van Deventer JSJ. Understanding the relationship between geopolymer composition, microstructure and mechanical properties. Colloids Surf A 2005;269:47–58.

[25] Škvára F, Kopecký L, Šmilauer V, Bittnar Z. Material and structural characterization of alkali activated low-calcium brown coal fly ash. J Hazard Mater 2009;168:711–20.

[26] Provis JL, Myers RJ, White CE, Rose V, van Deventer JSJ. X-ray microtomography shows pore structure and tortuosity in alkali-activated binders. Cem Concr Res 2012;42:855–64.

[27] Criado M, Fernández-Jiménez A, de la Torre AG, Aranda MAG, Palomo A. An XRD study of the effect of the $SiO_2/Na_2O$ ratio on the alkali activation of fly ash. Cem Concr Res 2007;37:671–9.

[28] Provis JL, Lukey GC, van Deventer JSJ. Do geopolymers actually contain nanocrystalline zeolites? A reexamination of existing results. Chem Mater 2005;17:3075–85.

[29] Zaharaki D, Komnitsas K, Perdikatsis V. Use of analytical techniques for identification of inorganic polymer gel composition. J Mater Sci 2010;45:2715–24.

[30] Álvarez-Ayuso E, Querol X, Plana F, Alastuey A, Moreno N, Izquierdo M, et al. Environmental, physical and structural characterization of geopolymer matrixes synthesized from coal (co-)combustion fly ashes. J Hazard Mater 2008;154:175–83.

[31] Lee WKW, van Deventer JSJ. Structural reorganisation of class F fly ash in alkaline silicate solutions. Colloids Surf A 2002;211:49–66.

[32] Kumar S, Kumar R, Mehrotra SP. Influence of granulated blast furnace slag on the reaction, structure and properties of fly ash based geopolymer. J Mater Sci 2010;45:607–15.

[33] Bernal SA, Provis JL, Rose V, de Gutierrez RM. Evolution of binder structure in sodium silicate-activated slag-metakaolin blends. Cem Concr Compos 2011;33:46–54.

[34] Xu L, Deng M. Dolomite used as raw material to produce MgO-based expansive agent. Cem Concr Res 2005;35:1480–5.

[35] Neto AAM, Cincotto MA, Repette W. Drying and autogenous shrinkage of pastes and mortars with activated slag cement. Cem Concr Res 2008;38:565–74.

[36] M.A. Cincotto, A.A.M. Neto, W. Repette. Effect of different activators type and dosages and relation with autogenous shrinkage of activated blast furnace slag cement. In: International congress on the chemistry of cement. 2003. p. 1878–1888.

[37] Fawer M, Concannon M, Rieber W. Life cycle inventories for the production of sodium silicates. Int J Life Cycle Assess 1999;4:207–12.

# CHAPTER 14

# Bond Between Steel Reinforcement and Geopolymer Concrete

## A. Castel
The University of New South Wales, Sydney, NSW, Australia

## 14.1 INTRODUCTION

Since the 1990s, geopolymer concretes (GPCs) have emerged as novel engineering materials with the potential to become a substantial element in an environmentally sustainable construction and building products industry [1–6]. GPC is the result of the reaction of materials containing aluminosilicate with alkalis to produce an inorganic polymer binder. Industrial waste materials, such as fly ash (FA) and blast furnace slag, are commonly used as the source of aluminosilicate for the manufacture of GPC due to the low cost and wide availability of these materials. With efficient use of other industrial byproducts, geopolymer binder can reduce embodied $CO_2$ by up to 80%, compared to ordinary Portland cement (OPC).

GPCs exhibit many of the characteristics of traditional concretes, despite their vastly different chemical composition and reactions [7,8]. The mixing process, the workability of freshly mixed geopolymers, the mechanical characteristics of the hardened material and durability appear to be similar to those for traditional OPC concretes. An important property of hardened concrete is its bond with reinforcing steel bars. However, only a few attempts to assess steel–GPC bond are reported in the literature. Some works, carried out using the pullout test [8,9], have shown that geopolymer mortars generally perform well. More recently, Songpiriyakij et al. [10] studied the bond between steel reinforcing bars and a GPC including FA, rice husk bark and silica fume. Results showed that the bond strengths of the GPC and traditional concrete were similar. Sarker [11] investigated bond strength of low-calcium FA-based GPC with deformed reinforcing steel bars using the ASTM A944 [12] beam-end test. The results

*Handbook of Low Carbon Concrete.*
DOI: http://dx.doi.org/10.1016/B978-0-12-804524-4.00014-2

were compared to the performance of equivalent traditional OPC concrete. The influence of different parameters, such as concrete compressive strength, diameter of the reinforcing bar and concrete cover, were analyzed. Generally, GPC showed higher bond strength than OPC concrete for the same test parameters. This higher bond strength was attributed to the higher splitting tensile strength of GPC than OPC concrete of the same compressive strength. Indeed, all pullout specimens failed in a brittle manner by splitting of the concrete along the bonded length of the reinforcing bars.

In this chapter, GPC bond with deformed reinforcing steel bars is studied using the standard RILEM pullout test [13]. This standard test method allows plotting the bond stress–slip diagram for GPCs. Two GPCs are used: a heat-cured low-calcium FA GPC and an ambient-cured ground granulated blast furnace slag (GGBFS) GPC. The specimens were tested at various ages ranging from 1 to 28 days. Results are compared to the performance of a reference OPC-based concrete. The experimental results obtained are used to recalibrate the existing model adopted by fib Committee [14] allowing correlating the steel–concrete bond strength to the mean compressive strength of concrete.

## 14.2 EXPERIMENTAL PROGRAM

### 14.2.1 GPC Mixes and Curing Regime

Two GPC mixes were used for this study. They were designed using the outcomes from both literature [15–21] and laboratory trials where different aluminosilicate materials proportions (FA and GGBFS), various activator concentration (8–14 M), and diverse activator-to-aluminosilicate source ratio (0.42–0.6) were tested.

Three different sources of aluminosilicate materials have been used in this study: a low-calcium type (ASTM C 618 Class F) FA, sourced by Eraring Power Station in New South Wales, Australia; a special-grade (ultrafine) FA branded as Kaolite High-Performance Ash (HPA) sourced by Callide Power Station in Queensland, Australia; and a GGBFS supplied by Blue Circle Southern Cement Australia. All details related to those three aluminosilicate materials such as oxide compositions and grading curves are available in Ref. [22]. The alkaline solution was made from a mixture of 12 molar (M) sodium hydroxide (NaOH) solution and sodium silicate solution with $Na_2O$. The mass ratio of alkaline solution to aluminosilicate material was 0.55.

The two GPC mixes are presented in Table 14.1. The first GPC mix (labeled GPC-FA) contains only 15% GGBFS. It is a low-calcium GPC. The second GPC mix (labeled GPC-S) contains 75% GGBFS. It is a high-calcium GPC. All GPCs were compacted using a poker vibrator and demoded 24 h after casting. The low-calcium GPC, GPC-FA, required an intense heat curing to achieve an acceptable performance. Two types of heat curing conditions were adopted:

1. 2D-curing: After casting, specimens were sealed to prevent excessive loss of moisture, stored in an 80°C oven for 1 day, and then cured in an 80°C water bath for a further 1 day. Then, all specimens where transferred to a controlled room at 23°C and 65% relative humidity until the day of the test.

2. 7D-curing: After casting, specimens were sealed to prevent excessive loss of moisture, stored in a 40°C oven for 1 day, and then cured in an 80°C water bath for a further 7 days. All specimens where then transferred to a controlled room at 23°C and 65% relative humidity until the day of the test.

The high-calcium geopolymer concrete GPC-S was ambient cured in a controlled environment ($T = 23°C, RH = 65\%$).

## 14.2.2 Reference OPC-Based Concrete

The performances of the GPC are compared to those of a traditional OPC-based concrete (labeled OPC40). The OPC concrete mix is

**Table 14.1** Geopolymer concrete mixes

|  | GPC-FA (kg/m³) | GPC-S (kg/m³) |
|---|---|---|
| FA | 193.5 | 80 |
| Kaolite HPA | 51.9 | — |
| GGBFS | 42.5 | 240 |
| Crushed coarse aggregate 1/10 kg/m³ | 1144.6 | 1215.2 |
| Sand 0/1 kg/m³ | 710.4 | 714.8 |
| Free water, kg/m³ | 59 | 25.5 |
| Sodium hydroxide solution (NaOH) | 45.2 | 54.9 |
| Sodium silicate solution (Na2SiO3) | 112.9 | 137.1 |

**Table 14.2** OPC based concrete mix

|  | OPC40 (kg/m³) |
|---|---|
| Cement CEM I 52.5 N CE CP2 NF | 400 |
| Sand 0/4 | 710 |
| Rolled gravel 4/10 | 532.5 |
| Crushed gravel 10/14 | 532.5 |
| Total water | 185 |

presented in Table 14.2. This concrete has been previously used as a reference structural concrete to assess the bond strength with carbon fiber–reinforced polymer rods or the bond strength between steel reinforcing bars and self-compacting concretes [23,24]. All concrete specimens were cured at 23°C and 65% relative humidity.

## 14.2.3 Testing Methods

Standard compressive and bond strength tests were conducted respectively in accordance with the European Standard NF EN 12390-3 and the RILEM recommendations [13]. Compressive strengths were conducted on cylindrical specimens of 110 mm (4.334 in.) diameter and 220 mm (8.668 in.) height. Automatic machines with 600 kN of capacity have been used for compression, and the rate loading was fixed at 5 kN/s (1.124 kip/s). For all concretes, tensile strength was measured using the standard splitting test.

The specimens prepared for the pullout tests were $100 \times 100$ mm in cross section. The ribbed bars used were 12 mm diameter. All bars had the same elastic limit (500 MPa). The total length of the rebar embedded in the concrete was 10 times the bar diameter, which was also the total length of the concrete specimens. To avoid an unplanned force transfer between the reinforcement and the concrete, the rebar were encased, at half of the length, in plastic tubes and sealed with silicone material as shown in Fig. 14.1. So the rebar had a bond length of five times the bar diameter. The concrete casting direction was perpendicular to the orientation of the rebar. The concrete cover was about 4.5 times the bar

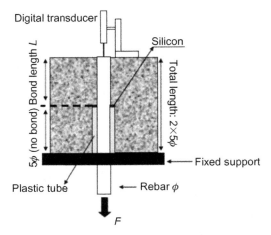

**Figure 14.1** Pullout test according to RILEM recommendations.

diameter and no confining reinforcement was provided. The pullout specimens were tested by applying the load to the bar at a rate of 0.1 kN/s
(0.02248 kip/s). The slip of the free end of the reinforcement was recorded
using a digital transducer with an accuracy of 0.001 mm.

To calculate the bond strength, a uniform distribution of bond stresses
along the bond length was assumed. It was calculated from the ultimate
pullout load using Eq. (14.1):

$$\tau_u = \frac{F_u}{\pi \cdot \phi \cdot L}$$

(14.1)

where $\tau_u$ is the ultimate bond stress (MPa), $\phi$ is the rebar diameter (mm),
$L$ is the bond length (mm), and $F_u$ is the ultimate pullout load (N).

For both pullout and compression tests, specimens were tested at different ages ranging from 1 to 28 days. Twenty-seven concretes batches
were produced in total. For each concrete mixture and age, at least three
specimens were used for both pullout test and compressive test. To determine the compressive strength of concrete of each pullout specimen at
the time of testing, the same numbers of compressive specimens, from the
same concrete batch, were tested following the pullout tests.

## 14.3 EXPERIMENTAL RESULTS

### 14.3.1 Mechanical Characteristics

Considering all tests regardless of the age of the concrete, the compressive
strength ranged between 10 and 50 MPa and between 11 and 72.4 MPa
for OPC concrete and GPC, respectively. The 28-day average compressive
strength, tensile strength, and elastic modulus of all concretes are presented
in Table 14.3. The FA GPC cured 2 days has a similar compressive strength
as the OPC concrete. The 28-day compressive strength of GPC-FA cured

**Table 14.3** Mechanical properties of the concretes

|  | OPC40 | GPC-FA (cured 2 days) | GPC-FA (cured 7 days) | GPC-S |
|---|---|---|---|---|
| Compressive strength, MPa | 44 | 46.5 | 58.5 | 72.4 |
| Elastic modulus, GPa | 32.0 | 24.8 | 25.3 | 32.9 |
| Tensile strength, MPa | 3.5 | 3.9 | 4.4 | N/A |

7 days and particularly GPC-S are significantly higher that of the OPC concrete.

Figs. 14.2, 14.3, and 14.4 show the increase in average compressive strength, elastic modulus and tensile strength, respectively, versus the duration of the heat-curing period for the low–calcium FA GPC. For the compressive and tensile strengths, three tests were performed for each curing condition. The elastic modulus was measured using one specimen only.

Figs. 14.2–14.4 show the sensitivity of Class F FA-based GPC to the curing condition. Up to 2 days, the increase in compressive and tensile strength is almost proportional to the duration of the heat curing. A minimum of 48 h (2D-curing) is required in order to reach a compressive

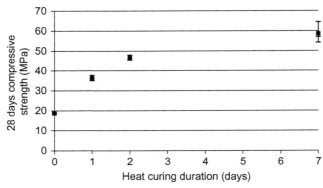

**Figure 14.2** Increase in 28 days fly ash GPC compressive strength versus the duration of the heat-curing period.

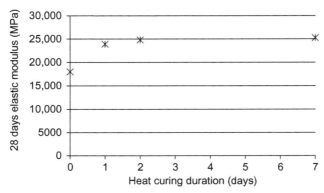

**Figure 14.3** Increase in 28 days fly ash GPC elastic modulus versus the duration of the heat-curing period.

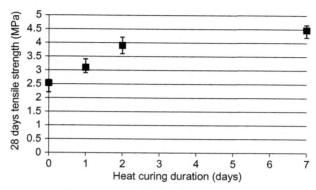

**Figure 14.4** Increase in 28 days fly ash GPC tensile strength versus the duration of the heat-curing period.

strength similar to the one of the reference structural OPC concrete. Fig. 14.3 shows that the elastic modulus is close to its maximum value of 25 GPa after 1 day of heat curing and does not change significantly with further, extended, heat curing.

In summary, the extension of the duration of curing from 2 to 7 days gave little increase in the elastic modulus and tensile strength; however, a 25% increase was observed in the compressive strength. The increase in energy consumption between 2 and 7 days of heat curing is significant, the economics of which needs to be considered in practice, especially when determining the overall carbon emission produced.

## 14.3.2 Low-Calcium FA GPC Bond Test Results

In this section, the steel–low-calcium FA GPC bond is compared to the reference steel–OPC concrete bond. The GGBFS geopolymer (GPC-S) concrete is not considered here because of its significantly higher compressive strength at 28 days.

Fig. 14.5A and B show the comparison between the bond stress–slip and normalized bond stress–slip curves, respectively, obtained for the 2D-cured geopolymer and the OPC concretes after 28 days using ribbed bars. The results show that, for an equivalent compressive strength, GPC performs better than OPC reference concrete. The normalized bond strengths of both types of concrete are equivalent (Fig. 14.5B), showing that the higher bond strength of the GPC corresponds to its higher tensile strength. The failure mode for all specimens was due to the pullout of the steel bar, leading the concrete to split after a slip of at least 1 mm.

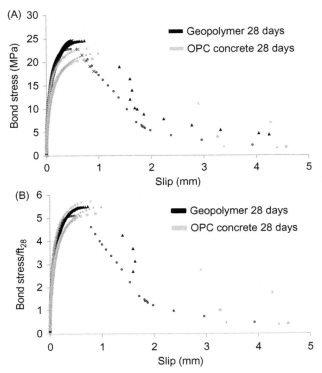

**Figure 14.5** Comparison between the bond stress–slip curves obtained for the 2 days heat-cured fly ash GPC and the OPC concrete after 28 days.

The overall bond stress–slip curves observed on both types of concrete are very similar. The same behavior was observed for the GGBFS GPC as well. Thus, existing models [14,25,26] allowing the prediction of the bond behavior of traditional concretes can be used or if necessary recalibrated for both Class F FA and GGBFS GPC.

Fig. 14.6 shows the bond stress–slip curves obtained after 28 days for the FA GPC pullout specimens heat cured for 7 days. When compared to the 2D-curing mode results, a 25% increase in bond strength is observed. This significant increase of the bond strength led to the yielding of the steel bar followed by concrete splitting at failure. As for both compressive and tensile strengths, increasing the duration of the heat curing from 2 to 7 days improves the bond strength.

Palomo et al. [21] and Fernandez-Jimenez et al. [27] investigated the microstructure development of heat-cured geopolymer binder using

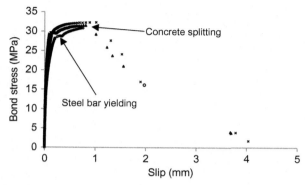

**Figure 14.6** Bond stress–slip curves obtained after 28 days for the fly ash GPC heat cured 7 days.

a low-calcium FA and alkaline solution, similar to the one used in this study. In these studies, heat curing at 85°C was maintained over a period of 90 days and the products of the activation reaction were mechanically and mineralogically characterized. The development of the geopolymer microstructure was assessed by using X-ray diffraction, Fourier transform infrared spectroscopy, and scanning electron microscope/energy dispersive X-ray spectroscopy. They concluded that the degree of nucleation (sub-stage denominated polymerization) of the FA continuously increases with time, especially during the early period of the thermal curing, with 70% of the total polymerization reached after just 2 days. It was demonstrated that the longer the curing time, the higher the average strength.

Kovalchuk et al. [28] studied the effect of thermal curing conditions on pore structure (total porosity and average pore diameter), down to a minimum pore diameter of $0.0067\,\mu m$, using a micrometrics autopore II 9220 porosimeter. They found that curing conditions, particularly the relative humidity, play an essential role in the development of the material's microstructural characteristics (such as porosity and phase composition), kinetics, and degree of reaction and their respective macroscopic properties. Large pores $(10{-}50\,\mu m)$ were observed on dried cured specimens, which caused a lowering of the compressive strength. Dry heat curing is not recommended for low-calcium FA systems. In contrast, when the specimens were wet cured, the resulting material developed a dense structure and a good correlation was observed between mechanical strength and total porosity.

The results reported in this chapter are consistent with the observations of Refs. [21,27–28]; the bond between ribbed steel bars and concrete,

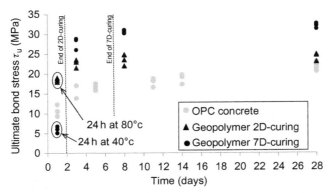

**Figure 14.7** Increase in bond strength for all fly ash GPC and OPC concrete specimens versus time.

together with both compressive and tensile strengths, increase with the increasing duration of the thermal curing in wet conditions. However, the increase in energy consumption needs to be considered and it may not be economic in practice.

Fig. 14.7 shows the increase in bond strength for all specimens tested versus time. The performance after 24 h of specimens stored at 40°C (after day 1 of the 7D-curing regime) is about 6 MPa, on average, which is low compared to the reference concrete. The bond strength of the specimens heat cured at 80°C for 24 h (after day 1 of the 2D-curing regime) reached an average of 18 MPa, which is better than the performance of the reference concrete at the same age and slightly superior to the reference concrete at the age of 28 days. Thus, high early bond strength can be achieved by providing an intensive period of heat curing. For the GPC, the average bond strength after 3 days was observed to be similar to that after 28 days. The polymerization leading to the bond development happens mostly during the heat-curing period. The results show that Class F FA GPC is well suited for precast applications. A 2015 study has shown that the optimum curing condition for this FA GPC is a temperature around 75°C for 18 h [29].

## 14.4 MODEL FOR BOND STRENGTH PREDICTION OF GPC

It is widely accepted that the bond strength is proportional to the square root of the average compressive strength. According to the fib model code [14], the bond strength of OPC concrete with ribbed steel reinforcement

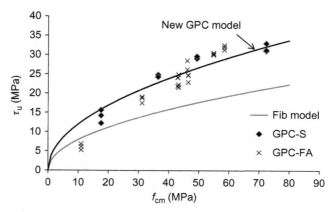

**Figure 14.8** Comparison between the fib model predictions, the new model predictions, and experimental results for all GPCs.

can be calculated from the average compressive strength of the concrete using Eq. (14.2):

$$\tau_{bmax} = 2.5\sqrt{f_{cm}} \tag{14.2}$$

where $\tau_{bmax}$ is the ultimate bond strength and $f_{cm}$ is the average compressive strength.

The GPC database offers 33 pairs of average compressive strengths $f_{cm}$ and bond strengths $\tau_u$. Fig. 14.8 shows the comparison between the fib model predictions and experimental results for all GPCs. The underestimation of the GPC bond strength by fib model is obvious especially for compressive strengths greater than 30 MPa. As a result, the fib model has been recalibrated for GPC by regression analysis. The model providing the best correlation with experiments ($R^2 = 0.83$) is as follows:

$$\tau_{u\text{-GPC}} = 3.83\sqrt{f_{cm}} \tag{14.3}$$

where $\tau_{u\text{-GPC}}$ is the ultimate bond strength and $f_{cm}$ is the average compressive strength. The predictions by the new model proposed are plotted in Fig. 14.8.

## 14.5 CONCLUSIONS

The results of this study show that the mechanical characteristics and bond strength of GPC composed mostly of Class F FA depend on the heat–curing conditions. For an equivalent compressive strength, both class

F FA and GGBFS GPC bond strength was observed to be better than that of the OPC reference concrete.

As expected, experimental results show that, for all concrete compressive strengths considered, the bond strength between GPC and reinforcing bars is significantly greater than the one calculated using the fib model code 2010 calibrated for OPC concrete. Results showed that traditional approach considering that the bond strength is proportional to the square roots of the average compressive strength of the concrete works very well for GPC. As a result, the fib model code 2010 has been recalibrated, providing accurate results for both high- and low-calcium GPCs.

Moreover, the overall bond stress–slip curves observed for GPC and OPC concretes are very similar. Thus, existing models allowing for the prediction of the bond behavior of traditional concretes can be used or if necessary recalibrated for both class F FA and GGBFS GPC.

## ACKNOWLEDGMENTS

This research was supported by the Australian Cooperative Research Center for Low Carbon Living. The experimental work was carried in the Structures Laboratory of the School of Civil and Environmental Engineering at University of New South Wales, Australia.

## REFERENCES

[1] J. Davidovits. Chemistry of geopolymeric systems, terminology. In: Proceedings of second international conference on geopolymers. 1999. pp. 9–40.

[2] Davidovits J. Geopolymer chemistry and applications, 2nd ed. France: Institut Geopolymere; 2008.

[3] Provis JL, van Deventer JSJ. Geopolymers: structures, processing, properties, and industrial applications. Cambridge: Woodhead Publishing Limited; 2009.

[4] Aly T, Sanjayan JG. Effect of pore-size distribution on shrinkage of concretes. J Mater Civil Eng 2010;22(5):525–32.

[5] Duxson P, Fernández-Jiménez A, Provis JL, Lukey GC, Palomo A, Van Deventer JSJ. Geopolymer technology: the current state of the art. J Mater Sci 2007;42(9):2917–33.

[6] N.A. Lloyd, B.V. Rangan. Geopolymer concrete with fly ash. In: 2nd International conference on sustainable construction materials and technologies. 2010. p. 1493–1504.

[7] Sofi M, van Deventer JSJ, Mendis PA. Engineering properties of inorganic polymer concretes (IPCs). Cem Concr Res 2007;37:251–7.

[8] Fernandez-Jimenez AM, Palomo A, Lopez-Hombrados C. Engineering properties of alkali-activated fly ash concrete. ACI Mater J 2006;103(2):106–12.

[9] Sofi M, van Deventer JSJ, Mendis PA. Bond performance of reinforcing bars in inorganic polymer concretes (IPCs). J Mater Sci 2007;42:3107–16.

[10] Songpiriyakij S, Pulngern T, Pungpremtrakul P, Jaturapitakkul C. Anchorage of steel bars in concrete by geopolymer paste. Mater Design 2011;32(5):3021–8.

[11] Sarker PK. Bond strength of reinforcing steel embedded in fly ash-based geopolymer concrete. Mater Struct 2001;44(5):1021–30.

[12] ASTM A 944-99. Standard test method for comparing bond strength of steel reinforcing bars to concrete beam-end specimens. West Conshohocken, USA: American Society for Testing and Materials Standard; 1999.

[13] RILEM Essai portant sur l'adhérence des armatures du béton: essai par traction. Mater Struct 1970;3(3):175–8.

[14] CEB-FIP, fib model code for concrete structures. 2010.

[15] Yip CK, Lukey GC, van Deventer JSJ. The coexistence of geopolymeric gel and calcium silicate hydrate at the early stage of alkaline activation. Cem Concr Res 2005;35(9):1688–97.

[16] Xie Z, Xi Y. Hardening mechanisms of an alkaline-activated class F fly ash. Cem Concr Res 2001;31(9):1245–9.

[17] Sun P. Fly ash based inorganic polymeric building materials. Michigan, USA: Wayne State University; 2005.

[18] Provis JL, et al. Correlating mechanical and thermal properties of sodium silicate-fly ash geopolymers. Colloids Surf A 2009;336(1–3):57–63.

[19] Hardjito D, Rangan BV. Development and properties of low calcium fly ash based geopolymer concrete. Perth, Australia; 2005.

[20] Ng TS. Development of a mix design methodology for high-performance geopolymer mortars. Struct Concr 2012;14(2):148–56.

[21] Palomo A, Grutzeck MW, Blanco MT. Alkali-activated fly ashes A cement for the future. Cem Concr Res 1999;29(8):1323–9.

[22] Castel A, Foster SJ. Bond strength between blended slag and Class F fly ash geopolymer concrete with steel reinforcement. Cem Concr Res June 2015;72:48–53.

[23] Al-Mahmoud F, Castel A, François R, Tourneur C. Effect of surface pre-conditioning on bond of carbon fiber reinforced polymer rods to concrete. Cem Concr Comp 2007;29(9):677–89.

[24] Castel A, Vidal T, François R. Bond and cracking properties of self-consolidating concrete. Constr Build Mater 2010;24:1222–31.

[25] Eligehausen R, Popov E, Bertero VV. Local bond stress–slip relationships of deformed bars under generalized excitations, Report No. 83/23, EERC. Berkeley, CA: University of California-Berkeley; 1983:162.

[26] Dahou Z, Sbartaï M, Castel A, Ghomari F. Artificial neural network model for steel-concrete bond prediction. Eng Struct 2009;31(8):1724–33.

[27] Fernandez-Jimenez A, Palomo A, Criado M. Microstructure development of alkali-activated fly ash cement: a descriptive model. Cem Concr Res 2005;35:1204–9.

[28] Kovalchuk G, Fernandez-Jimenez A, Palomo A. Alkali-activated fly ash: Effect of thermal curing conditions on mechanical and microstructural development—Part II. Fuel 2007;86:315–22.

[29] Noushini A, Babaee M, Castel A. Suitability of heat-cured low-calcium fly ash-based geopolymer concrete for precast applications. Mag Concr Res Sep. 2015:1–15.

# CHAPTER 15

# Boroaluminosilicate Geopolymers: Current Development and Future Potentials

**A. Nazari[1], A. Maghsoudpour[2] and J.G. Sanjayan[1]**
[1]Swinburne University of Technology, Hawthorn, VIC, Australia
[2]WorldTech Scientific Research Center (WT-SRC), Tehran, Iran

## 15.1 INTRODUCTION

Alkali-activated binders (geopolymers) are approximately a new class of construction materials that are produced from a silica-rich raw source material. The source, which must have some extent of alumina, is called an aluminosilicate source. Fly ash [1–3], metakaolin [4], and different types of slags [5,6] are the most commonly used aluminosilicate sources. Alkali activation of the starter material is performed by a mixture of alkali solution (sodium hydroxide (NaOH) or potassium hydroxide (KOH)) and a silica-rich source (sodium or potassium silicate). Its mechanism consists of dissolution of Al and Si in an alkali environment, orientation of dissolved species, and then polycondensation. This process results in the formation of a 3D network of silicoaluminate structures that appear in three sorts: poly(sialate) (–Si–O–Al–O–), poly(sialate–siloxo) (Si–O–Al–O–Si–O), and poly(sialate–disiloxo) (Si–O–Al–O–Si–O–Si–O) [7]. At low Si/Al ratio (=1), poly(sialate) structures are most likely to form while in higher ratios (>1), the formation of poly(sialate–siloxo) and poly(sialate–disiloxo) are most probable. It is reported that higher Si/Al ratio (but up to an optimum value) causes more condensed microstructures and higher values of compressive strength [7].

Some attempts have been made to introduce other types of starter mixtures in which alumina or silica parts are substituted entirely or partially. These include aluminogermanate geopolymer [8], phosphoric acid–based geopolymers with Si–O–Al–O–P–O structures [9], and borosilicate geopolymers [10]. Among them, borosilicate geopolymers revealed compressive strength as high as 57 MPa, which is excellent for construction

*Handbook of Low Carbon Concrete.*
DOI: http://dx.doi.org/10.1016/B978-0-12-804524-4.00015-4

technology. Formation of B–O bonds in geopolymer structures is supposed to be the cause of strengthening of these types of geopolymers.

In borosilicate geopolymers, anhydrous borax with the chemical composition of $Na_2B_4O_7$ is used for substituting silica-rich part of alkali activator (i.e., it is used instead of sodium or potassium silicate). Anhydrous borax is dehydroxylated by heating borax decahydrate ($Na_2B_4O_7 \cdot 10H_2O$) at 300°C. Then, an alkali activator is introduced into a high-purity silica source such as silica fume [10].

Despite its high strength, borosilicate geopolymer has not been developed and many features of this high-strength geopolymer are unknown. Silica fume used in Ref. [10] is not a common source for production geopolymers in construction technology. In the present work, the effects of borax content on compressive strength of boroaluminosilicate geopolymers at different ages of curing have been studied. The source material used is class F fly ash in which only a part of Al–O bonds are replaced by B–O ones. Microstructure changes were evaluated through Fourier transform infrared spectroscopy (FTIR) and scanning electron microscopy (SEM) analyses.

## 15.2 EXPERIMENTAL PROCEDURE

Fly ash used in this study was class F with the chemical composition 62.7 wt% $SiO_2$, 22.1 wt% $Al_2O_3$, 2.50 wt% $Fe_2O_3$, 3.10 wt% CaO, 0.52 wt% $SO_3$, and 0.40 wt% $Na_2O$. Its loss on ignition (LOI) was 2.6 wt%. Particle-size distribution of this fly ash is shown in Fig. 15.1. The chemical composition of fly ash was achieved by an X-ray fluorescence apparatus and its particle-size distribution was acquired by the ASTM C115 standard. Average particle size and Blain surface area of fly ash particles were 8 μm and 35.8 m$^2$/g, respectively.

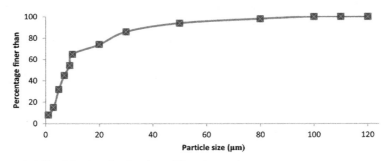

**Figure 15.1** Particle size distribution of fly ash.

Anhydrous borax was produced by heating borax decahydrate ($Na_2B_4O_7 \cdot 10H_2O$) at 150°C for 30 min and additional heating at 300°C for 15 h [10]. Anhydrous borax was then mixed with NaOH solution.

Fly ash was mixed with alkali activator (anhydrous borax + NaOH) for 5 min . To keep the workability of all mixtures in an acceptable range (between 60 and 65 mm), 1 wt% of water was replaced by polycarboxylate superplasticizer with a polyethylene condensate defoamed based admixture (Glenium C303 SCC). Although successful usage of superplasticizer in geopolymers is still in doubt, a previous study [11] showed that it is possible to use the appropriate mixture for increasing the fluidity of geopolymer paste. In this work, based on the experience, superplasticizer was used to keep the flow of all mixtures in an acceptable range. Mixture proportions of geopolymer specimens have been illustrated in Table 15.1.

Compressive specimens were prepared by pouring geopolymer pastes into 50 mm polypropylene cubic molds. Pouring of specimens was performed in two layers. The mold was half filled with paste and then vibrated for 45 s followed by filling the other half and vibrating for the same amount of time. Molds were covered by wet polythene sheets for 24 h to decrease carbonation. Demolded specimens were cured for 24 h at 70°C. Compressive strength tests were conducted using a hydraulic pressure jack with 100-ton capacity at 3, 7, 28, and 90 days of curing according to the ASTM C109 standard. For each mixture, three samples were tested and the average value was reported as the corresponding compressive strength.

SEM analysis was conducted on the fracture surface of different geopolymer specimens using a VEGA TESCAN microscope in secondary electron mode. FTIR analysis was done by a Bruker TENSOR27 FTIR apparatus. Specimens for FTIR analysis were prepared as KBr discs by mixing the powder samples (and those samples cured for 90 days) with KBr at a concentration of 0.2–1.0 wt%.

## 15.3 RESULTS AND DISCUSSION

### 15.3.1 Compressive Strength

Fig. 15.2 shows the compressive strength of all specimens. For all series of specimens, compressive strength increases at later ages. However, the differences between compressive strengths at early and later ages are not too high, indicating completion of most geopolymer reactions during the initial days of curing. The greatest strength is related to the G16 specimen

Table 15.1 Mixture proportions of the specimens

| Sample designation | Borax-to-NaOH weight ratio | Alkali-activator-to-fly-ash weight ratio | Content of fly ash (kg/m$^3$) | Content of borax (kg/m$^3$) | Content of NaOH flakes (kg/m$^3$) | Content of water (kg/m$^3$) | Content of superplasticizer (kg/m$^3$) |
|---|---|---|---|---|---|---|---|
| G1 | 0.593 | 0.75 | 1312 | 366 | 198 | 416 | 4.20 |
| G2 | 0.593 | 0.80 | 1276 | 380 | 205 | 431 | 4.35 |
| G3 | 0.593 | 0.85 | 1241 | 393 | 212 | 446 | 4.50 |
| G4 | 0.593 | 0.90 | 1208 | 405 | 219 | 459 | 4.64 |
| G5 | 0.700 | 0.75 | 1312 | 405 | 185 | 390 | 3.94 |
| G6 | 0.700 | 0.80 | 1276 | 420 | 192 | 404 | 4.08 |
| G7 | 0.700 | 0.85 | 1241 | 434 | 199 | 418 | 4.22 |
| G8 | 0.700 | 0.90 | 1208 | 448 | 205 | 431 | 4.35 |
| G9 | 0.806 | 0.75 | 1312 | 439 | 174 | 367 | 3.71 |
| G10 | 0.806 | 0.80 | 1276 | 455 | 181 | 380 | 3.84 |
| G11 | 0.806 | 0.85 | 1241 | 471 | 187 | 393 | 3.97 |
| G12 | 0.806 | 0.90 | 1208 | 486 | 193 | 405 | 4.09 |
| G13 | 0.912 | 0.75 | 1312 | 469 | 165 | 347 | 3.50 |
| G14 | 0.912 | 0.80 | 1276 | 487 | 171 | 358 | 3.62 |
| G15 | 0.912 | 0.85 | 1241 | 503 | 177 | 371 | 3.75 |
| G16 | 0.912 | 0.90 | 1208 | 519 | 182 | 383 | 3.87 |

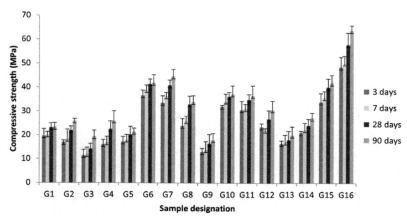

**Figure 15.2** Compressive strength of specimens.

at all ages while the smallest one is observed in G3 and G9 specimens. Since evaluating the parameters' effect on strength variations cannot be analyzed by only Fig. 15.2, it is followed by contour analysis. It should be mentioned the borax-to-fly-ash weight ratio is a symbol of the amount of boron, and alkali-activator-to-fly-ash represents the amount of $Na^+$ ions and water. Therefore, from now on in this study, when a discussion is made on these parameters, one can assume that main factors of boroaluminosilicate geopolymerization including boron, $Na^+$ ions, and water are investigated.

Fig. 15.3 shows the contour for the effects of alkali-activator-to-fly-ash weight ratios on compressive strength of geopolymers at different ages of curing and different borax-to-NaOH weight ratios. Fig. 15.3A illustrates that by using low borax-to-NaOH weight ratio (here 0.593), there is not any normal relationship between alkali-activator-to-fly-ash weight ratio and age of curing. The highest strength is achieved at later ages and alkali-activator-to-fly-ash weight ratio of 0.900 indicating that more boron is required to make high-strength boroaluminosilicate geopolymer pastes. Fig. 15.3B and C show approximately the same patterns for the considered dependence. The interesting point is that the highest strength occurs by using medium alkali-activator-to-fly-ash weight ratios. Although the amount of boron in the mixtures increases, at higher alkali-activator-to-fly-ash weight ratios, increasing water content and $Na^+$ ions has harmful effects on strength evolution even at later ages. This phenomenon seems strange and needs more evaluation. In the FTIR section

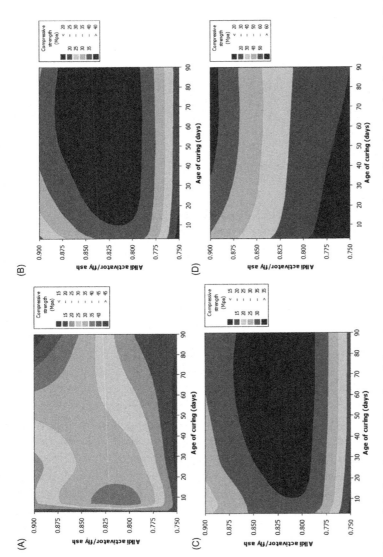

**Figure 15.3** Effects of alkali-activator-to-fly-ash weight ratios on compressive strength at different ages of curing; borax-to-NaOH weight ratios of (A), (B), (C), and (D) are 0.593, 0.700, 0.806, and 0.912, respectively.

of the present study, an attempt will be made to discuss this effect more. Finally, Fig. 15.3D shows that at the highest borax-to-fly-ash ratio (0.912), compressive strength increases by rising alkali–activator-to-fly-ash weight ratio. This indicates that by increasing boron in the mixtures, more water and $Na^+$ ions are required to achieve more boroaluminosilicate bonds. Additionally, it is evident that age of curing for a specific alkali–activator-to-fly-ash weight ratio has less effect on strength evolution. On the whole, higher strength in boroaluminosilicate is achieved at high boron concentration. However, the amount of boroaluminosilicate bond strongly depends on the amount of water.

## 15.3.2 Microstructure

Fig. 15.4 illustrates the SEM microstructure of boroaluminosilicate geopolymers for some selected specimens. As it is evident, various types of microstructures are totally different from those obtained in aluminosilicate geopolymers. A typical aluminosilicate geopolymer microstructure (with no special additive) includes a paste and some amount of unreacted fly ash (see, e.g., Ref. [12]). Even by adding other additives such as ordinary Portland cement (OPC) [13], iron-making slags [14], clays [15], and nanoparticles [16], microstructures do not alter completely and one cannot conclude anything easily just based on the microstructures achieved. Also, geopolymers made from other sources of aluminosilicates such as red mud [17] and metakaolin [18,19] have approximately the same microstructure in different conditions.

Another difference between the considered geopolymers and traditional ones is that some cracks appeared after fracture. In aluminosilicate geopolymers, there are many more cracks in fractured surfaces of geopolymers even at lower magnifications than those used in this study (see, e.g., Ref. [20]). Compared to them, as illustrated in Fig. 15.4, in boroaluminosilicate geopolymers, cracks rarely appear in microstructures. Only in Fig. 15.4G does one observe some cracks with the maximum depth of 200 nm. Although the appearance of cracks does not represent compressive strength, this indicates a different mechanism occurring during crack propagation in boroaluminosilicate geopolymers rather than aluminosilicate ones. In aluminosilicate geopolymers, during crack propagation, several microcracks nucleate and may affect energy dissipation. In boroaluminosilicate geopolymers, it is highly unlikely to observe crack branching during propagation. All the energy is consumed here for separating the two surfaces.

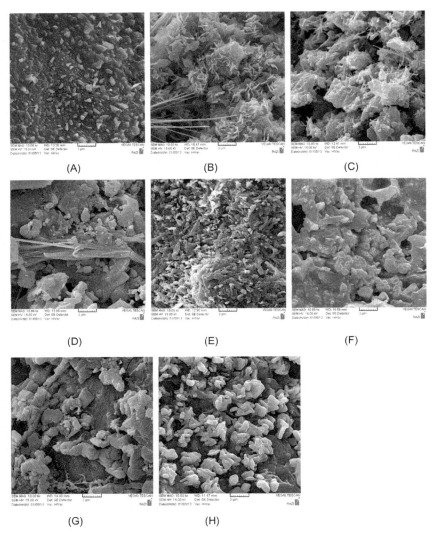

**Figure 15.4** SEM micrographs of (A) G1, (B) G3, (C) G7, (D) G8, (E) G11, (F) G12, (G) G15, and (H) G16 specimens.

As indicated, lots of unreacted fly ash particles are found in the microstructure of aluminosilicate geopolymers. Fig. 15.5 shows a typical microstructure of an aluminosilicate geopolymer we prepared. Unreacted fly ash, although weakening the strength of the paste, in a structure with high porosity (such as OPC concrete and geopolymer concrete) may act as a

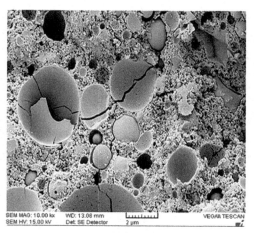

SEM MAG: 10.00 kx    WD: 13.08 mm                                          VEGAII TESCAN
SEM HV: 15.00 kV    Det: SE Detector    2 μm

**Figure 15.5** SEM microstructure of a typical aluminosilicate geopolymer.

barrier for crack nucleation. Unreacted fly ash particles have no adhesion to the paste and it may occur during shrinkage of the paste or crack propagation. Weak adhesion occurs whether as a result of inappropriate wetting of the particles due to the small amount of alkali activator, or because of crystalline nature of a specific fly ash, which retards its reaction. The holes appearing in the microstructure of Fig. 15.5 are the interface between unreacted fly ash particles and paste. It is evident a crack propagating toward unreacted fly ash must change its path or separate fly ash from the paste. Both of these possibilities require energy consumption and hence unreacted fly ash particles cause bridging in the paths of cracks. As Fig. 15.4 illustrates, there are fewer unreacted fly ash particles in any boroaluminosilicate geopolymers. Therefore, due to the mechanism of separation of the crack plane, various fracture surfaces occur.

Figs. 15.3A and 15.4B illustrate the microstructure of two of the weakest geopolymer pastes. These microstructures are entirely different. In Fig. 15.4A, related to the G1 specimen, a smooth surface with some geopolymer products is observed. This specimen has the lowest content of borax. By increasing the amount of water and borax in the G3 specimen, microstructure changes and some starlike needles with a maximum depth of 2 nm and some microscale-long needles again with a depth of maximum 2 nm are occurred. The presence of starlike needles cannot be considered as the reason of strength loss because in Fig. 15.4C, which is the microstructure of the second-ranked high-strength specimen (G7) among all of the 16 examined groups, these starlike needles are observed. Fig. 15.4D–F illustrate the microstructure of medium-strength geopolymers (G8, G11, and G12). All of them have higher boron content with respect to the G1, G3, and G7 specimens. The G8 and G12 specimens (Fig. 15.4D and F respectively) have completely large cleavage surfaces indicating their brittle fracture. The G11 specimen (Fig. 15.4E) has a rough surface, which signifies its abrupt separation under compression. However, the compressive strength of the G11 specimens is higher than those of the G8 and G12 specimens. Fig. 15.4G shows the microstructure of the G15 specimen, which has a compressive strength close to the G7 specimen. The G15 specimen has higher boron but less water with respect to the G7 specimen. This may be why some cracks occur in the microstructure of the G15 specimen (Fig. 15.4G). However, fracture surfaces of the G7 and G15 specimens are not similar at all, indicating different fracture mechanisms and strength gain when the amount of boron in the mixture changes. Finally Fig. 15.4H illustrates the microstructure of the G16 specimen with the highest strength. Its microstructure is to some extent similar to the G11 specimen (Fig. 15.4E). One can imagine that Fig. 15.4H is same as Fig. 15.4E but in higher magnification. In other words, abrupt separation has occurred here, but more energy is required since the surface is much rougher than that observed in Fig. 15.4E. In total, it seems that specimens with rough surfaces have higher strength.

## 15.3.3 FTIR Analysis Results

Aluminosilicate geopolymers have Si–O tetrahedrons, connected via corner-sharing bridging oxygen, and those tetrahedrons with n bridging oxygens are denoted as $Qn$ ($n = 0, 1, 2, 3,$ or $4$). Therefore, Si in Q3 configuration, for example, is surrounded by three bridging oxygens and a nonbridging oxygen. Amorphous $SiO_2$ is supposed to consist of only Q4 species forming a continuous random network [21]. The position of the first Si–O–X stretching bond (X = Si, Na, or H), which occurs at approximately $1100 \, cm^{-1}$ in amorphous geopolymers, is an indication of length and angle of the bonds in a silicate network. A shift of this band to lower wave numbers (i.e., decreasing its depth) indicates a lengthening of the Si–O–X bond, a reduction in the bond angle, and thus a decrease of the molecular vibrational force constant [21]. In the present boroaluminosilicate geopolymers, as a result of using borax, the formation of B–O bonds is anticipated. Main B–O stretching bonds appear between 1380 and $1310 \, cm^{-1}$ [22]. The signal at $2380 \, cm^{-1}$ can be assigned to $CO_2$, which reacts with extra amounts of NaOH and causes the formation of $HCO_{3-}$ [21]. FTIR spectra of all of 16 considered specimens have been illustrated in Figs. 15.6–15.9. Si–O bands in Figs. 15.6–15.9 consist of Si–O stretching in the range of $1000–1200\,cm^{-1}$, Si–O bending bond at $800\,cm^{-1}$ and between 890 and $975\,cm^{-1}$. Q4 and Q3 bonds can be assigned to those appeared at approximately 1100 and $1050\,cm^{-1}$, respectively [21].

Fig. 15.6 shows FTIR spectra for the specimens with borax-to-NaOH weight ratio of 0.593. These specimens have the lowest amount of borax between all the groups and the interesting point is that the only specimen with no B–O bond in this group (the G3 specimen with FTIR spectrum shown in Fig. 15.6C) has the lowest compressive strength. It is evident that the B–O bond has a vital role in increasing compressive strength since the G3 specimens with no B–O bond have the lowest strength. Additionally, the G3 specimen has the highest Si–O bending bond between these four specimens without any other Si–O stretching bonds. The G1, G2, and G4 specimens with a slightly more compressive strength have all of these primary bonds with approximately the same pattern.

Fig. 15.7 illustrates FTIR spectra for the specimens with borax-to-NaOH weight ratio of 0.700. Between these samples, G7 has the highest strength and its FTIR spectrum is shown in Fig. 15.7C. This may be as a

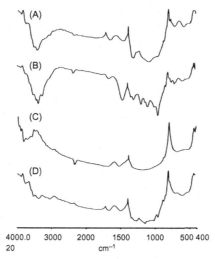

**Figure 15.6** FTIR analysis results for borax-to-NaOH weight ratio of 0.593; alkali-activator-to-fly-ash weight ratio for (A), (B), (C), and (D) are 0.75 (G1), 0.80 (G2), 0.85 (G3), and 0.90 (G4), respectively.

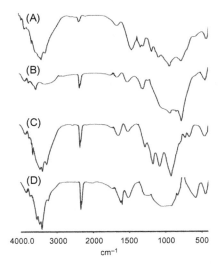

**Figure 15.7** FTIR analysis results for borax-to-NaOH weight ratio of 0.700; alkali-activator-to-fly-ash weight ratio for (A), (B), (C), and (D) are 0.75 (G5), 0.80 (G6), 0.85 (G7), and 0.90 (G8), respectively.

result of its sharper Si–O bonds. The compressive strength of this group increases gradually from G5 to G7 and then it decreases. The main reason that the G8 specimen may be related to its Si–O stretching bonds, which are weaker than those of the G6 and G7 specimens. Additionally, it has the biggest $CO_2$ peak. This indicates that the amount of NaOH is high here to contribute to boroaluminosilicate reaction, and hence the microstructure weakens.

Between G9 and G12, the specimens with borax-to-NaOH weight ratio of 0.806 and FTIR spectra illustrated in Fig. 15.8, G10 and G11 have the highest strengths. G9, as one of the weakest specimens, revealed no B–O bond. The lower strength of G12 on G10 and G12 may be related to the appearance of a sharp peak at approximately $800\,cm^{-1}$ which can be assigned to a Q2 stretching bond [21]. This means that lower Q4 bonds are formed and hence compressive strength decreases.

Fig. 15.9 shows FTIR spectra for the specimens with borax-to-NaOH weight ratio of 0.912. The highest amount of borax is used in this series of specimens. This group contained samples with the greatest strength (G16) and one with the lowest strength (G13). In the G13 specimen no B–O bond has formed indicating the importance of the formation of this bond on strength evolution. Instead, in G15 and G16 specimens, active B–O bonds are formed.

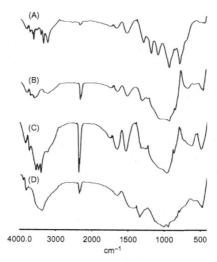

**Figure 15.8** FTIR analysis results for borax-to-NaOH weight ratio of 0.806; alkali-activator-to-fly-ash weight ratio for (A), (B), (C), and (D) are 0.75 (G9), 0.80 (G10), 0.85 (G11), and 0.90 (G12), respectively.

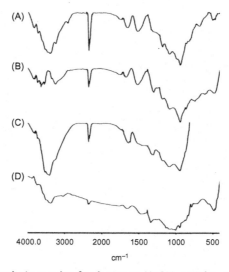

**Figure 15.9** FTIR analysis results for borax-to-NaOH weight ratio of 0.912; alkali-activator-to-fly-ash weight ratio for (A), (B), (C), and (D) are 0.75 (G13), 0.80 (G14), 0.85 (G15), and 0.90 (G16), respectively.

On the whole, the formation of B–O bonds in boroaluminosilicate geopolymers has a major effect on compressive strength and in specimens with no B–O bonds, compressive strength is the lowest.

## 15.4 CONCLUSIONS

In the current study, compressive strength and microstructure of boroaluminosilicate geopolymers were examined. From the results obtained, the following concluding remarks can be presented:

1. The best strength was related to the specimen with borax-to-NaOH weight ratio of 0.912 and alkali-activator-to-fly-ash weight ratio of 0.9. By increasing these ratios, the amount of borax in the mixture increases and the chance of formation of boroaluminosilicate bonds is increased as well.

2. The microstructure of boroaluminosilicate geopolymers was entirely different from that of traditional aluminosilicate geopolymers. Different fracture surfaces were also observed. Additionally, no microcracks were observed in boroaluminosilicate geopolymers.

3. FTIR spectra of boroaluminosilicate geopolymers consisted of all of the central aluminosilicate geopolymer bonds as well as B–O bonds. It was shown that B–O bonds have an important effect on strength gain where G3, G9, and G13 specimens with no B–O bonds have the lowest strengths.

## 15.5 FUTURE POTENTIAL STUDIES

Selection and justification of the use of starter materials are of foremost importance in this field of study and hence, these criteria are presented in detail. Additionally, conducting appropriate testing and analyzing procedures is of significant impact on identifying hidden features of this class of materials.

### 15.5.1 Materials

#### 15.5.1.1 Aluminosilicate Source

Primary aluminosilicate sources for the production of geopolymers are slags, fly ash, and metakaolin. Slags are highly porous materials with high calcium content and due to their appropriate reaction kinetic, slag-based geopolymers are cured at room temperatures. The strength of this class of geopolymers is high enough for use in many construction applications.

However, all the slag produced worldwide is now consumed by the concrete industry as a partial replacement of OPC and hence, there is little incentive to find use for these industrial byproducts. Metakaolin is another aluminosilicate source and although it is considered as an appropriate source material, it is more expensive than slag and fly ash since metakaolin is not an industrial byproduct. Fly ash on the other hand is a coal combustion byproduct that is abundantly available in Australia and worldwide. Depending on the source and makeup of the coal being burned, the components of fly ash vary considerably, but all fly ash includes substantial amounts of $SiO_2$ (both amorphous and crystalline) and CaO. In aluminosilicate geopolymers, calcium-(sodium)-aluminosilicate-hydrate [C-(N)-A-S-H] and sodium-aluminosilicate-hydrate (N-A-S-H) are the most likely amorphous gels formed during geopolymerization of class C and class F fly ash, respectively. Fly ash is naturally a little reactive material and in most cases, fly ash–based geopolymers are produced by oven curing. Oven-cured geopolymers gain their maximum strength at early ages [23]. To achieve fly ash–based geopolymer with reasonable strength by curing at room temperature, some attempts have been made to partially substitute fly ash by slag [24]. Our recent studies [25] showed the possibility of attaining boroaluminosilicate geopolymer with appropriate strength even by curing at room temperature. The effect of borax on the reaction kinetic of fly ash is unclear and more studies are required to discover its real nature. Attempts could be made to produce room temperature–cured fly ash-based boroaluminosilicate geopolymers through suitable mixture proportions and where required, fly ash could be substituted by small amounts of slags.

About 14 million tons of fly ash is being produced per annum in Australia while only 2–3 million tons are being used. In Victoria, massive deposits of brown coal, and their usage for the production of electricity, leave a considerable amount of brown coal ash. According to Energy and Earth Resources of Victoria State Government, "In 2010–11, total production of brown coal in Victoria amounted to 66.7 Mt. The 430 billion tons of brown coal located in Victoria represents a significant proportion of the world's brown coal endowment (http://www.australianminesatlas.gov.au/aimr/commodity/brown_coal.html)." Previous efforts to use brown coal ash in various applications have not been so lucky so far. We examined two types of fly ash in our previous works. In this work, the quality of fly ash was such that both aluminosilicate and boroaluminosilicate geopolymers had high strengths. In another work [25], the quality of fly ash was not suitable for the production of aluminosilicate geopolymers and all specimens had little compressive strength. Therefore, we called it "low-quality fly ash." Brown coal ash as a starter material could be used since the previous work [25] showed the ability of production of boroaluminosilicate geopolymers with reasonable strength through using a low-quality fly ash. Although our low-quality fly ash was not a brown coal one, the unknown features of reaction between borate and brown coal fly ash may cause reasonable strength. Our recent unpublished studies have created some potential for using brown coal fly ash as the starter material. However, in the case that using brown coal ash would be unsuccessful, other fly ash sources could be considered as a partial or total substitute for brown coal fly ash.

### 15.5.1.2 Alkali Activator (Borax + NaOH)

In aluminosilicate geopolymers, alkali activator is a mixture of a silicate-rich source such as sodium silicate or potassium silicate and a high-alkali solution such as NaOH or potassium hydroxide. Potassium-based activators are less common and are slightly more expensive. Sodium silicate partly helps the alkali reaction and provides sufficient $Si^{4+}$ ions. Higher contents of sodium silicate solution provide higher-strength specimens. However, increasing its amount causes a higher price and in some cases, it remains unchanged and a gluelike material covers the surface of the fly ash [26]. In boroaluminosilicate geopolymers, sodium silicate solution is entirely substituted by borax. Borax is an alkaline mineral material, one of the salts of boric acid, and is an eco-friendly material. It is widely used as a detergent in environmentally friendly cleansing powders. While critics of geopolymers are concerned about negative impacts of the production of highly concentrated sodium silicate solutions on the environment, borax can be a possible alternative. It is one of the main parts of boroaluminosilicate glasses. In these glasses, borax is used as a glazing agent (12–18 wt%). It is also used in porcelain cookware.

Borax, which was first used in geopolymer concrete by Williams and van Riessen [10], usually refers to decahydrate borax. Chemically sold borax is partially dehydrated and may be found as pentahydrate borax. Totally dehydrated borax can be attained by heating pentahydrate borax. However, all types of hydrated and dehydrated borax are commercially available and can be used to produce boroaluminosilicate geopolymers. Although dehydrated borax has been employed in Williams and van Riessen [10] and our previous work [25], it is possible to examine all types of borax to find the best results.

Our previous work [25] showed the possibility of geopolymers synthesis at room temperature in the presence of borax. Eliminating the oven curing of fly ash–based geopolymers can make a significant reduction in the final production cost.

### 15.5.1.3 Other Materials

Other relevant materials are aggregates, superplasticizers, and fibers. Justification of using these materials can be summarized as below.

- Aggregates

  In our current and previous work [25], only boroaluminosilicate geopolymer pastes were considered. For construction applications, it is crucial to evaluate properties of concrete rather than paste. After introducing aggregates into boroaluminosilicate pastes, concrete specimens are fabricated and tested.

- Superplasticizers

  Despite the widespread usage of superplasticizers in OPC concrete, superplasticizers have not been proven to have positive effects on the workability of geopolymers. However, some attempts were made at Swinburne University of Technology to determine the possibility of using a suitable type of superplasticizer based on mixture proportions of geopolymers [11]. The results of our current and previous work [25] show the opportunity of using superplasticizers correctly. Hence, further investigations can be carried on to evaluate the real effect of superplasticizers on boroaluminosilicate geopolymers.

- Fibers

  Steel fibers were suitably introduced to boroaluminosilicate geopolymer pastes in our previous study [27]. However, many factors such as diameter, length aspect ratio, and surface roughness of fibers, as well as mixture proportion and curing condition of geopolymer, require more in-depth studies. Consequently, a comprehensive program is proposed to evaluate the effect of steel fibers on the demanded properties of boroaluminosilicate geopolymer concrete. Also, our recently unpublished experiments show a good compatibility between polymer fibers and boroaluminosilicate paste. Polymer fibers and their role in reinforcing specimens could be carefully examined.

## 15.5.2 Experiments

Experiments can be conducted on unreinforced and fiber-reinforced geopolymers to determine their microstructure and physical, rheological, chemical, thermal, and mechanical properties. Tests can be categorized as following:

### 15.5.2.1 Experiments to Determine Physical and Rheological Properties

These properties include setting time, flowability, workability, bleeding rate, and pore structure of concrete. The first three sets of experiments are straightforward and essential properties of any particular mixture could be taken into consideration for all mixtures. The effect of type and amount of superplasticizers on these properties would be of high interest in both fiber-reinforced and unreinforced specimens.

Approximately 80% of the total concrete in Australia is for the construction of footpaths, slab-on-ground and warehouse and other industrial floors. Concrete cracks in these types of constructions are ubiquitous and the cost of repair and maintenance is a major budget item for asset owners such as councils and road authorities as these cracks often present trip hazards and result in significant public liability risks. It is widely known that the bleeding and evaporation rates cause plastic cracks. Unlike the conventional OPC concrete, the rheology of fresh geopolymer concrete is thixotropic due to the use of activators in the mix [28]. The thixotropic nature of geopolymer is likely to cause little bleeding in concrete resulting in a more smooth near-surface concrete than conventional Portland cement concrete. During evaluating the microstructure of the previously studied boroaluminosilicate geopolymers [25] and the current work, we found very few cracks in various magnifications. The relationship between plastic shrinkage, bleeding, and pore structure of boroaluminosilicate geopolymer and comparison of the obtained results for aluminosilicate geopolymer and OPC concrete will provide valuable knowledge on the exact reasons for cracking in concrete sections.

### 15.5.2.2 Experiments to Determine Chemical Properties

It is important to evaluate the chemical properties of boroaluminosilicate geopolymers very carefully. This part of the experiment could be conducted on unreinforced specimens. The principal acceptable chemical analyses to determine characteristics of geopolymers are X-ray diffraction (XRD) and nuclear magnetic resonance (NMR).

The XRD technique is widely used to determine crystalline and amorphous phases in many types of geopolymers. By comparing the spectrum of various specimens, effects of characteristics such as mixture proportions and curing conditions on phase evolution of those samples are determined. However, the most acceptable chemical analyzing tool for geopolymers is NMR, an analytical chemistry technique utilized in quality control and research for determining the content and purity of a sample as well as its molecular structure. NMR can quantitatively analyze mixtures containing known compounds. For unknown compounds, NMR can either be used to match against spectral libraries or to infer the basic structure directly. Once the basic structure is known, NMR can be used to determine molecular conformation in solution as well as studying physical properties at the molecular level such as conformational exchange, phase changes, solubility, and diffusion. To achieve the desired results, a variety of NMR techniques are available (http://chem.ch.huji. ac.il/nmr/whatisnmr/whatisnmr.html). The most acceptable technique for analyzing geopolymers through NMR studies is magic-angle spinning (MAS), a technique often used to perform experiments in solid-state NMR spectroscopies. 27Al and 29Si MAS NMR spectra are among those that are widely employed to determine the evolution of most aluminosilicate compounds. Other spectra such as 43Ca are used to evaluate the effect of additional constituent materials on aluminosilicate complexes. In boroaluminosilicate geopolymers, all of these spectra, as well as 11B, could be utilized to identify the hidden aspects. 11B has been successfully used to evaluate boron sites in borosilicate zeolites. A comprehensive study by performing step-by-step analysis and comparison of the results of XRD and NMR and their relationship with other examined properties is proposed for this type of geopolymer.

### 15.5.2.3 Experiments to Determine Mechanical Properties

For these materials, mechanical properties including compressive strength, flexural strength, tensile strength, and fracture toughness of geopolymer concrete specimens in both unreinforced and fiber-reinforced conditions are proposed to be studied. In addition to these conventional testing methods, creep properties of some of the high-strength geopolymer specimens could be tested. At Swinburne University of Technology, a comprehensive study on creep of aluminosilicate geopolymers is in progress. Bažant et al. [29] found that creep in long-span bridges built in the last few decades are still ongoing and may cause serious problems in the future. The main problem is that the exact mechanism of creep is not clear. It is understood to be either due to moisture diffusion or dislocations in addition to the possible formation of microcracks. A study of creep in geopolymer, which is a similar material in many aspects to OPC concrete despite being chemically different, will shed light from another angle and provide a greater understanding of this phenomenon in OPC concrete as well. Comparison between the creep results of aluminosilicate and boro-aluminosilicate geopolymers and OPC concrete may deliver invaluable information about this phenomenon.

As mentioned in the previous sections, two different types of reinforcements including steel and polymeric fibers can be used for reinforcing boro-aluminosilicate geopolymers. The effect of volume fraction, length, diameter, and aspect ratio of fibers on mechanical properties of fiber-reinforced specimens, especially their flexural strength, is of high interest. On the other hand, the effect of mixture proportions on flexural strength of reinforced specimens regarding strength of the paste and bond strength between fibers and paste could be evaluated. The strength of pastes could be determined by testing unreinforced specimens and the bond strength between fibers and paste could be estimated by conducting pullout tests.

### 15.5.2.4 Experiments to Determine Thermal Properties

Thermal properties of geopolymers that are of interest for these materials include (1) thermogravimetric analysis (TGA), (2) differential thermal analysis (DTA), (3) conduction calorimetry, (4) dilatometry, and (5) high-temperature strength. The first three experimental procedures can be carried out to evaluate phase changes and evolution during continuous or isothermal heating and could be conducted to compare chemistry and physics of unheated and heated specimens. The combination of the last two sets of experiments is done to evaluate the high-temperature compressive strength and transitional creep of specimens [30]. Very few experiments have been conducted to assess the dynamic compressive strength of geopolymer concrete at high temperatures. Most of them deal with heating the specimens to specific temperatures and measuring their properties after oven- or air-cooling. However, understanding the relationship between phase changes and high-temperature strength may be of interest. Additionally, in reinforced specimens, the effect of fibers on properties at high temperatures can deliver state-of-the-art knowledge.

Polymeric fibers, which can be used as reinforcements, are sensitive to heat and their nature will change based on amount of time exposed to heat. Thickness and chemical composition of boroaluminosilicate geopolymers will determine their thermal resistance. Effect of heat on bond strength and thermal stability of polymeric fibers could be studied by means of all five suggested groups of thermal analysis methods. The most significant change may occur in steel fibers due to their possible phase transformation. If steel fibers are heated above 723°C, a phase transformation happens and the body-centered cubic (bcc) crystalline structure of plain carbon steel fibers transforms to a face-centered cubic (fcc) crystalline structure [31]. This temperature is indicative and can be altered upon chemical composition and impurities of steel fibers. The interesting result of this transformation is shrinkage of steel due to higher packing factor of the fcc structure. This behavior is the geopolymer paste where its expansion causes further debonding between the steel fibers and matrix. However, at that temperature, everything depends on the behavior of geopolymer paste, which is a glassy viscoelastic material. On the other hand, after cooling specimens in different environments, two possible phase transformations in steel fibers are followed. The first is the reverse transformation of the fcc structure to the bcc structure with expansion against geopolymer paste shrinkage, and as a result, subsequent debonding is anticipated. The other is the transformation of the fcc structure to a body-centered tetragonal (bct) crystalline structure, which is called martensitic transformation and takes places at very high cooling rates [31]. Martensitic transformation induces a massive expansion of 4% [32] to steel fibers. However, steel fibers, which are surrounded by geopolymer paste, may not be affected by even high cooling rates and again the whole thing depends on chemical composition and thickness of specimens. High cooling rate may help to discover two essential characteristics. The first is to evaluate post-heat treatment properties of specimens, which is similar to the fire-exposed structures extinguished by water or other appropriate materials. The second and more important one is that the microstructure and chemistry of quenched specimens are highly likely to show the properties of specimens at high temperatures. Phase evolution in geopolymers is time-dependent and by rapid cooling of the structure, it is supposed that the microstructure of the specimen remains unchanged. A comprehensive comparison between phase evolution results at high temperature, and that could be acquired through conducting XRD and NMR on heat-affected specimens to verify this.

### 15.5.2.5 Evaluation of Microstructure

Microstructure evaluation of boroaluminosilicate geopolymer in our current and previous work [25] was one of the most interesting sections. Our previous studies showed the appearance of a wide variety of microstructures based on mixture proportions and curing conditions. As we indicate in those works, in traditional aluminosilicate geopolymers made from different sources, the resultant fracture surfaces are approximately the same. Although some differences are observed among various sources of aluminosilicates, for a particular aluminosilicate source, even by changing parameters such as NaOH concentration, microstructures are very similar, containing a paste with some unreacted fly ash particles. In boroaluminosilicate geopolymers, there are tiny unreacted particles, and a condensed paste with different fracture surfaces and crystals is observed. It would be worthwhile to understand the mechanisms of formation of these structures through an exact experimental procedure using various techniques.

Another interesting feature of these boroaluminosilicate geopolymers is their fracture surface, where little cracks are seen. In traditional aluminosilicate geopolymers, unreacted fly ash particles are a barrier to propagating cracks and therefore, they introduce tortuosities in crack shape. In boroaluminosilicate geopolymers, because of formation of a condensed paste, cracks are likely to be straight. However, branching of cracks or continuing in one direction only indicates the mechanism of crack propagation and does not reveal the strength of geopolymers. Although branching of cracks may dissipate the applied stress, the final strength of a brittle cement matrix structure depends on another aspect, namely the strength of the paste. We supposed that in the studied boroaluminosilicate geopolymers, the probable formation of functional crystals with the aim of sodium perborate functional groups may be the reason for the formation of high-strength pastes [25]. However, the real reasons require further study, with a focus on the microstructure and other characteristics of boroaluminosilicate geopolymers.

Both the microstructure and fracture surfaces of boroaluminosilicate geopolymers will provide exciting features and help us to discover the real nature of very complicated boroaluminosilicate reactions. Besides SEM, the thin-section analysis technique can be utilized to evaluate the microstructure of different concrete specimens. It provides valuable information on the dispersion of aggregates and fibers in a cross-section of specimens using light microscopy. Additionally, a targeted microstructure of these thin sections is attainable using SEM.

# REFERENCES

[1] Provis JL, Yong CZ, Duxson P, van Deventer JS. Correlating mechanical and thermal properties of sodium silicate-fly ash geopolymers. Colloids Surf A 2009;336(1):57–63.

[2] Sun P, Wu HC. Chemical and freeze–thaw resistance of fly ash-based inorganic mortars. Fuel 2013;111:740–5.

[3] Rashad AM. A comprehensive overview about the influence of different admixtures and additives on the properties of alkali-activated fly ash. Mater Des 2014;53:1005–25.

[4] Rovnaník P. Effect of curing temperature on the development of hard structure of metakaolin-based geopolymer. Constr Build Mater 2010;24(7):1176–83.

[5] Yusuf MO, Johari MAM, Ahmad ZA, Maslehuddin M. Strength and microstructure of alkali-activated binary blended binder containing palm oil fuel ash and ground blast-furnace slag. Constr Build Mater 2014;52:504–10.

[6] Puertas F, Torres-Carrasco M. Use of glass waste as an activator in the preparation of alkali-activated slag. Mechanical strength and paste characterisation. Cem Concr Res 2014;57:95–104.

[7] De Silva P, Sagoe-Crenstil K, Sirivivatnanon V. Kinetics of geopolymerization: role of $Al_2O_3$ and $SiO_2$. Cem Concr Res 2007;37(4):512–8.

[8] Durant AT, MacKenzie KJ. Synthesis of sodium and potassium aluminogermanate inorganic polymers. Mater Lett 2011;65(13):2086–8.

[9] Liu LP, Cui XM, He Y, Liu SD, Gong SY. The phase evolution of phosphoric acid-based geopolymers at elevated temperatures. Mater Lett 2012;66(1):10–12.

[10] Williams RP, van Riessen A. Development of alkali activated borosilicate inorganic polymers (AABSIP). J Euro Ceram Soc 2011;31(8):1513–6.

[11] Nematollahi B, Sanjayan J. Effect of different superplasticizers and activator combinations on workability and strength of fly ash based geopolymer. Mater Des 2014;57:667–72.

[12] Li Q, Xu H, Li F, Li P, Shen L, Zhai J. Synthesis of geopolymer composites from blends of CFBC fly and bottom ashes. Fuel 2012;97:366–72.

[13] Pangdaeng S, Phoo-ngernkham T, Sata V, Chindaprasirt P. Influence of curing conditions on properties of high calcium fly ash geopolymer containing Portland cement as additive. Mater Des 2014;53:269–74.

[14] Nath SK, Kumar S. Influence of iron making slags on strength and microstructure of fly ash geopolymer. Constr Build Mater 2013;38:924–30.

[15] Sukmak P, Horpibulsuk S, Shen SL. Strength development in clay–fly ash geopolymer. Constr Build Mater 2013;40:566–74.

[16] Phoo-ngernkham T, Chindaprasirt P, Sata V, Hanjitsuwan S, Hatanaka S. The effect of adding nano-$SiO_2$ and nano-$Al_2O_3$ on properties of high calcium fly ash geopolymer cured at ambient temperature. Mater Des 2014;55:58–65.

[17] Ye N, Yang J, Ke X, Zhu J, Li Y, Xiang C, et al. Synthesis and characterization of geopolymer from Bayer red mud with thermal pretreatment. J Am Ceram Soc 2014;97(5):1652–60.

[18] Bell JL, Driemeyer PE, Kriven WM. Formation of ceramics from metakaolin-based geopolymers: Part I—Cs-based geopolymer. J Am Ceram Soc 2009;92(1):1–8.

[19] Bell JL, Driemeyer PE, Kriven WM. Formation of ceramics from metakaolin-based geopolymers. Part II: K-based geopolymer. J Am Ceram Soc 2009;92(3):607–15.

[20] Hanjitsuwan S, Hunpratub S, Thongbai P, Maensiri S, Sata V, Chindaprasirt P. Effects of NaOH concentrations on physical and electrical properties of high calcium fly ash geopolymer paste. Cem Concr Compos 2014;45:9–14.

[21] Simonsen ME, Sønderby C, Li Z, Søgaard EG. XPS and FT-IR investigation of silicate polymers. J Mater Sci 2009;44(8):2079–88.

[22] Karabacak M, Kose E, Atac A, Asiri AM, Kurt M. Monomeric and dimeric structures analysis and spectroscopic characterization of 3, 5-difluorophenylboronic acid with experimental (FT-IR, FT-Raman, $^1H$ and $^{13}C$ NMR, UV) techniques and quantum chemical calculations. J Mol Struct 2014;1058:79–96.

[23] Ryu GS, Lee YB, Koh KT, Chung YS. The mechanical properties of fly ash-based geopolymer concrete with alkaline activators. Constr Build Mater 2013;47:409–18.

[24] Xu H, Gong W, Syltebo L, Izzo K, Lutze W, Pegg IL. Effect of blast furnace slag grades on fly ash based geopolymer waste forms. Fuel 2014;133:332–40.

[25] Nazari A, Maghsoudpour A, Sanjayan JG. Boroaluminosilicate geopolymers: role of NaOH concentration and curing temperature. RSC Adv 2015;5:11973–9.

[26] Tennakoon C, Nazari A, Sanjayan JG, Sagoe-Crentsil K. Distribution of oxides in fly ash controls strength evolution of geopolymers. Constr Build Mater 2014;71:72–82.

[27] Nazari A, Maghsoudpour A, Sanjayan JG. Flexural strength of plain and fiber-reinforced boroaluminosilicate geopolymer. Constr Build Mater 2015;76:207–13.

[28] Laskar AI, Bhattacharjee R. Rheology of fly-ash-based geopolymer concrete. ACI Mater J 2011;108(5):536–42.

[29] Bažant ZP, Yu Q, Li GH, Klein GJ, Krístek V. Excessive deflections of record-span pre-stressed box girder-lessons learned from the collapse of the Koror-Babeldaob Bridge in Palau. Concr Int 2010;32(6):44.

[30] Pan Z, Sanjayan JG. Stress–strain behavior and abrupt loss of stiffness of geopolymer at elevated temperatures. Cem Concr Compos 2010;32(9):657–64.

[31] Cayron, C. (2012). One-step theory of fcc-bcc martensitic transformation. *arXiv preprint arXiv:1211.0495*.

[32] Totten GE, editor. Handbook of residual stress and deformation of steel. ASM international; 2002.

# INDEX

*Note*: Page numbers followed by "*f*" and "*t*" refer to figures and tables, respectively.

Printed in the United States
By Bookmasters